# Cave and Karst Systems of the World

**Series editor**

James W. LaMoreaux, Tuscaloosa, USA

More information about this series at http://www.springer.com/series/11987

Harold L. Dibble · Shannon J.P. McPherron
Paul Goldberg · Dennis M. Sandgathe
Editors

# The Middle Paleolithic Site of Pech de l'Azé IV

Springer

*Editors*
Harold L. Dibble
Department of Anthropology
University of Pennsylvania
Philadelphia, PA
USA

and

Department of Human Evolution
Max Planck Institute for Evolutionary Anthropology
Leipzig
Germany

and

Institute for Human Origins, School of Human Evolution and Social Change
Arizona State University
Tempe, AZ
USA

Shannon J.P. McPherron
Department of Human Evolution
Max Planck Institute for Evolutionary Anthropology
Leipzig
Germany

Paul Goldberg
School of Earth and Environmental Sciences
University of Wollongong
Wollongong, NSW
Australia

and

Institute for Archaeological Sciences
Eberhard Karls Universitat Tübingen
Tübingen
Germany

and

Department of Archaeology
Boston University
Boston, MA
USA

Dennis M. Sandgathe
Department of Archaeology and Human Evolution Studies Program
Simon Fraser University
Burnaby, BC
Canada

and

Museum of Archaeology and Anthropology
University of Pennsylvania
Philadelphia, PA
USA

ISSN 2364-4591 ISSN 2364-4605 (electronic)
Cave and Karst Systems of the World
ISBN 978-3-319-57522-3 ISBN 978-3-319-57524-7 (eBook)
https://doi.org/10.1007/978-3-319-57524-7

Library of Congress Control Number: 2017948211



Printed on acid-free paper

This Springer imprint is published by Springer Nature
The registered company is Springer International Publishing AG
The registered company address is: Gewerbestrasse 11, 6330 Cham, Switzerland

*This book is dedicated to the memory of François Bordes and Denise de Sonneville-Bordes*

# Foreword

Préfacer un ouvrage de qualité qui concerne un sujet auquel on est personnellement attaché est toujours une agréable tâche et c'est donc le cas pour cette monographie du gisement du Pech de l'Azé IV, qui compte parmi les grands gisements du Paléolithique moyen du Périgord et d'Aquitaine et appartient à un ensemble de gisements fouillés et étudiés par le regretté Professeur François Bordes, gisements du Pech de l'Azé I, Pech de l'Azé II et Pech de l'Azé IV dans le cas présent, mais aussi Combe-Grenal tout proche, gisement de référence mondialement connu pour son importante séquence archéostratigraphique .

Du souvenir de François Bordes immense personnalité scientifique, ne saurait être dissocié celui de son épouse Denise de Sonneville-Bordes dont les travaux concernant le Paléolithique supérieur firent, et font encore autorité et qui fut mon Directeur de recherches.

J'ai rencontré Harold L. Dibble dans les années 1970 lorsqu'il effectuait de fréquents séjours à l'Université de Bordeaux au laboratoire de Préhistoire que dirigeait mon maître, le professeur François Bordes, illustre préhistorien réputé notamment pour sa parfaite connaissance des industries moustériennes que nos collègues venus du monde entier venaient examiner à Bordeaux dans les locaux de l'Université, bénéficiant ainsi de ses conseils, de ses enseignements et de sa grande expérience en matière de typologie et de technologie lithiques. La plupart des séries conservées dans les tiroirs du laboratoire provenaient des fouilles qu'il effectua dans des sites majeurs du Paléolithique moyen en Aquitaine, dans la région sarladaise, notamment Combe-Grenal et les différents locus du Pech de l'Azé parmi lesquels le Pech de l'Azé IV aux fouilles duquel Harold L. Dibble alors jeune étudiant, participa.

Quelques années plus tard nous entreprîmes Harold et moi la fouille du gisement de Combe-Capelle Bas dans la vallée de la Couze en Dordogne et c'est là que je rencontrais pour la première fois Shannon J.P. McPherron, un étudiant de l'Université de Pennsylvanie. Nous avons codirigé ensuite plusieurs campagnes de fouilles dans le gisement du Haut de Combe-Capelle (baptisé abri Peyrony, du nom de Denis Peyrony qui le fouilla entre les deux guerres).

Les activités de mes deux collègues en Aquitaine ne se limitèrent pas à ces gisements puisque, outre la saisie informatisée des carnets des fouilles de François Bordes, la révision et le reconditionnement du matériel archéologique recueilli par lui dans les sites du Paléolithique moyen du Sarladais, ils s'attaquèrent à plusieurs opérations de terrain en Aquitaine-Poitou-Charentes après quelques recherches en collaboration avec Alain Tuffreau, à Cagny l'Epinette dans le nord de la France.

D'autres opérations outre plusieurs relevés topographiques de gisements clefs, furent effectuées avec André Débenath à Fontéchevade, d'autres, avec Alain Turq au Roc de Marsal et à la Ferrassie en Dordogne, sans compter bien sûr, les fouilles au Pech de l'Azé IV ainsi que des travaux plus limités avec André Morala dans le gisement de la Gane proche de Combe-Grenal, travaux auxquels j'eus le plaisir de participer.

Je ne m'attarderai pas ici sur de multiples activités de terrain et des recherches au Maroc, en Egypte et en en ce qui concerne Shannon J.P. McPherron en Afrique de l'Est et à Jonzac (Charente).

C'est dire le dynamisme, l'esprit d'entreprise et l'opiniâtreté de ces deux chercheurs parfaitement organisés, auxquels nous devons déjà plusieurs monographies et de nombreux articles scientifiques dont certains méthodologiques puisqu'ils furent notamment les premiers à mettre en œuvre en Aquitaine, des méthodes d'enregistrement et de traitement des données très sophistiquées et novatrices à l'époque où nous avons fouillé Combe-Capelle bas.

La monographie du Pech de l'Azé IV à laquelle ont collaboré différents spécialistes renommés qui ont formé une équipe scientifique performante et très bien structurée, est présentée dans un style clair et limpide. Bien documentée et sérieusement argumentée comme leurs précédents travaux, elle accorde une grande importance à l'étude taphonomique des dépôts et des vestiges archéologiques qui y sont associés et veille avec soin à ce que les résultats présentés soient facilement disponibles pour d'autres chercheurs pour des études à venir. Bien illustrée, riche en tableaux, listes, décomptes divers, elle constituera désormais un ouvrage incontournable pour la connaissance du Moustérien et des néandertaliens du Sud-Ouest de la France et je ne puis que remercier tous les auteurs de cet ouvrage et plus spécialement ses coordinateurs

S'inscrivant ainsi dans une approche naturaliste mais également orientée vers l'Anthropologie et les sciences humaines, elle complète fort bien les études conduites par François Bordes dans ce gisement qui fut parmi les derniers qu'il fouilla tout au long d'une riche et brillante carrière hélas trop tôt interrompue et elle constitue désormais un brillant hommage à la mémoire de celui qui pendant trente ans, compta parmi les plus grands préhistoriens du $XX^{\text{ème}}$ siècle et dont la forte personnalité et les qualités scientifiques et humaines ont marqué toute une génération de chercheurs.

J'exprime ma sincère gratitude à mes deux collègues qui m'ont fait l'honneur et l'amitié de me demander d'en rédiger la préface et qui partagent avec moi m'attrait de celle belle région périgourdine qui les a quasiment acculturés.

Cet important ouvrage désormais de référence, constitue un prolongement attendu des recherches de mon illustre maître François Bordes.

Chercheur C.N.R.S honoraire
Docteur d'Etat es Sciences

Michel Lenoir

# Preface

This is the final report of excavations carried out at the Middle Paleolithic site of Pech de l'Azé IV (Pech IV), located in the Department of the Dordogne in southwest France. This site was originally excavated by François Bordes from 1970–1977, and the excavations reported on here were carried out from 2000–2003. The more recent project was directed by Harold L. Dibble and Shannon J.P. McPherron.

Unfortunately, Bordes died not long after finishing his excavation, and most of his work at the site remained unpublished, except for a brief preliminary report (Bordes 1975). To a very large extent, the primary goal of the recent project was to complete his work there, which included analyzing the lithic and faunal material recovered by him, to develop a chronology of the site based on newer numerical dating techniques, and to undertake a comprehensive study of the formation processes that had affected the site during and subsequent to the deposition of the materials. To a very large degree, these goals were achieved and, as a result, the faunal and lithic collections from the two excavations can be combined, resulting in a large, well-provenienced, and well-dated series of assemblages spanning roughly 60,000 years of the early Late Pleistocene.

## How We Came to Excavate Pech IV?

In the early 90s, Shannon went to the Musée National de Préhistoire (MNP) in Les Eyzies, France, to study handaxes from Mousterian of Acheulian Tradition (MTA) contexts as a follow-up to his dissertation on Acheulian handaxes (McPherron 1994; Iovita and McPherron 2011). One of the sites that interested him was Pech de l'Azé IV where in Layer F, at the top of the sequence, Bordes had identified a progression of MTA Type A to B. It turned out, however, that the collection was not at the museum. After a few phone calls, it was eventually determined that the Pech IV collections were housed at (what was then called) the Institut de Préhistore et de Géologie du Quaternaire (IPGQ) at the University of Bordeaux I, where Bordes had been professor (subsequently they were transferred to the Musée National de Préhistoire). However, there was a certain degree of pessimism about the research value of the collections. Indeed, with the exception of a limited study of the faunal assemblage by Laquay (1981), not much had been done with this material since it was excavated.

A few weeks after Shannon's visit to Les Eyzies, he and Harold were together on the island of Jersey to study collections from the site of La Cotte de St. Brelade. Over pints one night in a Jersey pub, we discussed the collection and the status of the site itself. Harold knew the site well—as a student he had excavated there with Bordes in 1976 and 1977, the last two years of the project (Fig. 1), and remembered it as having significant potential. Much of the evening was spent recounting those times—camping in the backyard of Bordes' house in the village of Carsac, the primitive (and nearly full) outhouse, baths in the Dordogne, lunches delivered to the site by the Hotel Delpeyrat, and dinners in the hotel's restaurant. We were, however, concerned about the fact that many people were dismissive of the site's research value, but

**Fig. 1** François Bordes at Pech IV, in 1976 (**A, B, and C**), and during his last day of excavation at the site in 1977 (**D**). Dibble is pictured in A

decided to look into the status of the collection with the idea that it might make a good future project.

A little while later we visited the IPGQ and took a quick inventory of the Pech IV collection. It was a large collection—around 90,000 stone tools and 30,000 pieces of bone, plus bags of bulk-provenienced material. As discussed in Chap. 1, the collection was not yet ready to be studied. Bordes had processed all the material from the first 4 years of excavations and sorted the lithics into boxes labeled with his type number and stored in trays with label indicating from which layer they had come. The rest of the collection, however, was unsorted, and a substantial portion was unwashed, unlabeled, and seriously at risk of becoming effectively lost to science. Thus, we requested permission from Jean-Philippe Rigaud, the then director of the institute, to begin organizing, processing, and analyzing the collection. We were aided considerably by Dominique Armand, with whom we had worked previously at La Quina, and who was at this time responsible for the curation of the various collections housed at the IPGQ. As part of this initial project, we digitally scanned all of the field notebooks, entered the coordinates of each object into a database, and washed and labeled the unprocessed objects.

The processing of this material, along with detailed analysis of the lithics, took place over the course of several years, between 1995 and 1999, during our work at Fontéchevade. This was possible simply because Fontéchevade yielded very few artifacts, and our focus there was primarily on site formation, which left time to do other research projects. Thus, with permission from the IPGQ, we took portions of the Pech IV material to our dighouse in Orgedeuil, near Fontéchevade, and as the excavations there proceeded, processed, and analyzed it.

Fig. 2 Analyzing the Pech IV lithics at the Fontéchevade dighouse in Orgedeuil (1999)

This continued through 1999, and the work on both sites provided training opportunities for many graduate students (Fig. 2).

Early on in this process of working with the Pech IV collections we learned that the original notebooks—in fact, the only copies of the entire documentation of Bordes' work there—were still with Denise de Sonneville-Bordes, Bordes' widow and herself the former head of the IPGQ. Harold had known de Sonneville-Bordes since his graduate school days excavating at Pech IV, but also from the close relationship the Bordes had with his professor, Art Jelinek, at the University of Arizona. Subsequently, we both saw her regularly, since she often visited our previous excavations at Combe-Capelle Bas, Cagny, and Fontéchevade. Still, she had a reputation for fiercely guarding the legacy of her late husband and her relationship with the department that housed the collections was an uneasy one. At that time she was also dealing with requests from the MNP to move all of F. Bordes' collections there. So, it was with some apprehension that we arranged to visit her and request a copy of the Pech IV notebooks. In fact, the visit went well and she graciously allowed us access to them. With help from our colleagues at Bordeaux, we immediately found a student (Abdeljalil Bouzzougar) who made three photocopies of all of the notebooks, giving the originals and one copy back to de Sonneville-Bordes (Fig. 3).

Entering the data from the scanned notebooks resulted in an additional job in the Fontéchevade lab. Each day one student would sit at a computer running an older version of our data entry software and enter the square, id, type of object, X, Y, and Z coordinates (including the local Z used for that particular day of excavation), the layer (where available), the date when the objects were recovered, and the name of the excavator (Fig. 4). This was done for each of the roughly 120,000 provenienced artifacts. Because the XY coordinates were local to each

Fig. 3 Denise de Sonneville-Bordes during one of her annual visits to the site (2002) (also pictured are Shannon J.P. McPherron far left and Harold L. Dibble)

square (measured from the west and south square boundary, respectively), and because the Z measurements were relative to a temporary datum that changed periodically, we had to transform them to a grid system for the site as a whole. After some effort, we were able to convert the notebooks into the same database format we use for our own excavations, and with this notebook information we were able to assign levels to the rest of the collection.

## Why Re-excavate Pech IV?

In 2000, the same year we started excavating Pech IV, we published our study of Bordes' Pech IV lithic collections (McPherron and Dibble 2000), and in that paper we outlined a number of reasons why we thought it was necessary to go back to the site. First, it was important to date the sequence. With his knowledge of the Pech I, II, and IV sequences, their respective lithic and faunal collections, and the sedimentological work of Laville (1973), Bordes put together a chronology for the three major Pech sites, I, II, and IV (Bordes 1978). Based on correlations with Pech II, which had already been dated using electron spin resonance (Grün et al. 1991, 1999), most of the Pech IV sequence would date to between 87 and 54 ka with the top part of the sequence, the MTA, dating to between then and the end of the Mousterian in southwest France. In re-excavating the site we hoped to test these conclusions and to bring some further precision to the dating of the various layers. In the end, we were able to apply electron spin resonance, radiocarbon, thermoluminescence, and optically-stimulated

Fig. 4 Flint Dibble at Fontéchevade entering from Bordes' excavation notebooks

luminescence dating techniques to the entire sequence. The results of this work are reviewed in Chap. 3.

A second reason to go back to Pech IV was the fact that the geology of the site, as done by Laville, remained unpublished. In fact, the only information available at the time came from Bordes' (1975) publication of the sequence based on the first 4 years of his excavation. But while Laville's work focused on establishing chronologies among sites in the Perigord, there had been virtually nothing done on the formation processes that may have affected the deposits. Of course, at the time of the site's initial excavation, studies of site formation processes were not yet part of most Paleolithic fieldwork. This did not mean that Bordes was unaware of issues surrounding the integrity of the deposits, but at the same time he did not systematically study this aspect on the site. However, since then site formation had become quite an issue in Paleolithic archaeology, and this topic had already developed into a major focus our own work at Combe-Capelle Bas (Dibble and Lenoir 1995; McPherron et al. 2005), Cagny l'Epinette (Dibble et al. 1997), and Fontéchevade (Chase et al. 2007; Chase et al. 2009; Dibble et al. 2006; McPherron et al. 2012). Thus, we felt it important to make this one of the major foci of the new excavations as well.

There was another reason why a thorough study of site formation was important for Pech IV. We mentioned earlier that there was some pessimism expressed concerning the potential of Pech IV. Much of this was expressed as rumors circulating that Bordes had missed important lateral changes in the sequence that resulted in his mixing two different and independent depositional sequences. This meant that the site's formation may have been much more difficult that Bordes had realized, even though he had noted other problems, including differences in the horizontal extent of cryoturbation in some layers, lateral variation within some layers between deposits closer to the front of the site and those situated toward (what he thought was) the back wall of the shelter, and the difficulty of making stratigraphic distinctions in the lowest layers. Although some information was available based on analysis of the objects and their positions within the three-dimensional grid system, a full site formation study, based in part on applying techniques we developed earlier, could only be undertaken with a new excavation.

We were fortunate in this regard to enlist the aid of Paul Goldberg (Fig. 5). Paul had also previously worked at Pech de l'Azé, but at Pech II, not IV, as part of his post-doctoral work on developing a methodology for doing micromorphology. This has become a particularly useful tool for studying site formation. He also had spent many years working at sites in Israel, especially at Kebara and Hayonim, which, like the lower deposits of Pech, had abundant evidence of the use of fire. Like Harold, the idea of essentially going back to where he had started his career was appealing to him, and his experience in studying fire use ultimately proved to be a major contribution at Pech IV and our later work at other sites. His study of the Pech IV sediments is presented in Chap. 2, along with the results of our work on site formation based on both geological and lithic evidence.

A fourth reason for re-opening Pech IV had to do with the issue of excavator bias. Based on his own work at the site in the 1970s, Harold knew that Bordes did not screen the sediment, and, furthermore, there were substantial inter-excavator differences in collection protocols—what kinds of objects were point-provenienced, what ones were just put into bags by square and level, and even which ones were discarded were, for the most part, left to the discretion of each excavator, who at the same time had varied amounts of experience in archaeological fieldwork. Of course, such variation can have a significant effect on the resulting collections, as we had already learned at Combe-Capelle (see Dibble and Lenoir 1995). So, by adhering to

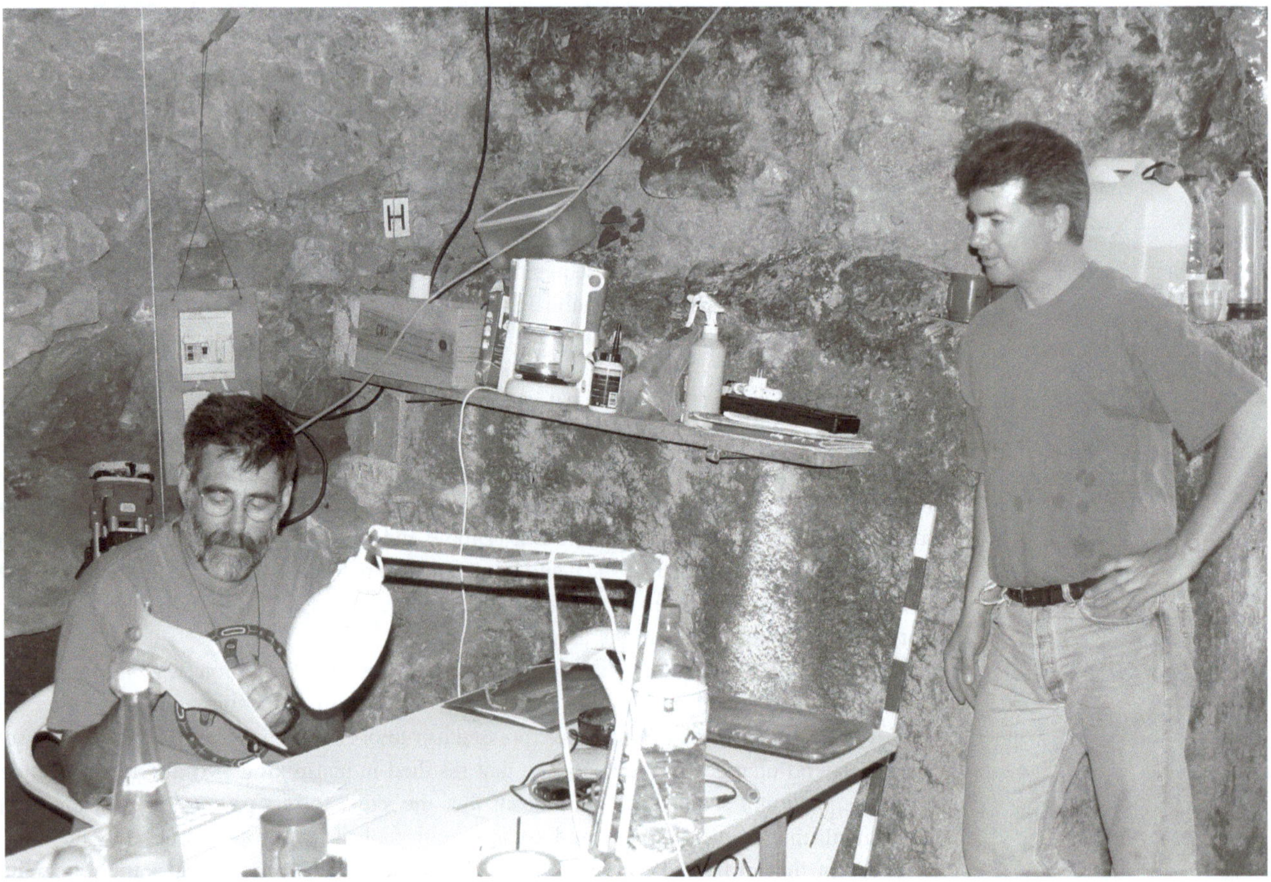

**Fig. 5** Paul Goldberg (*left*) and Dennis M. Sandgathe at Pech IV

strict excavation standards, as described in Chap. 1, and by digging some of Bordes' backdirt from Pech IV (as we had also done at Combe-Capelle), we could get an idea of the extent to which his collection had been affected (see Dibble et al. 2005).

Thus with these goals in mind, we received a 4-year, National Science Foundation grant, which when combined with funding from the Service Régional de l'Archéologie (Aquitaine) allowed us to excavated Pech IV from 2000 to 2003. Again, it proved helpful that Harold had worked at the site previously with Bordes during his final two seasons, since he was able to apply some of his initial understanding of the site, including some knowledge of the stratigraphic sequence as interpreted by Bordes. However, with Paul in the lead, we approached the site with fresh eyes and interpreted for ourselves the nature of the stratigraphic succession. We were also fortunate in locating Bordes' original site datum, allowing us to base the new excavation on the same coordinate system in all three dimensions and thus accurately integrate the new excavation with the former one.

Dennis M. Sandgathe joined our team in the first year of excavation at Pech IV. At that time, Dennis was a graduate student at Simon Frasier University, writing his dissertation on Levallois technology. Most of his previous archaeological fieldwork had been in the North American plains, at sites such as Head Smashed In Buffalo Jump, but he had also worked in Spain with Lawrence Straus at the Upper Paleolithic site of El Miron. Apart from his knowledge of lithics, we could see from the start that he had two qualities that made him stand out—he was an excellent excavator and he was able to work extremely well with the other student excavators. Thus, by the second season at Pech IV he became the site supervisor, a role that he continued to have at our later excavations at Roc de Marsal and La Ferrassie.

Most of the analysis was completed soon after the excavations stopped in 2003, but it took several more years to complete the final analysis. In 2005 and 2006 Daniel Richter placed additional dosimeters in the site for thermoluminescence dating, and in 2014 Zenobia Jacobs took samples for optically stimulated luminescence. In 2009, we worked at the site to clean some material that had fallen from the section, and in 2012 we went back to collect new samples for OSL dating (see Jacobs et al. (2016) and Chap. 3).

We have made a considerable effort to prepare the new collection for final curation. After having seen the excellent curation of the collections from La Cotte de St. Brelade in the Jersey Museum and Art Gallery at St. Helier, we decided to implement a similar system of individually bagging every provenienced artifact at Pech IV. In addition, we developed a barcode program for labeling these bags (see Dibble et al. 2007) and a computerized system for locating specific objects in storage. All of the excavated material is now housed in the MNP in Les Eyzies.

## A Preview of This Monograph

In our view a site report, such as this one, is a presentation of the excavation and the resulting data. The emphasis is on providing context of the collections and on providing basic descriptions of the material. A site report is not a place for detailed analyses of the collections; such publications belong in peer-reviewed journals (see list below of our previous publications on Pech IV), for no other reason than that specific analyses will change in focus through time as research questions change. To a very large degree, then, this volume presents the basic metadata essential for future studies by others of the objects themselves or their analyses of the analytical data that we have made publicly available.

The flow of this presentation is as follows. It begins, in Chap. 1, with an overview of Bordes' excavation of Pech IV and our subsequent work on his collection. In this chapter, we also present our own excavations of the site with our methodology and a narrative of the work. Chapter 2 covers the geology with an emphasis on the context of the site, description of the layers, and the site formation processes. For the latter, we include observations made on the artifacts themselves as well as observations of the sediments. An effort is made to unite these various lines of evidence, to see where the interpretations converge and where they may differ, something which potentially speaks to separate formation processes for the artifacts and the sediments. Chapter 3, by Nathan Jankowski, gives an overview of the dates for the sequence, all of which have been previously published elsewhere. The dating methods employed include radiocarbon dating of bone, thermoluminescence dating of heated flints, electron spin resonance dating of teeth, and optically stimulated luminescence dating of sediments. Chapters 4 and 5 present descriptions of the faunal assemblages: Chap. 4, by Jamie Hodgkins, presents a study of the fauna from two layers from Bordes' excavations from the perspective of zooarchaeology and taphonomy, and Chap. 5 is a report by Laura Niven (with contributions by Hélène Martin) of the fauna recovered during our own excavations. Chapter 6 describes the lithic assemblages. Bordes' collection (McPherron and Dibble 2000) and our collection (Turq et al. 2011) have already been published separately. Here we put the two collections together and examine changes through time. Chapter 7 provides a summary of all of these results.

Again, by no means do we consider this monograph to be the last word on Pech IV. First, this is a site that spans a long period of the late Upper Pleistocene, one that was excavated with modern techniques, and one that is well dated. There is undoubtedly much that can still be learned from its abundant collections about Neandertal behavior, and we encourage others to continue to make use of them in ways that go well beyond what we thought we might learn when this project started. In fact, the subject of Neandertal use of fire is a good example of how sites can continue to provide new evidence on questions that were not thought of when they were first excavated. Although we mentioned the presence of fire in the Pech IV sequence in our initial NSF proposal, we had no idea how interesting it would become. This was true in part because of the finds at Pech IV but also as a result of excavations we did subsequently, particularly at Roc de Marsal, France, where fire is also clearly evident at certain moments in the sequence. Explaining why fire comes and goes in late Mousterian sites in southwest France has become an important line of research for us (Dibble et al. 2009; Dibble et al. CA volume; Goldberg et al. 2012; Sandgathe et al. 2011), and given what we know now would have been one of the leading reasons for excavating the site. In fact, as of this writing we are planning a new project at the site, focusing on the lowermost layer, Layer 8, to further study how the numerous fires found in this level were used.

## Final Words

The core members of this team have continued to work together at many sites—Roc de Marsal, Abri Peyrony and La Ferrassie in France, Contrebandiers in Morocco, and on a survey project in Egypt, and we hope to continue to work together for as long as we are able. We look forward to continuing our collaborations with many of the senior researchers who worked with us at Pech IV and also to developing new collaborations with others. And not the least, we will continue to invite students from all over the world to join us in order to get experience and training in field archaeology and analytical methods (Fig. 6).

We would like to thank the following individuals for their assistance. Our biggest thanks go to the late Mme. de Sonneville-Bordes for having made this project possible. It has to be said that we were still apprehensive when she visited our excavations each year, but we will always be grateful for the kindness and trust she showed in us. We would also like to thank the Laplanche family for generously allowing us to work for years on their property, and the mayor of Carsac, who helped us find suitable housing in the village to serve as our dighouse and laboratory. A very special thanks goes to the entire Delpeyrat family for making us feel so at home in Carsac and who became a critical part of our Pech IV team over the years. Thanks also to Jean-Michel Geneste, Jean-Jacques Cleyet-Merle, and especially Alain Turq, who has been a close friend and colleague for over three decades. We thank Larry Bartram, Roland Nespoulet, Marie Soressi, and Laurent Chiotti for helping with our topographic surveys of the Pech de l'Azé sites.

Fig. 6 Crew photos from (clockwise from *upper left*) 2000, 2001, 2002, and 2003. Not all of the participants are included due to their schedules

Funding for the excavation and analysis of Pech IV came from the National Science Foundation, the Service Régional de l'Archéologie (SRA) for Aquitaine, the Leakey Foundation and the Max Planck Society. From the SRA we would like to especially thank Dany Baraud for his support of our work at Pech IV. Though she was not there for the initial excavation, many thanks go to Virginie Sinet-Mathiot who eventually became our lab manager and put the Pech IV collection into great shape to enter the museum. Finally, a project like this is not possible without the help of the more than 70 students who volunteered their time and resources to participate in the project. We are truly thankful for their efforts.

March, 2017

Harold L. Dibble
Shannon J.P. McPherron

## References

Bordes, F. (1975). Le gisement du Pech de l'Azé IV. Note Préliminaire. *Bulletin de la Société Préhistorique Française*, *72*, 293–308.

Bordes, F. (1978). Typological variability in the Mousterian layers at Pech de l'Azé I, II, and IV. *Journal of Anthropological Research*, *34*, 181–193.

Chase, P.G., Debénath, A., Dibble, H.L., & McPherron, S.P. (2009). *The Cave of Fontéchevade: A new investigation of the site and its paleoanthropological implications*. Cambridge: Cambridge University Press.

Chase, P.G., Debénath, A., Dibble, H.L., McPherron, S.P., Schwarcz, H.P., Stafford Jr., T.W., & Tournepiche, J.-. (2007). New dates for the Fontéchevade (Charente, France) Homo remains. *Journal of Human Evolution, 52*, 217–221.

Dibble, H., Marean, C., & McPherron, S. (2007). On the use of barcodes in excavation projects with examples from Mossel bay (South Africa) and Roc de Marsal (France). *The SAA Archaeological Record, 7*, 33–38.

Dibble, H.L., Berna, F., Goldberg, P., McPherron, S.P., Mentzer, S., Niven, L., Turq, A. (2009). A preliminary report on Pech de l'Azé IV, layer 8 (Middle Paleolithic, France). *PaleoAnthropology, 2009*, 182–219.

Dibble, H.L., Chase, P.G., McPherron, S.P., & Tuffreau, A. (1997). Testing the reality of a 'living floor' with archaeological data. *American Antiquity, 62*, 629–651.

Dibble, H.L., & Lenoir, M. (1995). *The middle paleolithic site of Combe-Capelle Bas (France)*. Philadelphia: The University Museum Press.

Dibble, H.L., McPherron, S.P., Chase, P., Farrand, W., & Debénath, A. (2006). Taphonomy and the concept of paleolithic cultures: The case of the Tayacian from Fontéchevade (Charente, France). *PaleoAnthropology*, 1–21.

Dibble, H.L., Raczek, T., & McPherron, S.P. (2005). Excavator bias at the site of Pech de l'Azé IV, France. *Journal of Field Archaeology, 30*, 317–328.

Goldberg, P., Dibble, H., Berna, F., Sandgathe, D., McPherron, S.J.P., & Turq, A. (2012). New evidence on Neandertal Bordes, F. (1975). Le gisement du Pech de l'Azé IV. Note Préliminaire. *Bulletin de la Société Préhistorique Française, 72*, 293–308.

Goldberg, P., Dibble, H., Berna, F., Sandgathe, D., McPherron, S.J.P., & Turq, A. (2012). New evidence on Neandertal use of fire: Examples from Roc de Marsal and Pech de l'Azé IV. *The Neanderthal Home: spatial and social behaviours, 247*, 325–340. doi:10.1016/j.quaint.2010.11.015

Grün, R., Mellars, P., & Laville, H. (1991). ESR chronology of a 100,000-year archaeological sequence at Pech de l'Azé II, France. *Antiquity, 65*, 544–551.

Grün, R., Yan, G., McCulloch, M.T., & Mortimer, G. (1999). Detailed mass spectrometric U-series analyses of two teeth from the archaeological site of Pech de l'Aze II: Implications for uranium migration and dating. *Journal of Archaeological Science, 26*, 1301–1310.

Iovita, R., & McPherron, S.P. (2011). The handaxe reloaded: A morphometric reassessment of Acheulian and Middle Paleolithic handaxes. *Journal of Human Evolution, 61*, 61–74. doi:10.1016/j.jhevol.2011.02.007

Jacobs, Z., Jankowski, N.R., Dibble, H.L., Goldberg, P., McPherron, S.J.P., Sandgathe, D., & Soressi, M. (2016). The age of three Middle Palaeolithic sites: Single-grain optically stimulated luminescence chronologies for Pech de l'Azé I, II and IV in France. *Journal of Human Evolution, 95*, 80–103. doi:10.1016/j.jhevol.2016.03.010

Laquay, G. (1981). Recherches sur les faunes du Würm I en Périgord. Bordeaux: Université de Bordeaux I.

Laville, H. (1973). *Climatologie et chronologie du Paléolithique en Périgord : étude sédimentologique de dépôts en grottes et sous abris*. Talence: Université de Bordeaux I.

McPherron, S.P. (1994). *A reduction model for variability in Acheulian biface morphology*. University of Pennsylvania.

McPherron, S.P., & Dibble, H. L. (2000). The lithic assemblages of Pech de l'Azé IV (Dordogne, France). *Préhistoire Européenne, 15*, 9–43.

McPherron, S.P., Dibble, H.L., Chiotti, L., Chase, P., Debénath, A., & Farrand, W.R. (2012). Processus de formation des sites et concept du Tayacien : l'exemple de Fontéchevade (Charente, France). *Acheuléen, industrie à bifaces, 116*, 321–347. doi:10.1016/j.anthro.2012.05.002

McPherron, S.P., Dibble, H.L., & Goldberg, P. (2005). Z. *Geoarchaeology, 20*, 243–262. doi:10.1002/gea.20048

Sandgathe, D.M., Dibble, H.L., Goldberg, P., McPherron, S.P., Turq, A., Niven, L., & Hodgkins, J. (2011). On the role of fire in Neandertal adaptations in western Europe: Evidence from Pech de l'Azé and Roc de Marsal, France. *PaleoAnthropology, 2011*, 216–242.

Turq, A., Dibble, H. L., Goldberg, P., McPherron, S. P., Sandgathe, D., Jones, H., Steenhuyse, A. (2011). Les Fouilles Récentes du Pech de l'Azé IV (Dordogne). *Gallia Préhistoire, 53*.

# Publications to Date Resulting from the Recent Excavations at Pech de l'Azé IV

There have been a number of publications based on the work done to date listed in chronological order of their publication date. The present volume relies heavily on these publications but is meant to be the complete record of our work at the site.


Dibble, H. L., Aldeias, V., Goldberg, P., McPherron, S. P., & Sandgathe, D. M. (2017). How did Hominins Adapt to Ice Age Europe without Fire? *Current Anthropology, 58.*

Dibble, H. L., Berna, F., Goldberg, P., McPherron, S. P., Mentzer, S., Niven, L., . . . Turq, A. (2009). A Preliminary Report on Pech de l'Azé IV, Layer 8 (Middle Paleolithic, France). *PaleoAnthropology, 2009*, 182–219.

Dibble, H. L., & McPherron, S. P. (2001). Pech de l'Azé IV *Bilan Scientifique 2000* (pp. 22–23): Direction Régionale des Affaires Culturelles de l'Aquitaine et Service Régional de l'Archéologie.

Dibble, H. L., & McPherron, S. P. (2003). Pech de l'Azé IV *Bilan Scientifique 2001* (pp. 24–30): Direction Régionale des Affaires Culturelles de l'Aquitaine et Service Régional de l'Archéologie.

Dibble, H. L., & McPherron, S. P. (2004). Pech de l'Azé IV *Bilan Scientifique 2002* (pp. 10–20): Direction Régionale des Affaires Culturelles de l'Aquitaine et Service Régional de l'Archéologie.

Dibble, H. L., & McPherron, S. P. (2006). The Missing Mousterian. *Current Anthropology, 47*, 777–803.

Dibble, H. L., & McPherron, S. P. (2007). Truncated-faceted Pieces: Hafting Modification, Retouch, or Cores? In S. P. McPherron (Ed.), *Tool v. Core: New Approaches in the Analysis of Stone Tool Assemblages*. Cambridge: Cambridge Scholars Publications.

Dibble, H. L., McPherron, S. P., & Goldberg, P. (2005). Pech de l'Azé IV *Bilan Scientifique 2003* (pp. 22–242): Direction Régionale des Affaires Culturelles de l'Aquitaine et Service Régional de l'Archéologie.

Dibble, H. L., Raczek, T., & McPherron, S. P. (2005). Excavator bias at the Site of Pech de l'Azé IV, France. *Journal of Field Archaeology, 30*, 317–328.

Hodgkins, J., Marean, C. W., Turq, A., Sandgathe, D., McPherron, S. J. P., & Dibble, H. (2016). Climate-mediated shifts in Neandertal subsistence behaviors at Pech de l'Azé IV and Roc de Marsal (Dordogne Valley, France). *Journal of Human Evolution, 96*, 1–18. doi:10.1016/j.jhevol.2016.03.009

Jacobs, Z., Jankowski, N. R., Dibble, H. L., Goldberg, P., McPherron, S. J. P., Sandgathe, D., & Soressi, M. (2016). The age of three Middle Palaeolithic sites: Single-grain optically stimulated luminescence chronologies for Pech de l'Azé I, II and IV in France. *Journal of Human Evolution, 95*, 80–103. doi:10.1016/j.jhevol.2016.03.010

Lin, S. C., McPherron, S. P., & Dibble, H. L. (2015). Establishing statistical confidence in Cortex Ratios within and among lithic assemblages: a case study of the Middle Paleolithic of southwestern France. *Journal of Archaeological Science, 59*, 89–109. doi:10.1016/j.jas.2015.04.004

Lin, S. C., Rezek, Z., Braun, D., & Dibble, H. L. (2013). On the Utility and Economization of Unretouched Flakes: The Effects of Exterior Platform Angle and Platform Depth. *American Antiquity, 78*, 724–745.

McPherron, S., Dibble, H., Goldberg, P., Lenoir, M., Sandgathe, D., & Turq, A. (2012). De Combe Grenal à Pech de l'Azé IV : L'Evolution des Méthodes de fouilles de François Bordes. In F. Delpech & J. Jaubert (Eds.), *François Bordes et La Préhistoire* (pp. 117–124). Bordeaux: Éditions du Comité des travaux historiques et scientifiques.

McPherron, S. J. P. (2005). Artifact orientations and site formation processes from total station proveniences. *Journal of Archaeological Science, 32*, 1003–1014. doi:10.1016/j.jas.2005.01.015

McPherron, S. P., & Dibble, H. L. (2000). The Lithic Assemblages of Pech de l'Azé IV (Dordogne, France). *Préhistoire Européenne, 15*, 9–43.

McPherron, S. P., Dibble, H. L., & Goldberg, P. (2005). Z. *Geoarchaeology, 20*, 243–262. doi:10.1002/gea.20048

McPherron, S. P., Soressi, M., & Dibble, H. L. (2001). Deux nouveaux projets de recherche á Pech de l'Azé (Dordogne, France). *Préhistoire du Sud-Ouest, 8*, 11–30.

McPherron, S. P., Talamo, S., Goldberg, P., Niven, L., Sandgathe, D., Richards, M. P., Dibble, H. L. (2012). Radiocarbon dates for the late Middle Palaeolithic at Pech de l'Azé IV, France. *Journal of Archaeological Science, 39*, 3436–3442. doi:10.1016/j.jas.2012.05.017

Niven, L. (2013). A diachronic evaluation of Neanderthal cervid exploitation and site use at Pech de l'Azé IV, France. In J. Clark & J. D. Speth (Eds.), *Zooarchaeology and Modern Human Origins: Human Hunting Behavior during the Later Pleistocene*. Dordrecht: Springer Verlag.

Richter, D., Dibble, H., Goldberg, P., McPherron, S. P., Niven, L., Sandgathe, D., . . . Turq, A. (2013). The late Middle Palaeolithic in Southwest France: New TL dates for the sequence of Pech de l'Azé IV. *Quaternary International, 294*, 160–167. doi:10.1016/j.quaint.2012.05.028

Sandgathe, D. M., Dibble, H. L., Goldberg, P., McPherron, S. P., Turq, A., Niven, L., & Hodgkins, J. (2011). On the Role of Fire in Neandertal Adaptations in Western Europe: Evidence from Pech de l'Azé and Roc de Marsal, France. *PaleoAnthropology, 2011*, 216–242.

Sandgathe, D. M., Dibble, H. L., Goldberg, P., McPherron, S. P., Turq, A., Niven, L., & Hodgkins, J. (2011). Timing of the appearance of habitual fire use. *Proceedings of the National Academy of Sciences, 108*, E298. doi:10.1073/pnas.1106759108

Turq, A., Dibble, H. L., Goldberg, P., McPherron, S. P., Sandgathe, D., Jones, H., . . . Steenhuyse, A. (2011). Les Fouilles Récentes du Pech de l'Azé IV (Dordogne). *Gallia Préhistoire, 53*.

# Contents

# Editors and Contributors

## About the Editors

**Harold L. Dibble** is the Francis E. Johnston Professor in the Department of Anthropology at the University of Pennsylvania, a Research Associate in the Department of Human Evolution at the Max Planck Institute for Evolutionary Anthropology (Leipzig, Germany), and a Research Affiliate, Institute of Human Origins, Arizona State University. He obtained his Ph.D. in Anthropology from the University of Arizona in 1981. He has excavated a number of Paleolithic sites in France (Combe-Capelle Bas, Cagny-l'Epinette, Fontéchevade, Pech de l'Azé IV, Roc de Marsal, La Gane, and La Ferrassie), Contrebandiers Cave in Morocco, and co-directed a survey for Paleolithic sites in Egypt. He is also director of the Laboratory for the Study of Ancient Technology at the University of Pennsylvania. His research primarily focuses on Middle Paleolithic, lithic analysis, lithic experimentation, and computer applications to archaeology.

**Shannon J.P. McPherron** is a senior researcher in the Department of Human Evolution at the Max Planck Institute for Evolutionary Anthropology in Leipzig, Germany. He obtained his B.A. in anthropology from the University of Arizona, Tucson, Arizona, and a Ph.D. degree in anthropology from The University of Pennsylvania. He is a Paleolithic archaeologist primarily interested in the evolution of hominin cultural abilities from the origins of stone tool use through to the dispersal of modern humans. He has excavated a number of Paleolithic sites main in France and North Africa. He currently involved in field projects in central Germany (Ranis), Bulgaria (Bacho Kiro), Morocco (Jebel Irhoud), and Ethiopia (Dikika and Mille-Logia Research Projects). He has also worked extensively on digital methods for excavation, artifact analysis, and the distribution of archaeological data.

**Paul Goldberg** is Professor Emeritus in the Department of Archaeology, Boston University. He obtained his B.A. in geology from the University of Colorado, Boulder, and M.S. and Ph.D. degrees in geology from The University of Michigan. He is currently Professorial Research Professor in the Centre for Archaeological Science at the University of Wollongong, and member of the Center for Archaeological Science at the University of Tübingen. His research interests focus on the application of micromorphology to the study of landscapes, soils, and site formation processes at archaeological sites. Most of his research is in the Old World, particularly cave sediments from China, France, Germany, Spain, Israel, and most recently Flores and Siberia.

**Dennis M. Sandgathe** is a full-time lecturer in the Department of Archaeology at Simon Fraser University in British Columbia and a Consulting Scholar with the University of Pennsylvania Museum of Archaeology and Anthropology. He received a B.A. Honours in Archaeology from the University of Calgary in 1995, an M.A. in Anthropology from the University of Alberta in 1999, and a Ph.D. in Archaeology from Simon Fraser University in 2005. He held a Social Sciences and Humanities Research Council of Canada postdoc at the University of Pennsylvania from 2006 to 2008. His early research was on the Early Prehistoric period of the Northern Plains of North America but subsequently has focussed primarily on the Paleolithic of Europe. He has excavated several Middle Paleolithic sites in France (Pech de l'Azé IV, Roc de Marsal, La Gane, and La Ferrassie) as well as El Miron in Northern Spain and Grotte de Contrebandiers in Morocco. His primary areas of interest are the role of stone tools in hominin adaptations, the evolution of the use of fire, Neandertal bio-cultural evolution, and other aspects of Paleolithic adaptations.

## Contributors

**Harold L. Dibble** Department of Anthropology, University of Pennsylvania, Philadelphia, USA; Department of Human Evolution, Max Planck Institute for Evolutionary Anthropology, Leipzig, Germany; School of Human Evolution and Social Change, Institute for Human Origins, Arizona State University, Tempe, AZ, USA

**Paul Goldberg** School of Earth and Environmental Sciences, Centre for Archaeological Science, University of Wollongong, Wollongong, NSW 2522, Australia; Institute for Archaeological Sciences, Eberhard Karls Universitat Tübingen, Tübingen, Germany; Department of Archaeology, Boston University, Boston, USA

**Jamie Hodgkins** Department of Anthropology, University of Colorado, Denver, CO, USA

**Nathan R. Jankowski** School of Earth and Environmental Sciences, Centre for Archaeological Science, University of Wollongong, Wollongong, NSW, Australia

**Sam C. Lin** School of Earth and Environmental Sciences, Centre for Archaeological Science, University of Wollongong, Wollongong, NSW, Australia

**Hélène Martin** INRAP-Centre Archéologique de Saint-Orens, St.-Orens de Gameville, France

**Shannon J.P. McPherron** Department of Human Evolution, Max Planck Institute for Evolutionary Anthropology, Leipzig, Germany

**Laura Niven** Annual Reviews, Palo Alto, CA, USA

**Dennis M. Sandgathe** Department of Archaeology and Human Evolution Studies Program, Simon Fraser University, Burnaby, BC, Canada; Museum of Archaeology and Anthropology, University of Pennsylvania, Philadelphia, USA

# Introduction

# 1

Dennis M. Sandgathe, Harold L. Dibble, Shannon J. P. McPherron, and Paul Goldberg

## The Pech de l'Azé Sites

The Middle Paleolithic site of Pech de l'Azé IV (Pech IV) is one of a cluster of four independent Lower and Middle Paleolithic sites located in the Perigord region of southwest France (Figs. 1.1, 1.2 and 1.3). They are situated about 50 m above the floor of a small, usually dry, valley formed by the Enéa, a small tributary of the Dordogne River. Pech I and II are opposite entrances of a single tunnel-like cave that cuts through a promontory in the limestone cliff. Pech III is a small cave in the same cliff located about 30 m west of the opening of Pech II. Pech IV, located roughly 80 m east of the mouth of Pech I, is a collapsed cave situated at the foot of the cliff.

The history of research at this complex of sites extends back to virtually the beginning of the discipline of Paleolithic archaeology. Pech I, or "Pey de l'Azé", as it was then spelled (this translates in English to 'Hill of the Donkey'), was initially excavated early in the nineteenth century by Jouannet and later by the Abbé Audierne (Bordes 1954), and was one of the sites described by Lartet and Christy (1864) in their seminal "Cavernes du Périgord". At some point during the nineteenth century most of the archaeological material inside the cave (identified as the Pech Ia locality; Bordes 1954) was removed by pothunters. However, in 1909, at the base of the cliff on the terrace just outside the cave, Peyrony discovered the cranium of a Neandertal child that had died around age five or six (Bordes 1954; Capitan and Peyrony 1909; Ferembach et al. 1970; Patte 1957; Maureille and Soressi 2000). In 1929–30 Vaufrey (1933) excavated in the terrace outside of the entrance to the cave (Pech Ib) and identified the sequence as containing only assemblages attributable to the Mousterian of Acheulian Tradition. Later, from 1948 until 1951, more excavations were carried out by Bordes (1954) in the same area. Most recently, Soressi excavated deposits in front of the site from 2004–5 (Soressi et al. 2002, 2007, 2008, 2013) (Fig. 1.4). These deposits were recently dated with optically-stimulated luminescence to roughly 51–48 ka (Jacobs et al. 2016).

Pech II was discovered by Bordes in 1948, thanks to the fact that some of the talus of the site had been cut away in the construction of a rail line that ran parallel to the cliff at this point. He excavated there from 1949 to 1951 and again from 1967 to 1969 (Fig. 1.5). Both outside the mouth of the cave (locality Pech IIb) and within the cave itself (Pech IIa), an occupational sequence began with the so-called Meridional Acheulian, followed by a variety of Mousterian industries (Bordes 1972). Schwarcz and Blackwell (1983) published two U-series dates from Pech II and two from Pech I, and Grün et al. (1991) published a series of ESR

D.M. Sandgathe (✉)
Department of Archaeology and Human Evolution Studies Program, Simon Fraser University, Burnaby, BC, Canada
e-mail: dms@sfu.ca

D.M. Sandgathe
Museum of Archaeology and Anthropology, University of Pennsylvania, Philadelphia, USA

H.L. Dibble
Department of Anthropology, University of Pennsylvania, Philadelphia, USA
e-mail: hdibble@sas.upenn.edu

H.L. Dibble · S.J.P. McPherron
Department of Human Evolution, Max Planck Institute for Evolutionary Anthropology, Leipzig, Germany
e-mail: mcpherron@eva.mpg.de

H.L. Dibble
School of Human Evolution and Social Change, Institute for Human Origins, Arizona State University, Tempe, AZ, USA

P. Goldberg
School of Earth and Environmental Sciences, Centre for Archaeological Science, University of Wollongong, Wollongong, NSW 2522, Australia
e-mail: paulberg@bu.edu

P. Goldberg
Institute for Archaeological Sciences, Eberhard Karls Universitat Tübingen, Tübingen, Germany

P. Goldberg
Department of Archaeology, Boston University, Boston, USA



H.L. Dibble et al. (eds.), *The Middle Paleolithic Site of Pech de l'Azé IV*,
Cave and Karst Systems of the World, https://doi.org/10.1007/978-3-319-57524-7_1

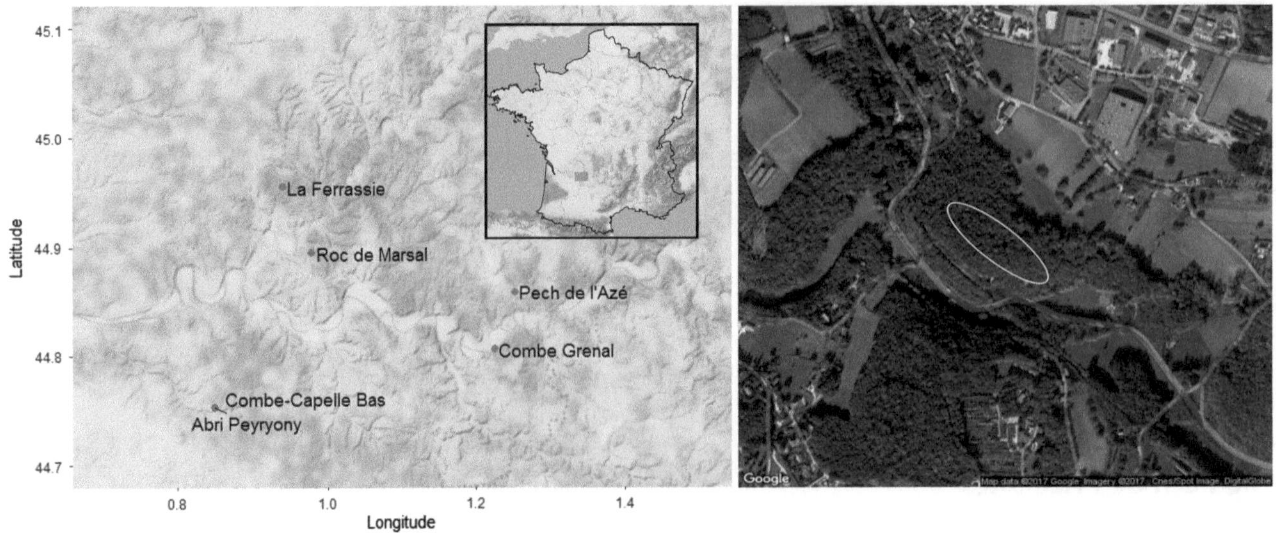


**Fig. 1.1** The location of Pech IV in southwest France (*left*) and satellite view of the hill containing the Pech sites (approximate location noted with ellipse)

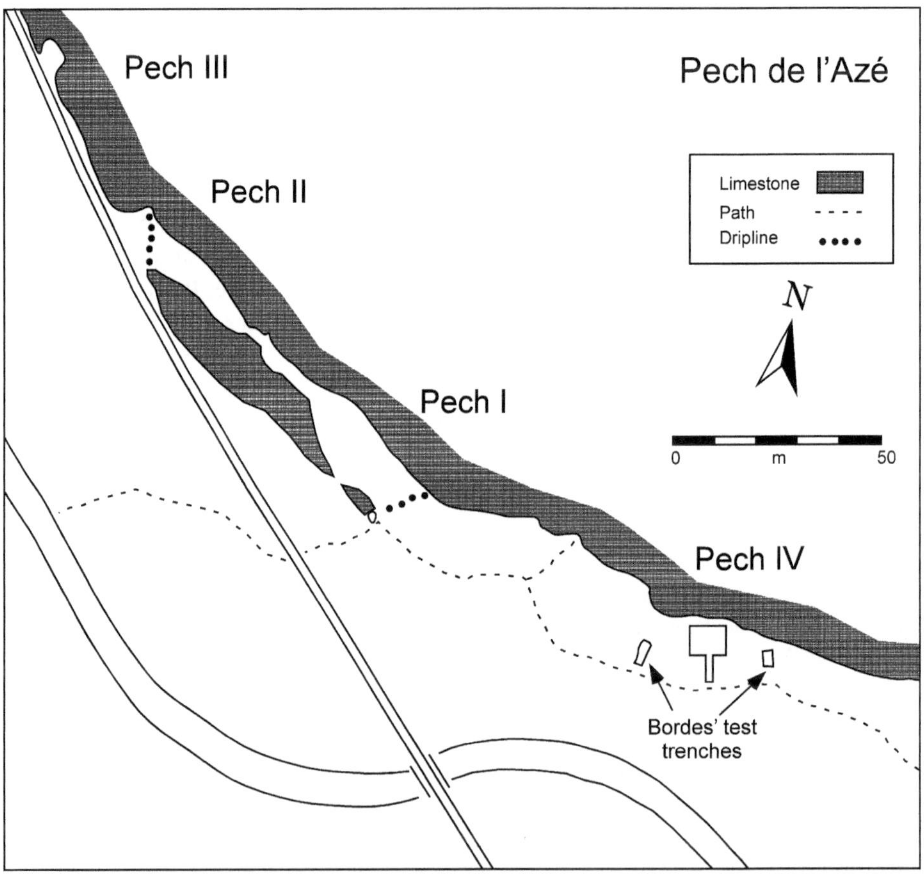


**Fig. 1.2** The Pech de l'Azé sites

dates based on 29 teeth from Pech II. Both sets of dates give a consistent picture, though the ESR dates provide more detail (see also Grün and Stringer 1991). More recently, OSL dates (Jacobs et al. 2016; see Chap. 3) suggest ranges from 100 to 55 ka for the upper ensemble (layers 2G1 to 4D) and roughly 180–140 ka for the lower ensemble (layers 7–9).

Pech III, discovered in 1951, is a very small cave that contained a sequence thought to correspond to the earlier

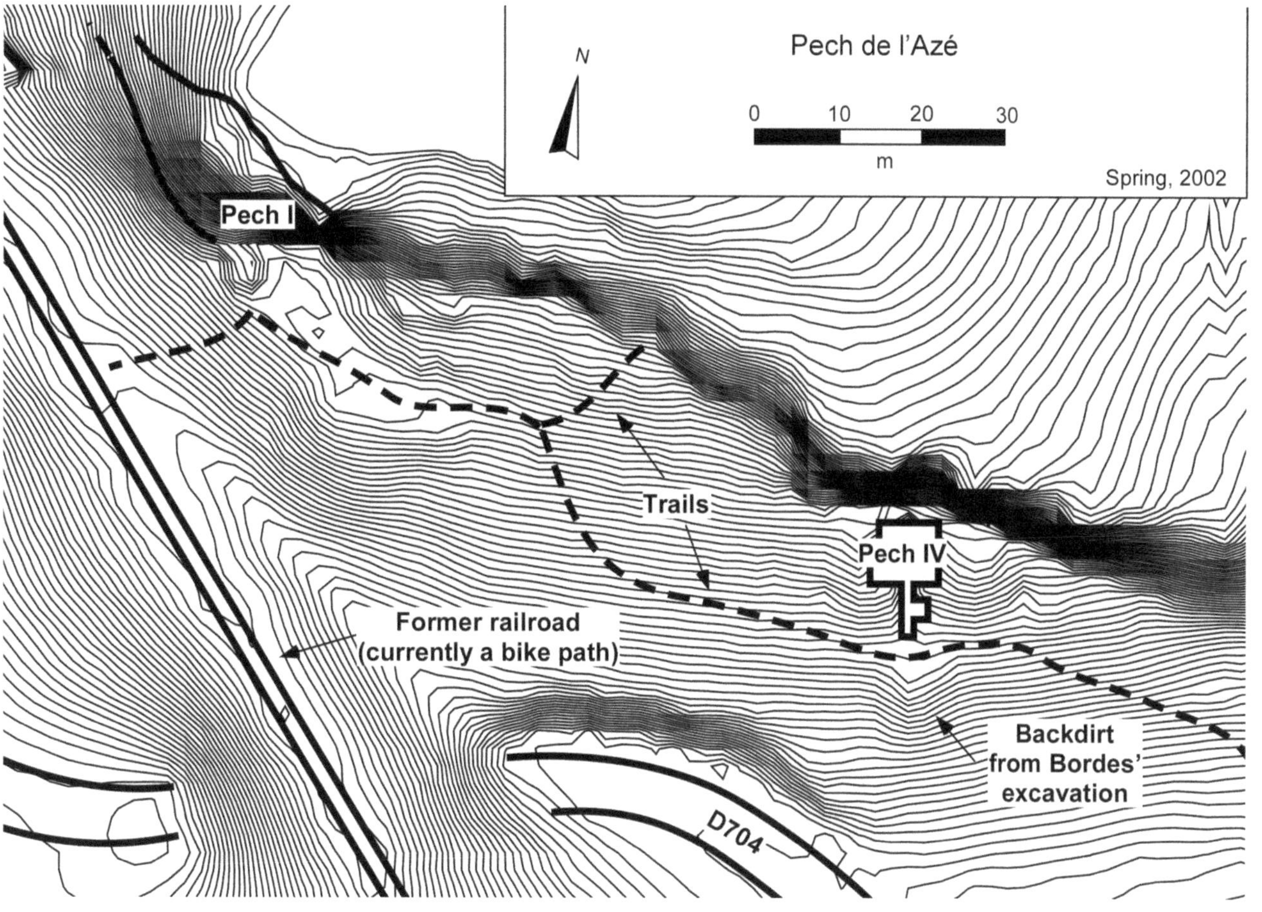


**Fig. 1.3** Topographic map between Pech I and IV

part of the Pech II sequence (Bordes and Bourgon 1951) (Figs. 1.5 and 1.6). The cave is now completely empty.

Pech IV was discovered and tested by Bordes in the spring of 1952 (Bordes 1954). In the following 4 years, 1953–1956, Mortureux, a dentist from the nearby town of Sarlat, excavated a larger trench, 1 by 9 m, into the site. He was stopped, however, by large blocks of roof fall and the demands of his practice. Because of Bordes' continued excavations at Pech I and II, among other sites, it was not until 1970 that he again began excavating Pech IV. He spent eight field seasons there, through 1977, and opened 52 $m^2$ (Figs. 1.7 and 1.8). In the first year, Bordes expanded Mortureux's trench into the site making it approximately 2 m wide and 11 m long through the slope deposits in front of the limestone cliff. In the following years, he opened a rectangular grid of 7 by 6 m against the cliff. Most of these squares were excavated to bedrock. At its maximum, against the cliff face, this meant a depth of roughly 4.5 m below surface, though a block of squares on the western side of the grid (C12–I13 and G14–H14) was only partially excavated leaving a series of steps (Fig. 1.6). Altogether he excavated just under 115 $m^3$. It is interesting to note that in terms of the investment of his time and amount of material that he recovered, Pech IV represents one of the largest excavations undertaken by Bordes during his career, second only to his work at Combe Grenal (McPherron et al. 2012; see Fig. 1.7). It was, however, the last site he excavated in France.

Unfortunately, the archaeological material Bordes excavated was never fully published. A preliminary note describing the stratigraphy, lithic industries, and fauna was published by him in 1975, based on analysis of material recovered through the 1973 season, and some of his interpretations of the industries were included in a later paper (Bordes 1981). The Mousterian industries included several examples of the named "facies": Typical Mousterian, Mousterian of Acheulian Tradition, and a new variant Bordes called the Asinipodian. Apart from these brief publications by Bordes himself, a dissertation was written on the fauna (Laquay 1981), and an early attempt at TL dating was made (Bowman et al. 1982). No other studies were made of these collections until the late 1990s when we started our project.

**Fig. 1.4** The site of Pech I. The area excavated most recently by Soressi is in the *lower left*

**Fig. 1.5** *Left* View of Pech II facing northwest. *Right* Entrance to Pech III

**Fig. 1.6** Former railroad bed that cuts in front of the Pech sites. Dibble (shown here surveying) is standing roughly opposite Pech I. Pech II is further along this roadbed, followed by Pech III

## Bordes' Excavations at Pech IV

There is relatively little documentation of Bordes' excavation methods at Pech IV. However, one of us (HLD) excavated there with Bordes in 1976 and 1977, and it is also possible to gain some insights into his methods through analysis of his field notebooks and of the archaeological materials themselves. There is no doubt that Bordes, during the course of his excavations at many sites, helped to usher in a new era of archaeological methods, including three-dimensional point proveniencing of most of the recovered objects and assigning unique numbers to them (see McPherron et al. 2005, 2012), along with careful attention paid to the geologic context of the finds. However, in other respects, his methods were lacking.

Bordes set up a 1 m grid system for the site, with east–west rows designated by letters and the north–south rows designated by numbers, much like a modern spreadsheet. Thus, each square meter, defined as the intersection of the two rows, was named by the combination of the letter and number, e.g., G12 or D18. Artifacts recovered from each square were given unique identifiers formed by the combination of the square name (or excavation Unit) and a sequential ID number, with the ID numbers going from 1 (the first object recovered from the square) to the last object recovered. So, the fifth artifact recovered from square G12 would be given the identifier G12–5, and so on through subsequent seasons until excavation in that unit stopped.

Typically, each meter square was excavated as a separate unit, in part to facilitate the recording of the *X* (east–west) and *Y* (north–south) coordinates of the objects. As each object was exposed, the *X* coordinate was measured from the western edge of the square to the middle of the object and the *Y* coordinate from southern edge. However, because the exposed walls at the periphery of each square were not often maintained to be truly vertical (due simply to human error typical of excavations carried out prior to the use of total stations), the actual surface area of a square varied as excavation descended. This of course introduced error in both the *X* and *Y* measurements, as illustrated in Fig. 1.9.

Although not as systematic, there were also problems with the measurement of the depth of objects or the

**Fig. 1.7** Looking grid west on Bordes' excavations in 1976. This portion of the site was not excavated completely to bedrock

*Z* coordinate. Bordes utilized a technique of positioning two horizontal strings at a known depth below datum, and to measure the depth of a particular object, the excavator would place a meter stick vertically on the object, line-up the strings visually (to solve the problem of parallax), and record the depth as read from the meter stick at the point where the two strings line-up. The actual *Z* coordinate would then be the sum of the vertical distance from the object to the strings and the vertical distance of the strings from the site datum. Although, in theory, this is an accurate system for recording Z, in practice the reliability of the measurement suffered from a number of problems, including not holding the meter stick vertically and the tendency for sagging in the strings due to fluctuations in humidity. As discussed in McPherron et al. (2005), the effects of all these problems resulted in the blurring of discrete zones of differing artifact densities that were much clearer when objects were provenienced with more reliable and precise methods (Fig. 1.10).

Although Bordes spent most of the excavation season on-site along with his student participants, he spent little effort in explaining various excavation protocols. At the beginning of each season, students were briefly instructed on how to measure the coordinates of the objects and to record certain observations in the field notebooks: the artifact ID number, its *X*, *Y*, and *Z* coordinates, the layer in which the object was found (occasionally these were later changed by Bordes himself based on the altitude of the object), a descriptive term for the object (scraper, bone, tooth, etc.), and a comment about the nature of the sediment (using the student's own system of description). Some excavators made hand-drawn maps as they worked, though many did not. He instructed students to record (i.e., provenience and number) most lithics (though without any particular attention to size cutoffs or other technological or typological criteria) and to provenience only "identifiable" pieces of bone (such as articular ends or teeth). Objects that were not provenienced or numbered were put in bags, which were in turn labeled by square, level, and the beginning and ending depths of the excavated volume of sediment. Thus, the decision of which objects were recorded, which ones were put into a bag labeled

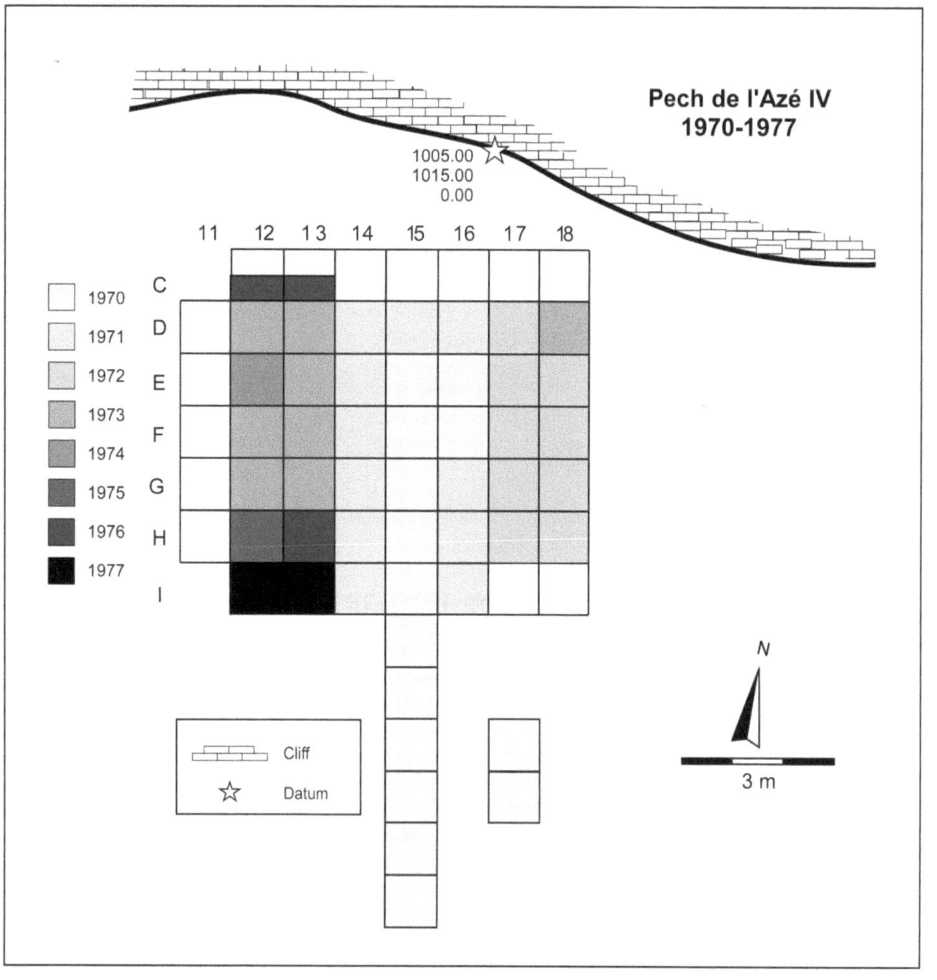


**Fig. 1.8** Bordes' excavation year by year

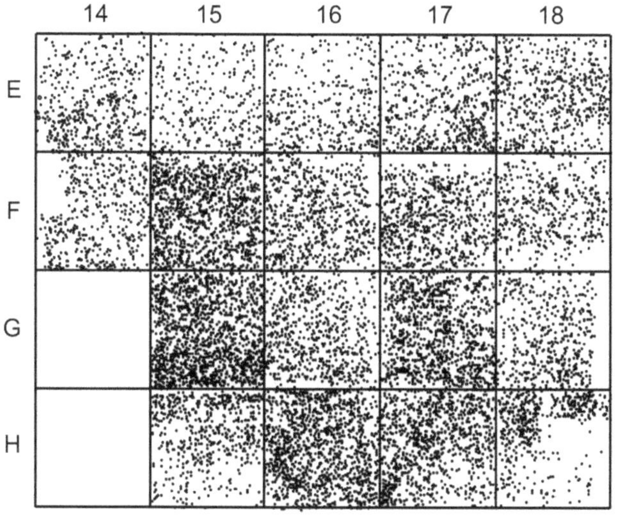


**Fig. 1.9** Plan view showing provenienced artifacts from Bordes' Layers X, Y, and Z in the central part of the excavation. Each grid cell is 1 $m^2$. Grid north is at the *top*. Gaps between squares are due to errors introduced by measuring from the edges of the unevenly excavated squares, while varying densities in adjacent squares reflect inter-excavator variation in deciding which objects to provenience

only by layer and depth, and which ones to be discarded was left primarily to the individual excavators, who, of course, had varying levels of background and training. Buckets that contained the excavated sediment and objects to be discarded were simply tossed downslope without screening. During the course of the new excavations we sampled Bordes' back dirt and analyzed the resulting lithics and fauna in a study designed to evaluate the nature of excavator bias during his excavation (Dibble et al. 2005).

A lack of clear standards on what to record led to a great deal of inter-personal variation. Some enthusiastic excavators had a tendency to record and number lithic objects quite literally as small as 2 mm in maximum size, whereas most others would either put such pieces in the small finds bags or simply discard them. Although most retouched pieces were provenienced, unretouched flakes were much more likely to be either retained in the bags or discarded. Few section drawings were made, and those that were are inconsistent in terms of terminology and description. Fortunately, as described below, it was possible to correct many of these deficiencies during our first phase of work with his collection.

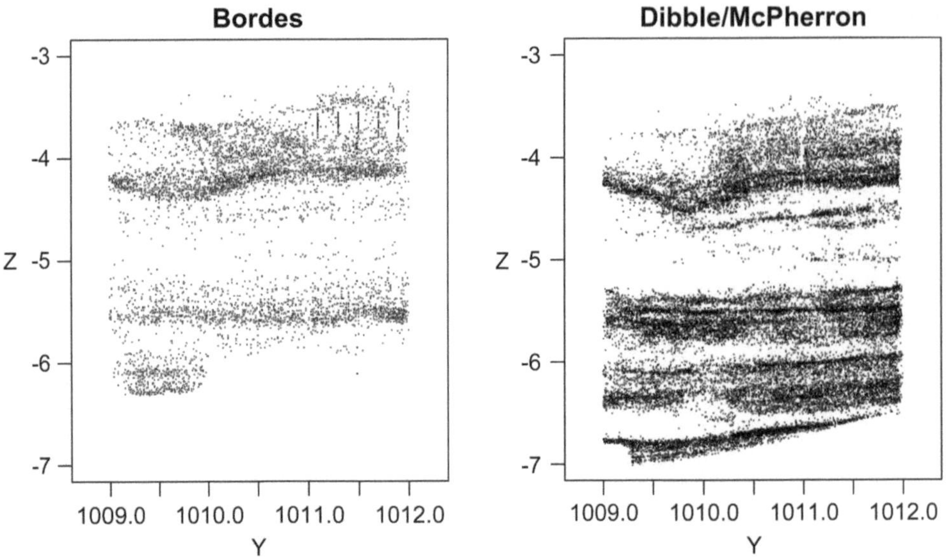


**Fig. 1.10** Comparison between Bordes' hand recording of artifact proveniences and those done with a total station. Note especially how the latter clarifies horizons of different artifact densities. Both projections represent artifact distributions through 1 m

## The Pech IV Sequence Based on the 1970–1977 Excavations

For the most part, Bordes' stratigraphic units (Fig. 1.11) were based on sedimentological variation, although in some cases he would also subdivide them either arbitrarily by depth or on the basis of changes in the composition of the assemblages. In his two preliminary reports on the Pech IV excavations (Bordes 1975, 1981), he provided the following descriptions of his stratigraphic succession and the Middle Paleolithic industries associated with each layer. In Chap. 2 we will discuss in greater detail the geological sequence as interpreted on the basis of the 2000–2003 excavations (as well as the correlation between his sequence and ours).

At the base of the sequence, resting on bedrock, Bordes distinguished three layers named, respectively, from bottom to top, Z, Y, and X, which consisted of multiple lenses that were sometimes difficult to distinguish from one another. They contained abundant traces of burning and apparently discrete fire features; in Layer Z, the burning appeared directly on the bedrock. Bordes considered the industries of these layers to be examples of Typical Mousterian.

Overlying Layers X, Y, and Z and separated from it by a layer of roof fall, Layer J was subdivided into several layers (from bottom to top J3–J1). All of the J layers were described as fairly pliable sand with rare *éboulis* (pieces of limestone from the cave roof or walls), a rich lithic industry, abundant fauna, and macroscopic evidence of fire. In terms of color, J3 was black toward its base, then grayer, and finally redder at the top.

At the bottom of the J layer, Layers J3C–J3A were thought by Bordes to represent an entirely new facies of Mousterian, what he termed the "Asinipodian." Indeed, there were many features of these assemblages that stand out. In particular, they contained a high number of truncated-faceted pieces (though Bordes did not recognize this type—see Debénath and Dibble 1994), Kombewa flakes and cores, and very small Levallois flakes and cores. Although the overall average size of Asinipodian tools and flakes was more or less similar to what was seen throughout the Pech IV sequence, the small size of some of the Levallois flakes led Bordes (1975: 298) initially to consider the term "Micro-mousterian" to describe this industry; he later coined the eponymous term Asinipodian (a rough Latin translation of Pech de l'Azé) to emphasize that Pech IV was the first site where it was recognized. Later studies have shown that these types co-occur in relatively high frequencies at a number of Paleolithic sites throughout western Eurasia (Dibble and McPherron 2006, 2007).

Layer J2, immediately above the J3 layers, was described by Bordes as having been affected by cryoturbation, with rounded limestone blocks (*éboulis*) and damaged flints in a sandy matrix. The effects of cryoturbation seemed to be more pronounced in the front of the site than in the rear. Above it, Layer J1 consisted of light red-brown sands with large blocks of limestone representing another partial collapse of the shelter. Both of these levels contained Mousterian artifacts.

Within Layer I, the stratigraphic distinction between Layers I1 and I2 is not well-marked. Layer I2 is characterized by numerous small limestone blocks, while I1 has fewer limestone blocks and fewer stone tools. Both layers are at times highly concreted. In Bordes' own words (1975: 298), the assemblages from I1 and I2 are "esthétiquement parlant, la plus belle [industrie] du site," and presented a very different kind of industry from the underlying Asinipodian. Here scrapers were the dominant tool, Levallois production was moderate, and the flakes and tools had the largest dimensions of any in the Pech IV sequence. Among the scrapers, there was a higher frequency of the more reduced

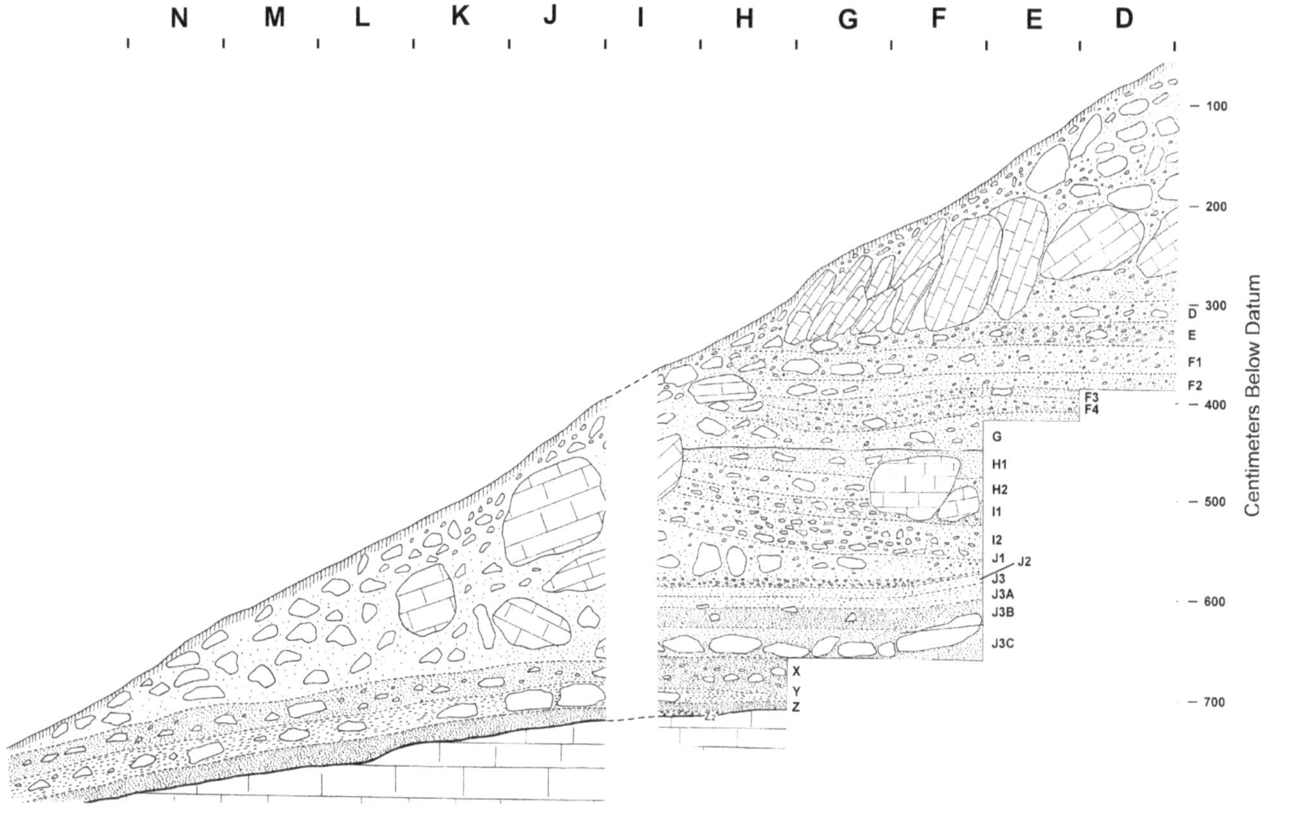


**Fig. 1.11** West profile of Pech IV, probably along the 14–15 square boundary (taken from Bordes' notes)

convergent and transverse types than in other assemblages of the site. Although the hiatus between the upper J layers and Layer I suggests some length of time between the deposition of these two units, the fact remains that it represents an extreme shift in tool production relative to flake production and in terms of artifact dimension.

Levels H1 and H2, described as sandy with scattered limestone blocks, contained very few lithics, though Bordes classified them as Typical Mousterian. Likewise, Layer G was nearly sterile. Bordes felt that some of the tools identified from this layer more likely represent pockets of material derived from Level F4 above, though our own analysis of the full assemblage suggests closer affinities, both typologically and technologically, to the underlying Layer I, and our new excavations suggest a shift at the top of Layer H toward a more Quina-type technology.

Layer F, again more or less arbitrarily subdivided by Bordes into four layers so that change through time in this thick deposit could be more easily detected (see McPherron et al. 2005), is the last substantial Mousterian deposit at Pech IV, though the assemblages, particularly Layer F4, were the richest at the site. All of the F layers were assigned to the Mousterian of Acheulian Tradition (MTA) industry with Bordes noting a shift from Type A MTA at the base to Type B MTA at the top of the sequence.

## Overview of the New Pech IV Project

The new research at Pech IV consisted of two distinct stages. The first, which took place between 1996 and 1999, is focused on the existing data and collections recovered by Bordes during his excavation. The second stage, from 2000 through 2003, was a renewed excavation at the site (McPherron and Dibble 2000; Dibble and McPherron 2007).

## Bordes' Collections

Bordes' collections, which consisted of the numbered objects (lithics and fauna) and several hundred bags of small finds, were initially stored at the Institut de Préhistoire et de Géologie du Quaternaire (IPGQ), Université de Bordeaux I, in Talence, France. In 2007 the material was transferred to the Musée National de Préhistoire (MNP) in Les Eyzies, France, where it is currently curated. Before this transfer, the collections and associated documents were in various states of curation. First, a portion of the lithic collection (approximately one-half) was washed, labeled (with the site name ["PA IV"], square, and sequential ID number) and organized by Bordes into layers and typological classes. For the material that had not been studied by Bordes (primarily the material excavated between 1973 and 1977), a portion was washed and labeled but was left unsorted either by layer or category.

There was also a significant portion of both the lithic and faunal collections, primarily from the last 2 years of excavations, that was not washed. The principal issue we encountered with this material was the deterioration of the artifact labels. At the time of excavation, each artifact was wrapped in foil and the ID number for the piece was written in pencil on masking tape that was then wrapped around the foil. This system was never intended to be permanent, and with the passage of time the tape sometimes lost its adhesive properties and became separated from the artifact. This had already happened with a small number of pieces (fewer than 50).

The small finds, lithics and fauna together, were stored in plastic bags labeled by square (or portion of square) and a depth range. For the most part, these bags and their contents had never been inventoried. By cross-checking the depth with the notebook data we were able to associate these finds with the numbered artifacts and thereby assign the proper stratigraphic unit to them. All of these materials were washed, put into new bags with permanent labels, and analyzed.

All of the basic provenience data, drawings and notes recorded during Bordes' excavation were retained at the Musée d'Aquitaine, Bordeaux, by D. de Sonneville-Bordes. Altogether, there are some 2500 pages of field notes and various plan and sections views. Over a period of 4 years the raw data contained in the field notebooks were entered into a computer database. These data consist of square, id number, *X* (relative to west edge of square), *Y* (relative to south edge of square), and *Z* (relative to the daily reference datum) coordinates, a code indicating type of artifact (retouched tool, flake, core, tooth, etc.), sediment description as recorded by the excavator, layer assignment, excavator name, and date of excavation. We modified the coordinate data to create a global grid system for the site as a whole relative to the original site datum (which we eventually located on the cliff above the site). In addition, each notebook page was scanned and saved in a high-resolution format, as a means both to archive the notebook information and to facilitate editing of the entered notebook data.

Altogether, approximately 92,000 lithic artifacts from Bordes' excavations were inventoried and analyzed. For complete flakes, tools, and cores a full set of descriptive and analytical observations were made, building on the system described in Dibble and Lenoir (1995; see also; Debénath and Dibble 1994; Chase et al. 2009). These include detailed observations of technology, typology, morphology, and raw material. For broken lithic artifacts, a more restricted set of observations was made depending on the nature of the object. The goal of this analysis was to provide a thorough typological and technological description of the industries.

The entry of both the notebook data and the analytical data for the lithic artifacts resulted in the creation of a large database, and a large part of our efforts focused on the organization and maintenance of it. Some of the problems we faced were those inherent in any large database, but many were also due to the fact that this collection had not been systematically processed and adequately curated. Some problems were simply the result of the fact that the material was excavated and processed before computers were in common use.

One issue that became apparent during our analysis is that there is a relatively high number of duplicate ID numbers in the Pech IV collection. This is a problem that is much more serious in archaeological work than is commonly realized and, in fact, it becomes apparent only with computerized inventorying of the entire collection (which allows for quick and accurate verification of identification numbers). Artifact labeling is often considered a trivial aspect of archaeological fieldwork, but it is fundamentally important since the only way to link a particular object with other data (such as its original provenience) is through the identifying number written on the object itself. In our system, which we developed during the course of work at several sites (Dibble and McPherron 1988; Dibble and Lenoir 1995; Chase et al. 2009; Dibble et al. 2006), each time an artifact is picked up and analyzed; its identifying number is the first thing entered in the computer. The computer then checks to see that (a) the number is a valid number in our system and (b) that it has not already been analyzed. In making these checks during the course of analysis of the Pech IV material, it became clear that there were a number of errors related to artifact labeling, and unfortunately we had no way to correct this problem after the fact (i.e., work out which artifact was correctly labeled and which had a duplicate label). Since duplicate ID numbers make it impossible to relate external data to a specific artifact, these cases were set aside (though not permanently deleted) in the main database.

As noted above, another problem in the Pech IV collections is the tremendous amount of inter-excavator variability in the minimum sizes of numbered and provenienced artifacts. The issue is not that small artifacts are not important, but for comparative studies it is essential to be consistent in terms of how different materials are collected and analyzed. Inter-excavator variability in terms of minimum size cutoffs can grossly affect a number of measures, including artifact densities, basic counts, artifact size calculations, and artifact class ratios. For this reason, we coded all lithic artifacts less than 2.5 cm in maximum dimension as such and, again, set them aside (though not permanently deleted) in the main database. This not only helps to minimize intra-excavator variability at Pech IV itself, but it also makes the remaining data sample more comparable to other sites we have excavated with the same controls.

Finally, we had to determine or verify the layer from which each of the artifacts came. There were three sources of information regarding the proper layer assignments of the excavated artifacts: Bordes' own assignments (whereby artifacts were stored together and the layer indicated on their container), notebook entries by the individual excavators (including Bordes) indicating the layer, and the artifacts' position in three-dimensional space. Only about one-half of the existing collection had been assigned by Bordes into layers, and there were clear problems even with these. Some such problems were relatively easy to spot and correct, thanks to our ability to plot the points on the computer, and by doing this (Fig. 1.12) we determined that at least two drawers of material had incorrect layer labels. Thus, it became clear that the storage of the existing collection by layer had serious errors and that any prior use of the collection without verification with the notebook data would lead to significant problems of interpretation. However, by utilizing all of these sources of information, the overwhelming bulk of the material has been assigned to their proper stratigraphic provenience. As discussed in Dibble et al. (2009), a similar problem exists with Bordes' collection from his excavation at Combe Grenal, though in this case, because only half of the objects were individually labeled, it is impossible to verify the original provenience for the entire collection.

During the analysis of the lithic materials from Bordes' collection, approximately ten percent of the objects were digitally photographed.

## The New Excavation

Following an initial topographic mapping of the site in 1998 (see Fig. 1.3), the renewed excavations at Pech IV began in 2000, co-directed by Dibble and McPherron. The excavation itself continued through 2003.

Deciding where to dig involved choosing among the three main, intact sections left by Bordes (see Fig. 1.8). Any further excavation along the north wall would have quickly removed sediment immediately adjacent to the cliff face, which in turn would have effectively destroyed the connection between the east and west stratigraphic profiles. Compared to the east wall, the deposits on the west side of the site were thicker, contained more abundant artifacts, had far fewer large rocks, and appeared to be closer to the center of the site. Furthermore, the west section was closer to the stratigraphic section that Bordes illustrated in his description of the site stratigraphy (see Fig. 1.11), making it easier for us to understand his interpretation of each layer as work proceeded. Therefore, we decided to excavate to bedrock the entire western wall, squares D11, E11, F11, G11, and H11, as well as adjacent squares (D12, D13, E12, E13, F12, and F13, as well as a thin edge of the 14 column through rows D to F) that he had partially excavated. The area thus excavated is shown in Fig. 1.13. Most of our squares were excavated to bedrock, with two exceptions: (1) the row of squares H11–H14 and the southern half of G11–G14, both of which stop within a layer of thick limestone blocks, and (2) the northern half of G11–G14, which is an untouched bench of sediment

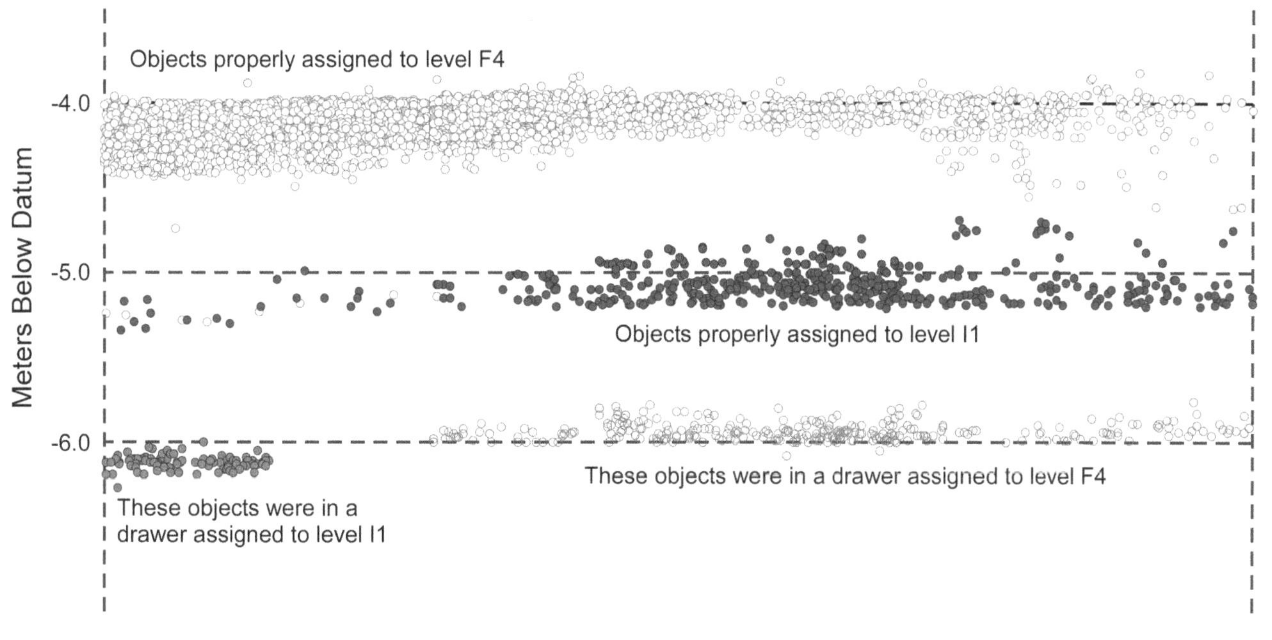


**Fig. 1.12** A profile view demonstrating some of the curation problems in Bordes' Pech IV collection. The coordinates used to draw the profile come from the square notebooks. The layer designations come from labels on the drawers containing the stone tools. In several instances, labels had apparently been inadvertently switched

from our Layer 8. Layer 8 proved very difficult to excavate because of its many combustion features, and so the decision was made to reserve a small portion of it for a future time when better methods could be developed and used. Altogether, based on the number of 7-L buckets of sediment collected, we removed just over 15 $m^3$ of deposit during the new excavations, which yielded approximately 19,500 provenienced lithics, around 23,000 faunal remains, and one hominin tooth (see Table 1.1).

At some point in the 1980s, following Bordes' death, the exposed Pech IV sections were protected with cement blocks and/or poured concrete (Fig. 1.14). Although it initially appeared to be a thin covering of concrete that could be removed fairly quickly, it turned out to be so difficult that only parts of the west and north walls could be removed in the first season (Figs. 1.13, 1.14 and 1.15). The rest of the walls were removed over the next two seasons. Because removal of the walls left the site exposed, a fence was erected around the site. Ultimately, a more permanent structure was erected to protect the site.

Fortunately, we were able to find Bordes' original site datum in the cliff overlooking the site. This datum enabled us to continue with the same grid he defined for his excavations. Bordes gave this datum a *Z* of 0, which means that all of the *Z* coordinates in the excavation are negative values. In order to avoid any negative numbers in the *X* (east–west) and *Y* (north–south) axes during our excavation, the *X* and *Y* coordinates for the datum were arbitrarily given large values. Because Bordes used a tape measure to layout his site grid, some error in defining square boundaries was inevitable and with this method errors accumulate in the placement of individual unit boundaries. Because of this, and perhaps also due to some post-excavation erosion, there is a 10–15 cm gap between the eastern most extent of the new excavations compared with the western most extension of Bordes' (Fig. 1.16).

**Fig. 1.13** The extent and depth of the new excavations

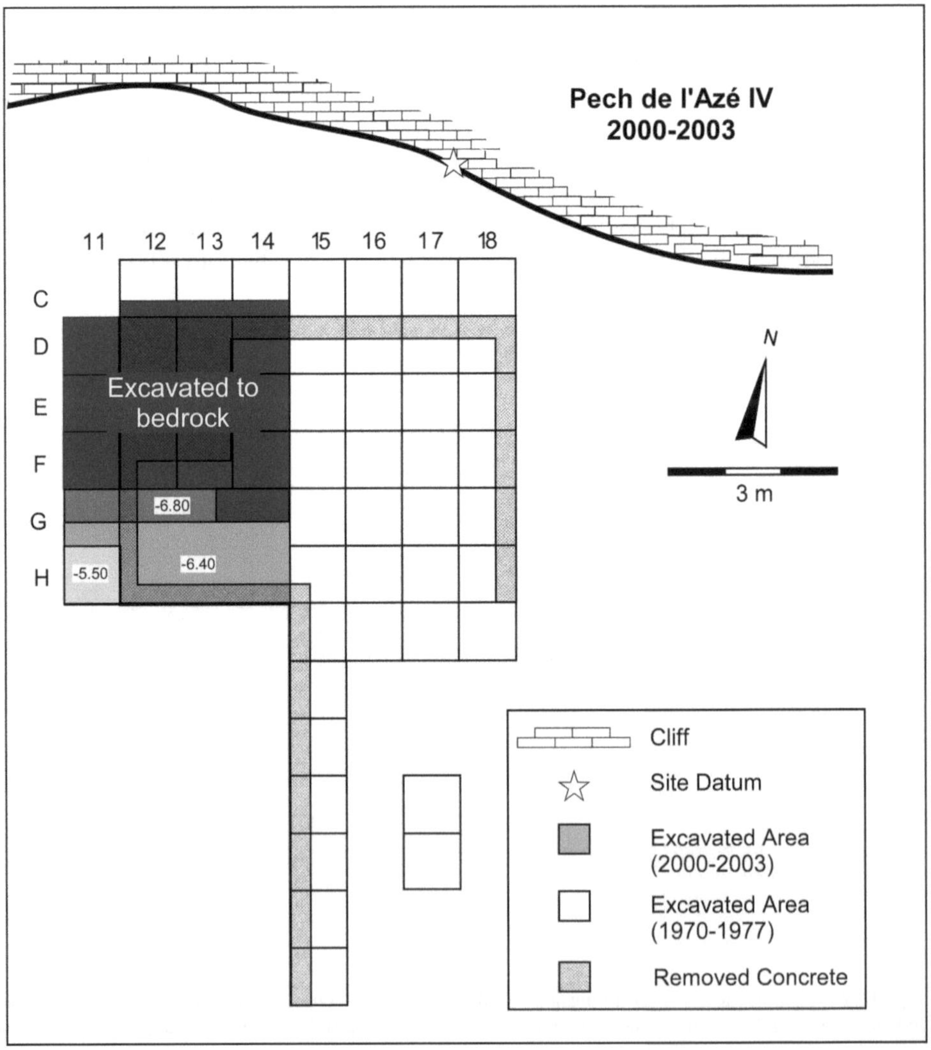

**Table 1.1** Counts of provenienced faunal and lithic remains, and volume excavated, from each layer

| Layer | Fauna | Lithics | $m^3$ |
|---|---|---|---|
| 3A | 746 | 1638 | 1.15 |
| 3B | 956 | 2938 | 0.79 |
| 4A | 1078 | 275 | 2.44 |
| 4B | 412 | 49 | 0.74 |
| 4C | 3827 | 652 | 0.74 |
| 5A | 2640 | 1453 | 0.71 |
| 5B | 1034 | 715 | 1.44 |
| 6A | 4895 | 2500 | 2.16 |
| 6B | 5346 | 2532 | 2.53 |
| 7 | 341 | 4214 | 0.79 |
| 8 | 1736 | 2597 | 1.86 |
| Total | 23,011 | 19,563 | 15.34 |

**Fig. 1.14** (*left*) Pech IV looking west prior to the demolition of the cement wall and (*right*) after partial removal of the wall. For reference, the flat bench on the *left* is the same as the flat bench still partially intact on the *right*

## Methods Used During the 2000–2003 Excavations

The excavation methodology and techniques employed at Pech IV were based on those developed in the course of work on several other French Paleolithic sites including La Quina, Combe-Capelle Bas, Cagny l'Epinette, and Fontéchevade (Dibble and Lenoir 1995; Chase et al. 2009) and subsequently employed in the course of work at the French Paleolithic sites of Roc de Marsal, and La Ferrassie (Dibble et al. 2006, 2008, 2012; Goldberg et al. 2012; Turq et al. 2008). The backbone of this methodology is the use of a total station connected to a small, hand-held computer (Dibble 1987; Dibble and McPherron 1988). The computer ran self-authored software designed to allow for rapid and accurate proveniencing of excavated items and greatly reduced error in the three-dimensional proveniences. Data collected from the field were then transferred to a central database.

Our methodology was based on a number of goals. First, it was designed to maximize the accuracy and reliability of the recording of the positions of objects and samples. The use of a total station itself had a precision of 1 mm, and the correct positioning of the instrument was verified continuously. Excavators were trained extensively on how to position the prism on the objects or samples being collected,

**Fig. 1.15** View of Pech IV at the end of excavation in 2003. The *black*, rectangular bench is Layer 8

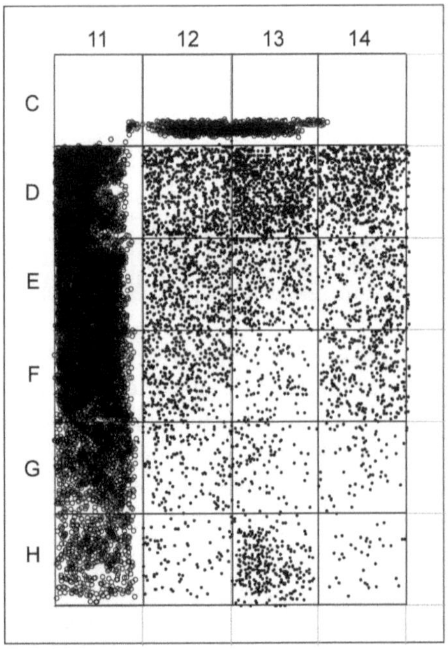


**Fig. 1.16** Lateral contact between Bordes' excavations (levels G, H, I1, and I2 in columns 12–14) and our excavations (Layers 4a–c, 5a in column 11)

which helped to minimize reliability issues. The second goal was to increase the efficiency of the recording process. By having the total station connected directly to a field computer, data were transferred quickly and without the error that can happen with hand entry of data. The same was true for entry of analytical data (see McPherron and Dibble 2002). The third goal was to maintain strict standards or protocols on what kinds and what size objects to be point-provenienced, and when objects should be placed in the buckets of sediment for later recovery after wet screening. This greatly reduced the degree of inter-excavator variability.

During excavation, all lithic artifacts and faunal remains equal to or greater than 2.5 cm in maximum dimension were point-provenienced. What this means is that such objects were given a unique identification number (following the same system of Square-ID as used by Bordes, or what is termed now Unit-ID) and their three-dimensional coordinates were recorded by the total station (in squares that were originally excavated by Bordes, our ID numbers picked up where his left off). Other relevant variables were also recorded at this time, including the layer in which the object was found, the name of the excavator, the date, and a general code indicating the kind of object (lithic, bone, mineral, etc.). Other objects, such as minerals (e.g., pieces of manganese oxide) and teeth (though not microfauna), were provenienced regardless of size. Objects smaller than the minimum cutoff were bulk provenienced with the sediments and recovered during the wet screening at the lab.

Most provenienced objects were recorded with a single *X*, *Y*, and *Z* coordinate at the center, base of the object (i.e., the surface the object rests on). There are, however, some important exceptions. For elongated objects that showed a clear orientation, two points were recorded, i.e., one at each of its ends (McPherron et al. 2005). These two points provide both the horizontal (bearing) and vertical orientation (plunge) of the object, which, as described in Chap. 2, are useful for inferring site formation processes. For rocks that were 30 cm or larger in maximum dimension, multiple shots were taken. Typically, one point was recorded at the center of its upper surface, and then 3–5 more points were recorded around the outline at its base. This provided some indication of the size, shape, volume, and orientation of each rock. Although these rocks were not saved, their recorded coordinates were assigned an identification number that consisted of the name of the excavation unit (i.e., square) and five random letters as the ID (e.g., D11–XIGFE). Smaller rocks (10–30 cm in maximum dimension) were provenienced with only a single point and also assigned a random 5-letter designation. All rocks less than 10 cm in maximum dimension went into the bucket of sediment and were thus counted as part of the volume of each bucket.

All lithic and faunal objects were processed in a similar manner. After they were point-provenienced, objects were placed in reclosable plastic bags with an affixed label and barcode (Dibble et al. 2007) indicating the identification number. Any fragile items, such as bone fragments, or items that may exhibit fragile cultural modification, like pieces of mineral, were wrapped in tinfoil before they were placed in their plastic, reclosable bags. In the lab, the durable objects were washed in water (without detergent), using only fingers or a soft brush to remove adhering sediments; some lithics with concretions were also soaked for several minutes in white vinegar. Almost all objects were large enough to be labeled with indelible ink, with the site name ("PA IV") and the Unit-ID. To prevent the kinds of labeling mistakes that plagued Bordes' collections, a system for artifact labeling was followed in which one person labels the artifact and a *second* person then reads the number from the artifact and verifies that it corresponds to the number on the accompanying tag. In this way, virtually all of the labeling errors were eliminated. Once washed and labeled, all objects were put into fresh plastic bags, with a new barcode label indicating the object ID number and type of object. Ultimately, all of the objects were organized into various boxes by class of object (for the lithics) or by ID number (for the fauna), and the boxes were grouped together by layer and put into plastic trays. Each box and each tray also had a unique identifier and has a barcode label. In this way, each curated object was put into a specific box and each box into a specific tray, and these locations were all put into the final database. Thus, by knowing the Unit-ID of a particular object, it is possible to find in which tray, and which box within that tray, the object is stored.

Excavators were instructed to work within a relatively small area, usually within an area of approximately a quarter-square. As they worked, they put all sediments and rocks smaller than 10 cm into a bucket along with all unprovenienced objects. When the bucket was filled with 7 L of sediment, or when a new layer was encountered, the excavator recorded with the total station a point on the excavation surface at the center of the area in which they worked, and thereby assigned a new Unit-ID for the bucket of sediment. The buckets of sediment were then wet screened through two mesh sizes (6 and 2 mm), and the objects recovered in this manner were sorted into lithics or fauna (and, where appropriate, other categories, such as minerals), and these aggregated bags of material were given the same Unit-ID and coordinates as the bucket itself. In this way it is possible to compute an accurate measure of the quantity of sediment removed from each layer and across the excavated area. Furthermore, any lithics or fauna found in a bucket that were greater than 2.5 cm in maximum dimension, which should have been point-provenienced during excavation, were assigned a new ID number with the coordinates of the bucket itself. In order to mark these records as having approximate coordinates, a special code of "LAB" was entered as the value for excavator.

### Sample Collection

In addition to artifacts, many kinds of samples were taken during the excavation and virtually all were point-provenienced in the manner described above.

Various kinds of materials were collected to be used specifically for dating purposes, including charcoal, sediment samples, burned flint, and large mammal teeth. When charcoal was encountered in potentially datable quantities (a few grams), it was provenienced, collected, and wrapped in foil to avoid any contamination. For burned flints and large teeth suitable for thermoluminescence and electron spin resonance the object was wrapped in foil, along with any sediments still adhering to it, and placed in a small plastic bag. This bag was then placed in a larger plastic bag along with a larger sample of sediments from within a 10 cm radius of the object. This provided the dating lab a means of recording the background radiation in the immediate vicinity of the objects. In addition, each object was also photographed in situ along with a small tag indicating the identification number (following the same convention of Unit-ID as described above) and the type of sample (e.g., TL or ESR). These photographs were to provide further information for the dating lab about the nature of the surrounding sediments and their potential concentration of background radiation. All of these samples were given normal object identifiers (Unit-ID) and special codes ("RC14", "TL", "ESR", "OSL"; see Chap. 3 for descriptions of these dating methods). In 2014, 30 OSL samples were collected at night under red light (see Jacobs et al. 2016); three more unconsolidated samples were collected as blocks and stabilized with plaster bandage. The positions of all samples were recorded with the total station, and all samples were sealed in black plastic bags to prevent light exposure.

Because many of the numerical chronological techniques require accurate measurements of background radiation, dosimeters were placed at least 30 cm into the unexcavated sediment and left for a minimum period of time (usually one year) (see Richter et al. 2013). While the exact position of the opening in which the dosimeter was placed was provenienced at the time the dosimeter was inserted, our normal practice was to position them in sediment that would be excavated in the following season; this allowed us to record the exact *X*, *Y*, and *Z* coordinates of the dosimeter when it was eventually recovered during subsequent excavation. Dosimeters were identified by a special Unit designation ("DOSIM") and numbered sequentially.

As a further source of data on background radiation in the site sediments, small (tablespoon-sized) samples were taken each morning from the surface of each of the excavation squares, usually at the center of each 50 × 50 cm quadrant. These were identified as normal objects (i.e., Unit-ID) with a code of "RINKDOSE" (named after Jack Rink, who originally suggested taking such samples). Nearly 1500 of such samples were collected during the course of excavation, representing a collection of relatively evenly distributed sediment samples for the entire volume of excavated sediment.

Nearly one hundred blocks of sediment for micromorphological analyses were taken, both from the western part of the site and from the east section left by Bordes. Each sample was recorded with two points: one at the upper extent of the sample and one at the lower. These samples were also identified by a special unit designation ("PDA4") and then numbered sequentially.

## Casts

With the assistance of Alain Dalis, a mold was made of northern and eastern faces of the Layer 8 deposits. The resulting cast is currently on display in the Musée National de Préhistoire, in Les Eyzies.

## Photography

A photographic record of the site during the course of the excavation was considered to be a very important component of the project methodology. The first task every morning was to take photos of each excavation square to record the nature of the deposits encountered and the progress of the excavation. These photos were generally taken looking directly down on each square with a photo board and north indicator (usually a trowel) in the square. Often camera settings were adjusted so as to have several different exposures as needed. General site photos were also frequently taken and included specific excavation areas as they changed with the removal of the deposits, stratigraphic sections, people at work at various tasks in the course of excavating, and visitors to the site. Many photos were also taken of special tasks, such as the removal of the brick and concrete walls, the placement of the dosimeters, and the collection of the sediment cores used for attempted DNA analysis. A series of time-lapse photos (that is, one photo taken every five minutes from one location throughout the working day) were also recorded.

In addition, all lithic objects were photographed in the lab during processing, usually of the exterior surface but sometimes of both surfaces. Though the exact procedure changed some over the course of the excavation, generally artifacts were photographed on a copy stand with a digital camera. A centimeter scale was placed to one side. In post-processing, we attempted to automatically replace the background with a solid shade of blue, to crop the photograph, and to write the Unit-ID into the photograph.

## Yearly Progress of the Excavations

For the most part, excavations took place over 6-week seasons during the summer.

2000: The first year of excavation was mostly confined to squares F11, G11, and H11 with some limited excavation in the lower bench left by Bordes. Unfortunately, the time available for excavation was severely reduced due to the need to remove a portion of the concrete walls and erect scaffolding. Thus, only the very top of the Mousterian deposits was exposed in the "11" squares before the season ended.

2001: In the second season most of the rest of the concrete walls were removed and excavation continued on the squares along the top of the west section (we also opened two new squares, D11 and E11) and on the lower bench. By the end of the season, all five of the "11" squares had been excavated to a depth between 2 and 3 m below datum. In addition, the six squares of the bench were taken down approximately 0.75–1 m to a thick, site-wide layer of roof fall, which in turn overlaid the basal Layers 7 and 8 (see Fig. 1.17).

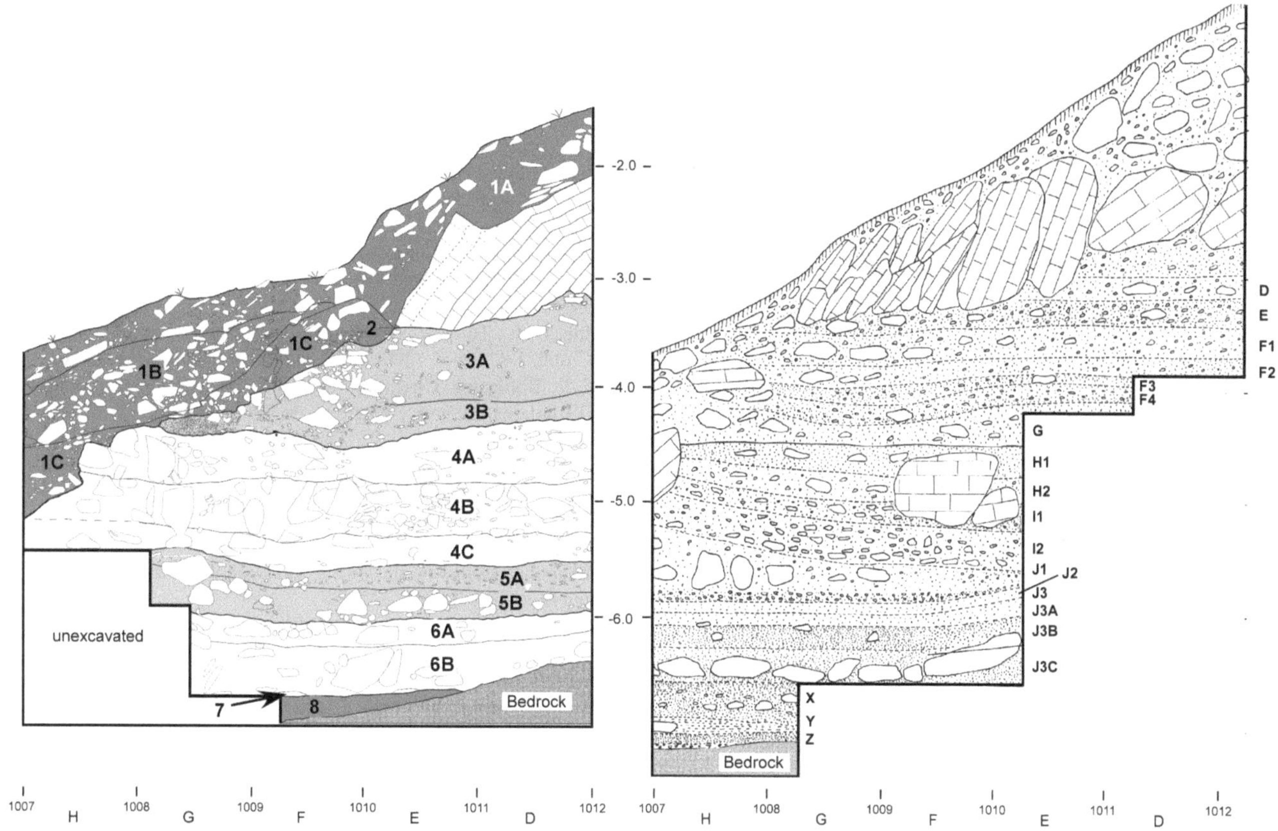


**Fig. 1.17** Comparison of Bordes' stratigraphic section at *right* (at the 14–15 column boundary) and that of the recent excavation at *left* (10–11 column boundary)

2002: Work was focused on squares D11, E11, and F11 where the cultural material concentrations were highest. By the end of the season, squares D11, E11, and F11 were taken down to a level between about 0.5 and 0.75 m above bedrock, square G11 to about 1.5 m above bedrock, and H11 about 2 m above bedrock. In addition, the entirety of the bench left by Bordes was excavated to bedrock.

2003: In this year all of the squares D11–F14 were excavated to bedrock. A small bench of the lowermost layer (Layer 8) was left in the northern half of squares G11–G13, and a higher bench left in the southern portion of squares G11–G13 and H11–H14. These benches were left primarily to preserve the deposits of Layer 8, which contained numerous combustion features. Although we had attempted to excavate this layer in such a way as to expose individual features—through the use of both *décapage* (gradually exposing the surface) and advancing from the side section, taking small (25 cm wide) slices—neither technique was entirely successful. Thus, the decision was made to preserve them for future excavations. Beyond this small bench (roughly 2.6 m by 70 cm), probably 20–30 $m^2$ of this layer remains.

2009: In the winter of 2009 a relatively large mass of limestone blocks and sediments had collapsed from the face of the north section, likely as a result of large tree roots growing into the Holocene deposits that cap the site. The slump included a significant quantity of large limestone blocks and poorly consolidated silts that included Pleistocene and Holocene components. Salvage work and clean up were carried out in July of 2009. Materials recovered included $\approx$100 lithics, 60 faunal pieces, and 11 ceramic sherds.

## The Current Status of the Pech IV Collections and the Site

In 2007, all of Bordes' collection from Pech IV was accessioned by the Musée National de Préhistoire in Les Eyzies, France, and all of the materials excavated during the new excavation were similarly transferred there after the end of the excavations. For Bordes' lithic material, a complete inventory was made of the contents of each box at the time it was analyzed by us. For the recent excavations, there is a similar inventory for all of the material. Aside from the objects themselves, all of the data collected during the first stage of this project, including scans of the field notebooks, all analytical data (lithic observations and measurements), and digital photos of approximately ten percent of Bordes' collection has been made available online. Our data and all of our artifact photographs have also been made available online.

## References


Bordes, F. (1954). Les gisements du Pech de l'Azé (Dordogne). I, Le Moustérien de tradition acheuléenne. *L'Anthropologie, 58,* 401–432.

Bordes, F. (1972). *A tale of two caves.* HarperCollins Publishers.

Bordes, F. (1975). Le gisement de Pech de l'Azé IV: Note préliminaire. *Bulletin de la Société Préhistorique Française, 72,* 293–308.

Bordes, F. (1981). Vingt-cinq ans après: le complexe moustérien revisité. *Bulletin de la Société Préhistorique Française, 78*(3), 77–87.

Bordes, F., & Bourgon, M. (1951). Le gisement du Pech de l'Azé-Nord. Campagnes 1950–1951. Les couches inférieures à *Rhinocéros mercki. Bulletin de la Société Préhistorique de France, 48* (11/12), 520–538.

Bowman, S. G. E., Loosemore, R. P. W., Sieveking, G. d. G., & Bordes, F. (1982). Preliminary dates for Pech de l'Azé 4. *Pact, 6,* 362–369.

Capitan, L., & Peyrony, D. (1909). Deux squelettes humains au milieu de foyers de l'époque moustérienne. *Comptes rendus des séances de l'Académie des Inscriptions et Belles-Lettres, 53*(11), 797–806.

Chase, P. G., Debenath, A., Dibble, H. L., & McPherron, S. P. (2009). *The cave of Fontéchevade: Recent excavations and their paleoanthropological implications.* Cambridge University Press.

Debénath, A., & Dibble, H. L. (1994). *Handbook of paleolithic typology: Lower and Middle Paleolithic of Europe.* Philadelphia: University of Pennsylvania Museum of Archaeology.

Dibble, H. L. (1987). Direct measurement of artifact provenience with an electronic theodolite. *Journal of Field Archaeology, 14,* 249–254.

Dibble, H. L., & Lenoir, M. (Eds.). (1995). *The Middle Paleolithic site of Combe-Capelle Bas (France)* (Vol. 91). UPenn Museum of Archaeology.

Dibble, H. L., & McPherron, S. P. (1988). On the computerization of archaeological projects. *Journal of Field Archaeology, 15,* 431–440.

Dibble, H. L., & McPherron, S. P. (2006). The missing Mousterian. *Current Anthropology, 47*(5), 777–803.

Dibble, H. L., & McPherron, S. P. (2007). Truncated-faceted pieces: Hafting modification, retouch, or cores? In S. P. McPherron (Ed.), *Tools versus cores alternative approaches to stone tool analysis* (pp. 75–90). Newcastle: Cambridge Scholars Publishing.

Dibble, H. L., Marean, C. W., & McPherron, S. P. (2007). On the use of barcodes in excavation projects: Examples from Mossel Bay (South Africa) and Roc de Marsal (France). *The SAA Archaeological Record, 7*(1), 33–38.

Dibble, H. L., Raczek, T. P., & McPherron, S. P. (2005). Excavator bias at the site of Pech de l'Azé IV, France. *Journal of Field Archaeology, 30*(3), 317–328.

Dibble, H. L., McPherron, S. P., Chase, P. G., Farrand, W. R., & Debénath, A. (2006). Taphonomy and the concept of Paleolithic cultures: The case of the Tayacian from Fontéchevade. *PaleoAnthropology,* 1–21.

Dibble, H. L., McPherron, S. P., Sandgathe, D., Goldberg, P., Turq, A., & Lenoir, M. (2009). Context, curation, and bias: An evaluation of the Middle Paleolithic collections of Combe-Grenal (France). *Journal of Archaeological Science, 36,* 3540–3550.

Dibble, H. L., Aldeias, V., Alvarez-Fernández, E., Blackwell, B., Hallet-Desguez, E., Jacobs, Z., et al. (2012). New excavations at the site of Contrebandiers Cave, Morocco. *PaleoAnthropology, 2012,* 145–201.

Dibble, H. L., Turq, A., McPherron, S. P., Sandgathe, D., Guibert, P., Castel, J.-C., et al. (2008). *Roc de Marsal. Bilan Scientifique 2006: 7–9.* Bordeaux: Direction Régionale des Affaires Culturelles de l'Aquitaine et Service Régional de l'Archéologie.

Ferembach, D., Legoux, P., Fenart, R., Empereur-Buisson, R., & Vlcek, E. (1970). L'Enfant du Pech-de-l'Azé. *Archives de l'Institut de Paléontologie Humaine, 33,* 1–180.

Goldberg, P., Dibble, H. L., Berna, F., Sandgathe, D., McPherron, S. P., & Turq, A. (2012). New evidence on Neandertal use of fire: Examples from Roc de Marsal and Pech de l'Azé IV. *Quaternary International, 247*(1), 325–340.

Grün, R., & Stringer, C. B. (1991). Electron spin resonance dating and the evolution of modern humans. *Archaeometry, 33,* 153–199.

Grün, R., Mellars, P., & Laville, H. (1991). ESR chronology of a 100,000-year archaeological sequence at Pech de 1'Aze II, France. *Antiquity, 65,* 544–551.

Jacobs, Z., Jankowski, N., Dibble, H. L., McPherron, S. P., Soressi, M., Sandgathe, D., et al. (2016). The age of three Middle Palaeolithic sites: single-grain optically stimulated luminescence chronologies for Pech de l'Azé I, II and IV in France. *Journal of Human Evolution, 95,* 80–103.

Laquay, G. (1981). *Recherches sur les faunes du Würm I en Périgord.* Talence: Université de Bordeaux I.

Lartet, E., & Christy, H. (1864). Cavernes du Perigord. *Revue archéologique, 1,* 233–267.

Maureille, B., & Soressi, M. (2000). A propos de la position chronostratigrahique de l'enfant du Pech-de-l'Azé I (commune de Carsac, Dordogne): la résurrection du fantôme. *Paleo, 12,* 339–352.

McPherron, S. P., & Dibble, H. L. (2002). *Using computers in archaeology: A practical guide.* Boston: McGraw-Hill Mayfield, xvi + 254 pp.

McPherron, S. P., & Dibble, H. L. (2000). The lithic assemblages of Pech de l'Azé IV (Dordogne, France). *Préhistoire Européenne, 15,* 9–43.

McPherron, S. P., Dibble, H. L., & Goldberg, P. (2005). Z. *Geoarchaeology, 20*(3), 243–262.

McPherron, S. P., Dibble, H. L., Goldberg, P., Lenoir, M., Sandgathe, D., & Turq, A. (2012). De Combe Grenal à Pech de l'Azé IV: L'évolution des Méthodes de Fouilles de François Bordes. *François Bordes et la Préhistoire. Actes du colloque de Bordeaux en 2009,* 65–77.

Patte, É. (1957). *L'enfant néanderthalien du Pech de l'Azé.* Paris: Masson.

Richter, D., Dibble, H. L, Goldberg, P., McPherron, S. P., Niven, L., Sandgathe, D., et al. (2013). The late Middle Palaeolithic in Southwest France: New TL dates for the sequence of Pech de l'Azé IV. *Quaternary International, 294,* 160–167.

Schwarcz, H. P., & Blackwell, B. (1983). $^{230}$Th/$^{234}$U age of a Mousterian site in France. *Nature, 301,* 236–237.

Soressi, M., Armand, D., d'Errico, F., Jones, H. L., Pubert, E., Rink, W. J., et al. (2002). Pech-de-l'Azé I (Carsac, Dordogne): nouveaux travaux sur le Moustérien de tradition acheuléenne. *Bulletin de la Société Préhistorique Française, 99,* 5–12.

Soressi, M., Jones, H. L., Rink, W. J., Maureille, B., & Tillier, A.-M. (2007). The Pech-de-l'Azé I Neandertal child: ESR, uranium series, and AMS 14C dating of its MTA type B context. *Journal of Human Evolution, 52,* 455–466.

Soressi, M., McPherron, S. P., Lenoire, M., Dogandzic, T., Goldberg, P., Jacobs, Z., et al. (2013). Neandertals made the first specialized bone tools in Europe. *PNAS, 110*(35), 14186–14190.

Soressi, M., Rendu, W., Texier, J.-P., Daulny, É. C. L., D'Errico, F., Laroulandie, V. … Tillier, A.-M. (2008). Pech-de-l'Azé I (Dordogne, France): nouveau regard sur un gisement moustérien de tradition acheuléenne connu depuis le XIXe siècle. *Les sociétés Paléolithiques d'un grand Sud-Ouest: nouveaux gisements, nouvelles méthodes, nouveaux résultats. -Actes des journées décentralisées de la SPF des 24–25 novembre 2006*, pp. 95–132.

Turq, A., Dibble, H. L., Faivre, J. P., Goldberg, P., McPherron, S. P., & Sandgathe, D. (2008). Le Moustérien Récent Du Périgord Noir: Quoi De Neuf? In J. Jaubert, J.-G. Bordes, & I. Ortega (Eds.), *Les Sociétés du Paléolithique dans un Grand Sud-ouest de la France: nouveaux gisements, nouveaux résultats, nouvelles méthodes*: *Mémoire de la Société Préhistorique Française* (Vol. 48, pp. 83–94).

Vaufrey, R. (1933). Le Moustérien de tradition acheuléenne au Pech-de-l'Azé (Dordogne). *L'Anthropologie (Paris), 93,* 125–127.

# Stratigraphy, Deposits, and Site Formation

2

Paul Goldberg, Shannon J.P. McPherron, Harold L. Dibble, and Dennis M. Sandgathe

## Introduction

From the outset of the Pech IV project, geoarchaeology played an integral role in the excavations. In addition to performing detailed field observations, much of the geoarchaeological analyses employed the technique of archaeological micromorphology. In addition, observations made on the archaeological material that have relevance to site formation processes are discussed in this chapter. These observations include artifact orientations, edge damage, burning and breakage of lithic artifacts, and an analysis of the small lithics fraction (5–25 mm) coming from the screens.

Principally, we focused on a number of aspects of the site including:

(1) Detailed microstratigraphic observation and recording of the deposits, particularly noting lateral variations through time;
(2) Correlation of the revised stratigraphy with that developed by F. Bordes from his excavations in the 1970s;
(3) Reconstruction of site formation history, documenting geogenic processes, anthropogenic activities, as well as post-depositional, and syn-depositional processes;
(4) Establish temporal relationships among Pech I, II, and IV based on the stratigraphic observations and analyses.

## Site Setting

As described in Chap. 1, the Pech sites are situated along the cliffs of a small dry valley (the Farge), about 5 km SE of the town of Sarlat-la-Canéda. The bedrock of Pech de l'Azé is part of Upper Coniacian yellow sandy bioclastic limestone (Unit C4b), composed of subangular quartz, muscovite, some glauconite, and rounded microfossils (Capdeville and Rigaud 1987) (http://infoterre.brgm.fr/). All of the Pech de l'Azé sites are associated with erosion of an anticlinal ridge of mostly Mesozoic limestones during the Oligocene, Miocene, and Pliocene, with karstic dissolution along joints during the Quaternary (Salomon and Astruc 1992; Turq et al. 2011); evidence of numerous solution cavities—some of which are exploited commercially—occur in the vicinity. Figure 2.1 shows the general alignment of Pech I, II, and IV, which reflects karstic development along the same structural axis; minor jointing that parallels this axis was observed in the bedrock floor of Pech IV.

Bordes originally described Pech IV as a collapsed rockshelter, presumably based on its morphology: a steep

P. Goldberg (✉)
School of Earth and Environmental Sciences, Centre for Archaeological Science, University of Wollongong, Wollongong, NSW 2522, Australia
e-mail: paulberg@bu.edu

P. Goldberg
Institute for Archaeological Sciences, Eberhard Karls Universitat Tübingen, Tübingen, Germany

P. Goldberg
Department of Archaeology, Boston University, Boston, USA

S.J.P. McPherron · H.L. Dibble
Department of Human Evolution, Max Planck Institute for Evolutionary Anthropology, Leipzig, Germany
e-mail: mcpherron@eva.mpg.de

H.L. Dibble
e-mail: hdibble@sas.upenn.edu

H.L. Dibble
Department of Anthropology, University of Pennsylvania, Philadelphia, PA, USA

H.L. Dibble
School of Human Evolution and Social Change, Institute for Human Origins, Arizona State University, Tempe, AZ, USA

D.M. Sandgathe
Department of Archaeology and Human Evolution Studies Program, Simon Fraser University, Burnaby, BC, Canada
e-mail: dms@sfu.ca

D.M. Sandgathe
Museum of Archaeology and Anthropology, University of Pennsylvania, Philadelphia, USA


H.L. Dibble et al. (eds.), *The Middle Paleolithic Site of Pech de l'Azé IV*,
Cave and Karst Systems of the World, https://doi.org/10.1007/978-3-319-57524-7_2

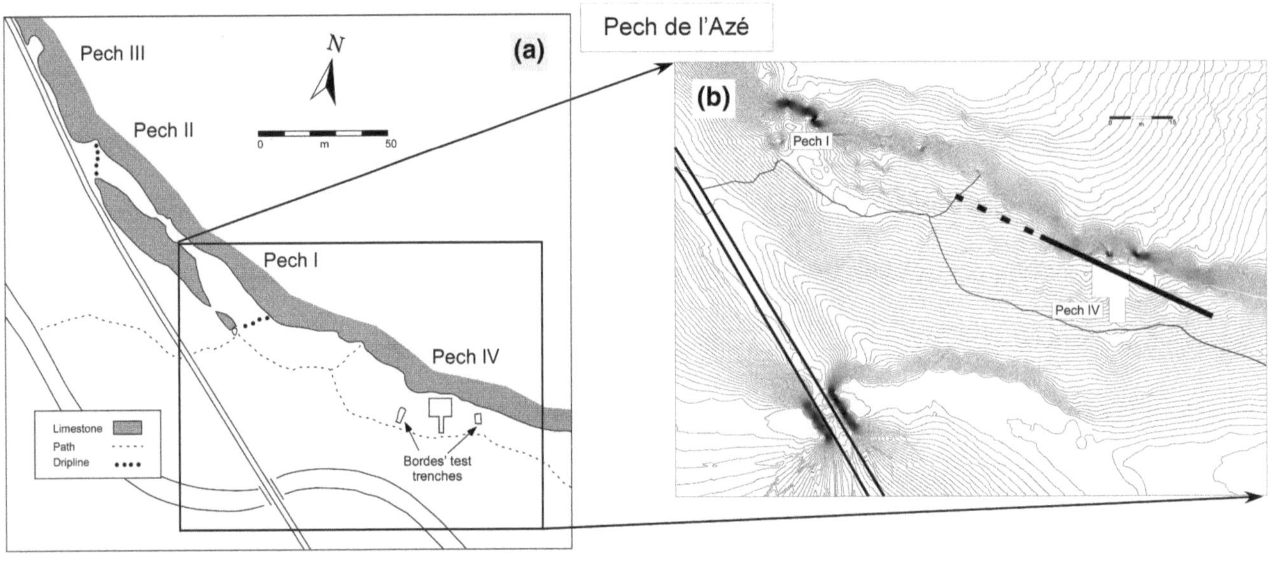


**Fig. 2.1** **a** Four sites of Pech de l'Azé. Note that Pech IV generally follows the orientation of the chamber connecting Pech I and II. **b** Topographic map of the Pech IV site before the recent excavations. The *line* depicts the limits of the north wall of the cave; *dashed lines* indicate the projection of the cave concealed under slope fill

bedrock wall at the back (north) with the remains of a small bedrock overhang just above the excavation. Our recent research, however, has revealed that Pech IV is most likely a collapsed cave connected to the karstic system associated with that of Pech I and II (Fig. 2.1). Supporting evidence for this inference is discussed below, but here it briefly includes the position of large blocks of roof collapse visible in the south, west, and east areas of the excavations, coupled with deposits that locally slope from north to south (i.e., from the inside of the inferred cave interior towards the entrance), and fabric (orientation) analyses (see below). Finally, fist-sized remnants of dripstone derived from the original cave roof were found within the lower deposits; these would not form in an aerated space such as that of an open rockshelter.

Thus, erosion of the valley—somewhat oblique to the axis of the Pech I, II, and IV (Fig. 2.1)—has caused the retreat of the cave mouth of Pech IV, resulting in its appearance as a rockshelter with coarse limestone block remnants representing a collapsed overhanging roof. A similar picture can be seen at Pech I today (Soressi et al. 2013), though in this case the southern portion of the cave wall is largely intact.

## Stratigraphy

Bordes' original description of the deposits encompassed a mixture of lithology and lithic industries within the different layers (Bordes 1975) (see Chap. 1; Table 2.1), although he never published a detailed and systematic lithological description of these units. The stratigraphic subdivision of the deposits from the current excavations consists of eight major geological units which are subdivided into a total of 13 layers, and their relationship to the levels identified by Bordes is shown in Table 2.2. The current stratigraphic divisions are based on lithostratigraphy using the criteria of color, composition, texture, grain morphology, internal structure, and lateral and vertical changes (Fig. 2.3).

As can be seen in Fig. 2.4a and Table 2.2, there is general similarity between our units and his, and a number of layers generally overlap or correspond to those of Bordes. However, since our sequence is based on different exposures from the ones seen by him (his section drawing is along the 15 column, or 4 m east of our West Section drawing), some localized mismatches are to be expected. Nevertheless, as can be seen in Fig. 2.4b, despite some mismatches and lensing/pinching out of layers, the stratigraphic units are traceable across the entire expanse of both excavations. Consequently, we are confident with Bordes' overall sequence, our correlations with it, and its lateral variations.

The deposits described below are based mostly on observations during our excavations (squares C11 through H13- best illustrated in the West and South Sections), as well as the witness sections left by Bordes: a corridor of excavated squares that act as access to the site (the 15 line) and a section in the eastern part of the excavations (Figs. 2.2 and 2.3). Below is a general summary of the stratigraphy at the end of the recent excavations (Fig. 2.4).

## Micromorphological Methods

As discussed below, our work at Pech IV concentrated on comprehensive examination of the deposits exposed in Bordes' and our sections in order to identify Bordes' original

**Table 2.1** Pech de l'Azé IV—Summary of Bordes' stratigraphy (Bordes 1975)

| Level | Lithology/geology | Industry |
|---|---|---|
| A1 | Brown sand and éboulis with modern vegetation | Poor in lithics. Medieval |
| A2 | Brown sand and éboulis | Iron age to Medieval |
| B | Brown sand and éboulis | Traces of Mousterian |
| C | Brown sand and éboulis | Traces of Mousterian |
| D | Reddish sand with éboulis | Richer in lithics |
| E | Reddish sand with éboulis | Similar to D but fewer lithics |
| • F1 | Rich in finer éboulis at the rear and medium-sized éboulis at the front. Subdivided in four subunits somewhat arbitrarily | • F1: MTA (B) |
| • F2 | | • F2: MTA (B) |
| • F3 | | • F3?: MTA (A-B) |
| • F4 | Erosional pockets at the base | • F4: MTA (A) |
| G | (not stated) | Mousterian |
| H | Sand with scattered éboulis | Few tools; Typical and Mousterian? |
| • I1<br>• I2 | Finer éboulis, locally strongly cemented. I2 becomes redder and softer toward the cliff at the rear of the cave | Typical Mousterian |
| • J1 | • J1: light reddish brown sand with large limestone blocks from roof collapse | Typical Mousterian? |
| • J2 | • J2: Éboulis rounded by cryoturbation, with broken flints in a sandy matrix. Toward the back, the éboulis is more angular and the lithics are almost intact | • J2: Typical Mousterian? |
| • J3 | • J3 (subdivided into a, b, c): loose sand poor in éboulis, rich in worked lithics; traces of hearths; it is red at the top, becoming gray and black toward its base. Rests on large slabs of limestone from the roof | • J3: Typical Mousterian |
| • J3a | | • J3a-c: Asinipodian |
| • J3b | | |
| • J3c | | |
| X | Underlies blocks of J3 and is black to gray; abundant broken and cryoturbated flint. Laterally, intercalated with dark brown layer containing traces of fireplaces | Typical Mousterian |
| Y | Reddish and found in the front of the cave; toward the rear it appears to be replaced by roof collapse | Typical Mousterian |
| Z | Rests on bedrock, which is locally reddened by fire | Typical Mousterian |

stratigraphic divisions (Bordes 1975) and bring them up to date. Nevertheless, there is only so much sedimentary and stratigraphic information that can be gleaned from field observations, no matter how diligent one tries to be. In order to supplement these observations in our interpretations of site formation processes, we employed micromorphology, which is the study of undisturbed sediments and soils—predominately in thin section—using a petrographic microscope (Courty et al. 1989). Since it is concerned with the microcontextual attributes of sedimentary and soil features and objects (e.g., bones, flints, charcoal) (Goldberg and Berna 2010), micromorphology is effective in reconstructing the formation of natural and archaeological layers and soils (Goldberg and Macphail 2006).

About 200 samples were collected as large intact blocks of sediment using the procedures outlined in Goldberg and Macphail (2003). These blocks were wrapped tightly in the field with tissue paper and packaging tape, and their coordinates were recorded using a total station and integrated into the site database (McPherron et al. 2005); loose, bulk sediment samples were also collected. The sample blocks were transported to Boston University, and then oven dried for one week at 60 °C. The samples were impregnated with a mixture of polyester resin and styrene (7:3) catalyzed with MEKP and allowed to harden for ~4–5 days. The hardened blocks were again heated in the oven overnight at 60 °C, then trimmed and sliced using a diamond blade rock saw. At this time, areas were identified for thin section analysis and

**Table 2.2** Approximate correlation between Bordes' levels and the geological layers used here

| Bordes level | Our layers |
|---|---|
| A-E | 1, 2 |
| $F_1$ | 3A |
| $F_2$ | |
| $F_3$ | |
| $F_4$ | 3B |
| G | 4A |
| $H_1/H_2$ | 4B |
| $I_1$ | |
| $I_2$ | 4C |
| $J_1$ | 5A |
| $J_2$ | 5B |
| $J_3$ | |
| $J_3A$ | 6A |
| $J_3B$ | |
| $J_3C$ | 6B |
| X | 7 |
| Y | 8 |
| Z | |

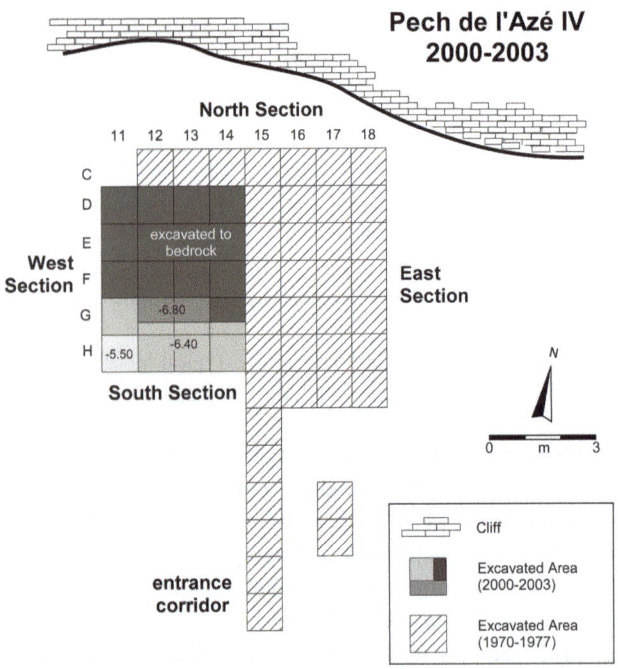


**Fig. 2.2** Plan of site, showing grid, location of excavated squares, depth to which different sections were excavated, and the main sections discussed in the text

preparation, and in the case of a large sample, multiple thin sections were made. The trimmed blocks were shipped to Spectrum Petrographics (Vancouver, WA, USA), where they were prepared into thin sections, 50 × 75 mm with a thickness of 30 μm.

About 130 thin sections were made from the samples collected in 2001, 2002, and 2003 (Table 2.3). The sections were scanned on a flatbed scanner (Arpin et al. 2002) and examined with stereoscopic and petrographic microscopes at varying magnifications and illuminations: plane-polarized light (PPL), cross-polarized light (XPL), and dark field illumination; in addition, some were observed using blue light epi-fluorescence. Oblique incident light was used in some cases to distinguish organic material from charcoal and burned bone from stained bone (Courty et al. 1989). Descriptive nomenclature follows that in (Stoops 2003) and (Courty et al. 1989).

## Results

### Layer 8 (≡Bordes' Levels YZ)

Bordes originally identified three mostly anthropogenic basal layers (X, Y, and Z) above the bedrock that were similar in archaeological composition. He described them as burned areas and stated that the "bedrock is strongly rubefied by fire (Bordes 1975: 294)." He also distinguished the layers on the basis of texture: the first, X (roughly equivalent to our Layer 7), was "blackish gray with cryoturbated layers at the top"; the second, Y (the upper part of our Layer 8), was a "reddish layer at the front of the shelter that grades into *éboulis* toward the interior"; and the third, Z (the base of our Layer 8), was described as "resting on bedrock and reddened by fire". In light of the fact that it is difficult to clearly differentiate among them, we have grouped the lower two layers (Y and Z) into one major unit, Layer 8.

Overall, Layer 8 consists of massive to coarsely bedded dark reddish brown, greasy clayey, organic-rich silty sand with burned bone (some calcined) and flint, with little rock fall. The darkest levels are exposed along the west side of the entrance section, particularly in squares G-H 14 (Figs. 2.3, 2.5 and 2.6). The base of the layer is locally cemented by carbonate (see below), whereas the top of the layer exhibits a marked increase in abundance of anthropogenic components (Fig. 2.7). This layer is especially prominent in the south-western portion of the site (squares G-H-I, 12-13-14; Figs. 2.3 and 2.5), where it is thickest (~45–50 cm); however, it thins laterally to the north and south: it pinches out to the north along the D squares, just at the location where the bedrock floor rises, and it does not extend to the rear wall of the cave (Figs. 2.3 and 2.4).

Along the E, D, and I rows of squares, the organic zones interfinger with lighter brown, noncalcareous sand, whereas organic-rich sands are well exposed to the west along the F/G line indicating that the intense use of fire continued along this direction but not much to the north and south.

**Fig. 2.3** **a** General view of Pech IV showing clockwise, the South, West, and North Sections, respectively. *Arrow* points to red circular stain on bedrock thought by Bordes to represent the effects of burning. The dark layer resting on bedrock (Sq. H12-14) is Layer 8, which pinches out to the north (*right*) as the bedrock floor rises. The remains of large blocks of excavated roof fall constitute the main part of the South wall and cover layers 7 and 8. **b** West wall of Pech IV showing stratigraphic units of the current excavations and those of Bordes. Note the increased proportion of sandy sediments toward the north (*right*), particularly in Layers 4 and 3. **c** Detailed view of lower part of West Section and part of the South Section at *lower left*. Note the pinching out of Layer 8 at the base, the slope of the bedrock floor to the *left* (south), and the angular rock fall in the base of Layer 6

They appear to be concentrated under the large blocks of unexcavated rock fall occurring in the block of squares defined from G12 to I14 and represent the locus of repeated burning activity, which took place just inside the cave beneath the former dripline/brow (Figs. 2.3a and 2.5d).

It is important to note that the overall dark color of Layer 8, particularly where it occurs as cm-thick black organic bands and stringers (Fig. 2.7), reflects not so much the presence of charcoal, but that of char (gelified organic matter or fat from bones and meat formed by burning) (Ligouis 2006; Villagran et al. 2013) (Figs. 2.8b, d and 2.9a–c); in fact, it is this char that provides the "greasy" feel of the deposits in the field. Moreover, some of the darker bands are capped by thin, cm-thick bands of ash (Fig. 2.9d), and a number of them (especially in Sq. E13) tend to be cemented (Fig. 2.9d and 2.10c–f). This punctuated induration of only the ashes suggests that they were cemented relatively rapidly after deposition (Berna and Goldberg 2008; Madella et al. 2002), as the other sediments between them are not cemented. In any case, most of these couplets of dark band/ash represent in situ hearths/fires/burning events. On the other hand, parts of Layer 8 that consist of more diffuse dark bands or an overall homogeneous "dark" deposit (e.g., Fig. 2.9) are products of hearth rake out, locally accompanied by trampling whereby previously combusted materials were spread out before the next burning event took place.

In addition to the above, other localized microfacies variations occur in Layer 8. Along the south face of square F13, for example, 3–4-cm-thick red (2.5YR5/6) bands occur (Figs. 2.8c and 2.9a–f). At first, these were thought to be rubefied layers resulting from heating by fire. However, one 5-cm-thick continuous band (Fig. 2.9a, b) appears to be too thick to be caused simply by heating, especially over such a wide lateral extent. Such colorations were also found on the bedrock floor of the cave (Figs. 2.3a and 2.9c) and were similarly noted by (Bordes 1975). Finally, the fact that red coloration was observed to cut across the natural bedding suggests that at least part of the reddening is a post-depositional effect, possibly tied to oxidation of the organic matter in these lowermost layers (Karkanas et al. 2002).

Lastly, the basal part of Layer 8, which rests on bedrock, is locally cemented with micrite and microsparite, possibly, or at least in part, representing recrystallized ashes. Moreover, the sediment in these cemented areas is locally porous, with vesicles and channels within a compact matrix (Fig. 2.11). Presumably, these phenomena are tied to water flow that was concentrated along floor of the cave, as the overlying deposits above these basal 10–15 cm are not

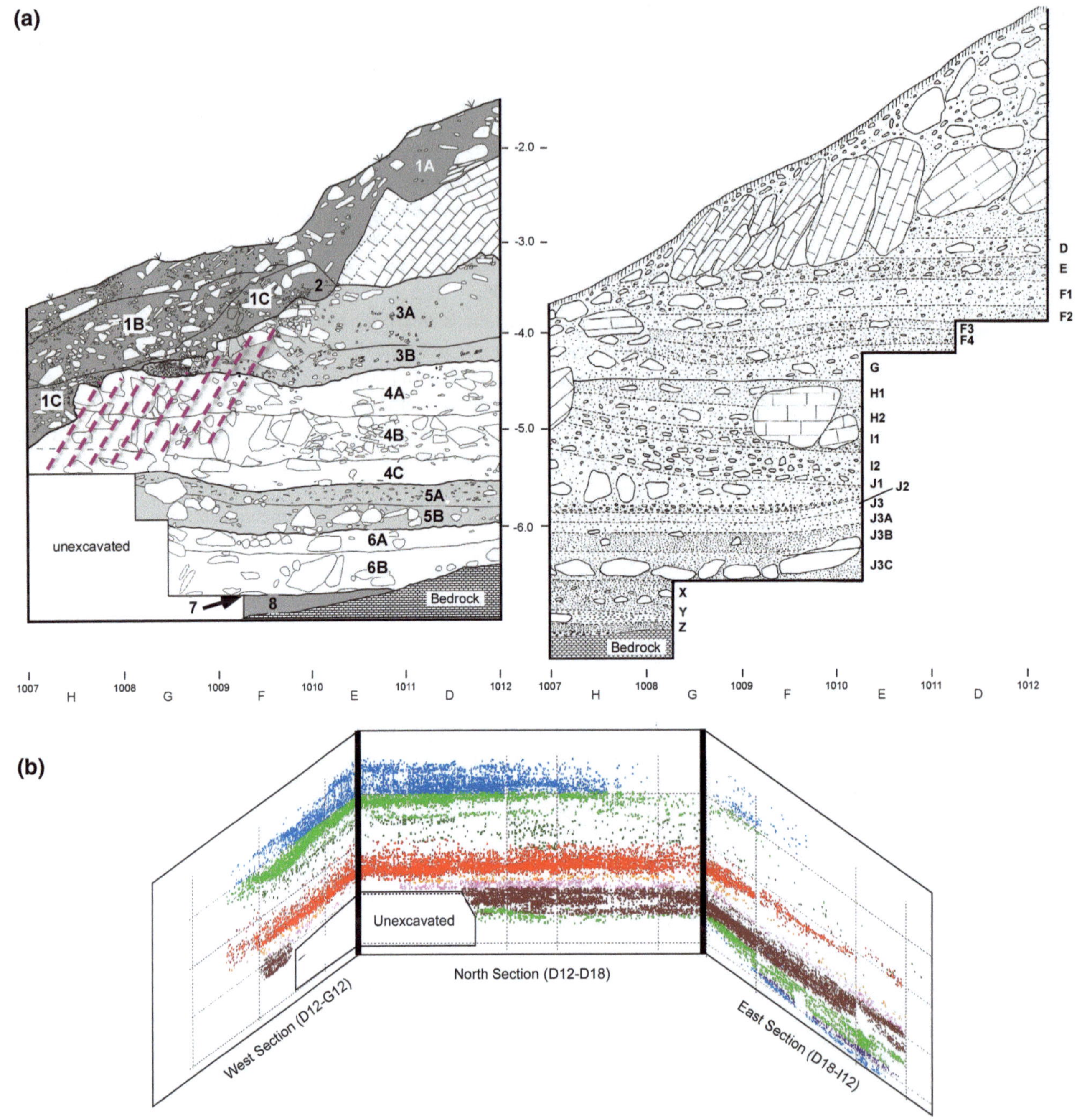


**Fig. 2.4** **a** *Left* Drawing of West section showing the stratigraphic units used in the current project. The *diagonal hatching* indicates strong secondary calcite cementation of the deposits, demarcating the former positions and retreat of the dripline (see text). *Right* Bordes' West section, which is approximately 3–4 m east of our final West Section. **b** *Panel diagram* showing lithic projections on three sections, colored by level, showing the lateral continuity across the site

cemented except for localized calcification of ashes (Fig. 2.11).

### Layer 7 (=The Uppermost Part of Bordes' Level X)

Bordes (1975) noted the presence of a ca. 5 cm unit of angular, worn, and abraded flints at the top of his XYZ sequence (Figs. 2.7e and 2.12). We decided to separate this thin, lithologically distinct unit from Bordes' Layer XYZ because of its generally geogenic character, lack of organic matter, and its lighter color. Overall, Layer 7 consists of cm-sized angular pieces of flint in a massive, strong brown (7.5YR5/8) sandy matrix; bones are relatively rare but also heavily rounded. Cryoturbated angular flint and bone-rich sand truncate Layer 8 in squares DEFG 13 and 14, but in

**Table 2.3** Location and field descriptions of thin sections

| Sample no. | Spectrum no. | Square | Layer | X | Y | Upper Z | Lower Z | Description |
|---|---|---|---|---|---|---|---|---|
| *2001* | | | | | | | | |
| PDA-4-1 | *LSI-1* | D11 | 3A | 999.721 | 1011.25 | −3.731 | −3.731 | Compact to weakly cemented clayey sand with flint chips and bone. Quite massive and homogeneous, with inclusions of rotten, rounded, cm-sized bedrock |
| PDA-4-2 | *LSI-2; LSI-3* | F19 | 8 | 1006.56 | 1009.99 | −6.812 | −6.998 | From very base of profile. Bottom ~3 cm weathered micaceous sandy limestone bedrock with some quartz that grades into and is overlain by dark reddish brown (5YR3/3 damp), noncalcareous, organic-rich "dirty" sand, somewhat washed. Much burned bone and flint. In one locality ca. 1 cm thick, locally concreted ash layer. Also, a few 1 cm diameter carbonate nodules/rhizoliths. Laterally, sediments are somewhat redder and look rubefied. In fact, on W entrance section, there are stringers of reddened sediment, charcoal and lighter zones that appear to be ash |
| PDA-4-3 | *LSI-4; LSI-5* | F19 | 8 | 1006.61 | 1010.01 | −6.802 | −6.681 | Organic-rich, greasy clayey/silty sand with abundant burnt flint and bone. Somewhat mottled and looks bioturbated, especially to S. Locally appears to be slightly rubefied (burrowed areas?). In all, appears to be burrowed hearth-like material. Organic-rich, greasy clayey/silty sand with abundant burnt flint and bone |
| PDA-4-4 | *LSI-6; LSI-7* | E19 | 6B | 1006.63 | 1010.16 | −6.461 | −6.655 | Bottom of Layer J, including remains of decayed roof fall at bottom. Above roof fall, sediment consists of crumbly organic-rich sand with some clay. Looks somewhat similar to sample 4-3, but lighter color. Many roots, and some cm in diameter root concretions but otherwise is noncalcareous |
| PDA-4-5 | *LSI-8* | F19 | 5B | 1006.77 | 1009.79 | −6.045 | −6.192 | Massive gritty clayey sand, locally with angular rockfall, bones, and flints, which are generally horizontal. Laterally quite extensive, although to N are larger rockfall. Possible clay translocation. Many pores and modern rootlets; some calcareous hypocoatings |
| PDA-4-6 | *LSI-9* | F19 | 6 | 1006.79 | 1009.77 | −6.218 | −6.359 | Similar to 4-5 but much stonier, with abundant bone and lithics. But overall, fine-grained matrix. Massive and laterally extensive but no bedding (trampled?) |
| PDA-4-7 | *MNJ-24* | G19 | 8 | 1006.46 | 1008.74 | −7.084 | −7.165 | Lightly stratigraphically lower than sample 4-2. About 4–5 cm thick of orangey sand in middle and black above and below. Black generally loose and rich in bone and roots. Appears burrowed |

(continued)

**Table 2.3** (continued)

| Sample no. | Spectrum no. | Square | Layer | X | Y | Upper Z | Lower Z | Description |
|---|---|---|---|---|---|---|---|---|
| PDA-4-8 | *LSI-10; MNJ-19* | D13 | 6B | 1001.5 | 1011.23 | −6.031 | −6.136 | From ledge of sediment on top of large tabular slabs of roof fall. Massive, soft, poorly sorted silty clay coarse sand with larger granule inclusions. Many flint chips which are commonly vertically oriented |
| PDA-4-9 | *LSI-11* | F19 | 5B | 1006.75 | 1009.68 | −5.902 | −5.983 | About 10–15 cm thick. Poorly sorted sandy gravel, consisting of rounded and spherical 1 cm in diameter limestone gravel and about 2 cm subangular to sub-round gravel. Clearly eroded into Layer J and corresponds to early stage of roof collapse of the overlying blocks in Layer I. Back plots along the 18 line show somewhat lower densities, possibly because of erosion. Seems to represent small, localized channels, although difficult to determine direction of flow |
| PDA-4-11 | *LSI-11* | E19 | 4C | 1006.72 | 1010.37 | −5.292 | −5.415 | Stony sand. Calcareous, massive and mixed with limestone blocks, varying from rounded cm-sized clasts to angular pieces about 1–5 cm across. Many roots. Locally cemented with calcite in broad zones. Many blackened bones; several flakes |
| PDA-4-12 | *LSI-13* | E19 | 5A | 1006.71 | 1010.6 | −5.572 | −5.641 | Similar to sample 4-11 and is stratigraphically above 4-10 but ~80 cm to N and ~30 cm below sample 4-11. More cemented than 4-11 and several mm-sized (modern?) root casts of carbonate. Several unburnt bones |
| PDA-4-13 | *LSI-14; LSI-15* | D13 | 4A | 1006.84 | 1011.74 | −4.758 | −4.851 | Weathering rind on bedrock consisting of |
| | | | | | | | | 1: Yellow (10YR7/6) soft, calcareous (?) sand. Hard in upper part |
| | | | | | | | | 2: Very pale brown (10YR8/4) chalky sand with many fine pores and roots. Looks like moon milk. Abundant fine roots at contact between 1 and 2 |
| | | | | | | | | 3: Similar to 1 and grades laterally from 2 |
| PDA-4-14 | *LSI-16* | D19 | 3 | 1006.82 | 1011.54 | −4.579 | −4.666 | Additional transition sample with part closer to wall being similar to 4-13 (3). The part away from the wall is more compact sand, which is typical of the finer sediment in the shelter |
| PDA-4-15 | *MNJ-20* | D19 | 3 | 1006.85 | 1011.18 | −4.654 | −4.745 | Compact clay sand with rhizolithic concretions |
| PDA-4-17 | *LSI-17* | E-19 | 3B | 1006.86 | 1010.87 | −4.686 | −4.6 | From gravelly pocket in the middle of E: compact mixture of weathered (phosphate rinds?) limestone clasts in a yellowish red (5YR 5/8) silty sand matrix. Matrix has fine pores and appears to be decalcifying. Limestone clasts vary from cm-sized rounded up to 8–10 cm flattish sub-rounded to subangular clasts, which are scattered throughout and are matrix supported. At base, some are |

(continued)

**Table 2.3** (continued)

| Sample no. | Spectrum no. | Square | Layer | X | Y | Upper Z | Lower Z | Description |
|---|---|---|---|---|---|---|---|---|
| | | | | | | | | horizontally bedded and perhaps in cryoturbation pockets. Entire feature is ~100 by 100 cm. Matrix on both sides of feature is similar red clayey sand |
| PDA-4-18 | *LSI-18* | D19 | 5B | 1006.75 | 1011.43 | −5.771 | −5.825 | Similar to sample 4-15. Soft clayey sand. Mottled with localized clay concentrations as well as lighter washed zones. A few rare rhizoliths. Many roots. Looks like percolation zone |
| PDA-4-19 | *LSI-19; LSI-45* | G19 | 4C | 1006.71 | 1008.53 | −5.399 | −5.458 | Just above large rock fall in Sq. F19; from stony sand just above this. Locally cemented. Collected between rock fragments |
| PDA-4-20 | *LSI-20; LSI-21* | G11 | 4A | 999.025 | 1008.8 | −4.39 | −3.96 | Top of cemented zone beneath layers 1A and 1B that unconformably sits on it. Many flattish to circular, cm-sized stones. Clasts are rounded, some with weathered/AP(?) thin patchy rinds. Seems almost to be a laminar crust at the very top of W face |
| PDA-4-22 | *LSI-23* | D11 | 3A | 999.853 | 1011.67 | −3.868 | −3.96 | About 35 cm below PDA-4-1, on N end of square. Here, sediments appear to be somewhat finer grained, i.e., greater proportion of matrix to weathered limestone debris. This matches the general trend of finer grained material closer to the N, rear wall |
| PDA-4-23 | *LSI-24* | D11 | 3B | 999.841 | 1011.73 | −4.023 | −4.124 | From finer unit ca. 15 cm below sample 4-22. Soft clay sand with remnants of weathered limestone pieces. Some bone, flint chips. Becomes coarser to the south |
| PDA-4-24 | *LSI-25; MNJ-21* | D11 | 4A | 999.842 | 1011.91 | −4.553 | −4.702 | Massive, dense clayey sand, with few artifacts |
| PDA-4-25 | *LSI-26* | D11 | 4A | 999.843 | 1010.95 | −4.421 | −4.368 | At zone of interfingering between fine sandy facies and stony sandy facies. Vertically in between samples 4-23 and 4-24. Poorly sorted silty sand with some carbonate roots. Looks calcareous. Gets stonier to S |
| PDA-4-26 | *NTO-002; NTO-003* | E11 | 3B | 999.836 | 1010.65 | −4.41 | −4.285 | Lateral equivalent to boundary across samples 4-23 and 4-25. Sample spans contact between finer sediment below and more angular limestone above. On freshly broken surface contact not clear but perhaps in thin section. Could be similar to sample 4-17 but fewer cryoturbation fragments appear. Angular bone and flint chips throughout. Limestone fragments appear weathered. Sample ~85–90 cm below I-2/I-3 contact |

(continued)

**Table 2.3** (continued)

| Sample no. | Spectrum no. | Square | Layer | X | Y | Upper Z | Lower Z | Description |
|---|---|---|---|---|---|---|---|---|
| PDA-4-27 | *LSI-27* | E11 | 3A | | | −3.768 | −3.836 | Appears to be a lobe-like mass that eroded into underlying 3. Seems to be oriented NW to SE. Sediment is crumbly, partly cemented calcareous silt sand with limestone clasts that are generally weathered and sub-rounded. Some roots. In hand lens looks partly decalcified and jumbled; possibly colluvium or solifluction |
| PDA-4-28 | *LSI-28* | D11 | 4B | 999.798 | 1011.53 | −4.876 | −4.967 | Continuation of soft clay sand below sample 4-23. Very bottom could be in top of J as it is somewhat stony and lighter in color. Equivalent unit to S is stonier but some weathered stones here as well |
| PDA-4-29 | *LSI-29* | D11 | 5A | 999.623 | 1011.73 | −5.575 | −5.681 | Soft crumbly sand with abundant bone and flint that seems clast supported. Somewhat darker than other samples but gives the same Munsell color of 5YR 5/8. Widespread across the site and rich in bone and flint but here it is burrowed either by animals or by *clandestins* |
| PDA-4-30 | *LSI-30* | D11 | 4B | 999.794 | 1011.01 | −5.199 | −5.266 | Small sample from between stones of soft reddish sand. Some bone; locally quite stony |
| PDA-4-31 | *LSI-31* | F11 | 5A | 999.878 | 1009.94 | −5.764 | −5.813 | Soft, crumbly bone-rich sand. Seems to be more calcareous than reddish sand but with same old color here. Bones are ~ cm in size, and some appear rounded and even polished (cryoturbation??). Black sediment on front part of sample could be terrier or related to wall construction |
| PDA-4-32 | *NTO-004; NTO-005* | F11 | 6A | 999.932 | 1009.79 | −6.039 | −6.158 | Just below layer of loose rockfall. Upper part is coarse and grades to more sandy at base. Limestone pebbles quite rounded [probably by cryoturbation]. Rich in lithics and bones. Asinipodian layers |
| PDA-4-33 | *LSI-32* | D11 | 5B | 999.89 | 1011.26 | −5.784 | −5.861 | Rocky sediment from cryoturbation level. Many well-rounded limestone pebbles and artifacts have worn appearance. Difficult to collect sample because of stoniness. The large sample with much padding might be somewhat disturbed. Also collected smaller sample of mostly fine clay sand between rockfall. This layer thickens to S and is below 4-31 and 4-32. Possibly equivalent layer of "gravel" in sample 4-9 |
| PDA-4-34 | *LSI-33* | H10 | 1A | 999.009 | 1007.51 | −3.777 | −3.78 | Clayey sand with many aggregates, roots, and tabular rock fragments with slabs measuring about 8 cm thick by about 30–40 cm across. Some more rounded cm-sized clasts |

(continued)

**Table 2.3** (continued)

| Sample no. | Spectrum no. | Square | Layer | X | Y | Upper Z | Lower Z | Description |
|---|---|---|---|---|---|---|---|---|
| PDA-4-36 | *LSI-34* | G10 | 4A | 1008.16 | −4.368 | −4.368 | −4.409 | From and above contact with 3B. Two subsamples: upper (4-36) is composed of 1–2 cm in diameter, angular to sub-rounded limestone fragments produced by cryoturbation/solifluction. Subsample 36 bis is just below this and consists of cemented reddish sand with many fine pores and carbonate hypocoatings. Seems to be eroded by cryoturbation gravels |
| PDA-4-37 | *LSI-35; LSI-36* | D18 | 8 | 1006.79 | 1011.77 | −6.167 | −6.327 | From just above bedrock. "Greasy", moist, organic sand. Many flints and bones, some of which are burnt. Some rockfall. Roughly equivalent to sample 4-4. To S gets slightly lighter color and looks bioturbated |
| PDA-4-38 | *LSI-37* | Pech I | – | – | – | – | – | Cross-bedded Middle to Upper Coniacian sandstone collected from the wall made by Bordes in Pech I |
| PDA-4-39 | *LSI-38* | Pech I | – | – | – | – | – | Lower Coniacian marly limestone from next to bike path below Pech I. Possibly bioturbated |
| PDA-4-40 | *LSI-39; LSI-40* | D18 | 4A | 1006.8 | 1011.51 | −4.952 | −5.076 | Massive sand with some carbonate rhizoliths. Uppermost is coterminous with bottom of sample 4-13 |
| PDA-4-42 | *NTO-006; NTO-007* | D18 | 4C | 1006.77 | 1011.58 | −5.288 | −5.397 | Similar to 4-40 and -41 but perhaps richer in clay. In fact, appears to be broad domains of perhaps translocated clay. Perhaps darker (5YR5/6). Some flint at bottom, fewer rhizoliths. Perhaps sandy zone next to wall is produced by decalcification of sandy bedrock followed by translocation of clay |
| PDA-4-43 | *LSI-41* | D18 | 5A | 1006.78 | 1011.45 | −5.516 | −5.516 | Same as -41 and -42 but, more modern roots; more flints, larger rhizoliths; clayey domains appear more prominent and larger |
| PDA-4-44 | *NTO-008; NTO-009* | D18 | 5A | 1006.79 | 1011.44 | −5.618 | −5.741 | Between samples 4-34 and 4-18. Similar to others with abundant fine roots. Flints at an angle here and in sample -42 (bioturbation?). Clear clay band at bottom and clayey domains |
| PDA-4-45 | *LSI-42* | E11 | 3A | 999.831 | 1010.74 | −3.613 | −3.708 | Soft, poorly sorted massive calcareous silty sand with modern roots; 1- to 4-cm-sized fragments of angular to rounded limestone. Flint and bone chips, many burnt. Could be soliflucted level and some cryoturbation |
| PDA-4-46 | *MNJ-22* | E11 | 3A | 999.822 | 1010.73 | −4.076 | −4.076 | Generally similar to sample 4-45 but fewer roots, more compact and massive |
| PDA-4-47 | *LSI-43; LSI-44* | E11 | 4A | 999.825 | 1010.58 | −4.504 | −4.608 | Transition between red sandy facies and stony sand facies. Some bone. A few cm-sized stones. Locally cemented. More stony to S |

(continued)

**Table 2.3** (continued)

| Sample no. | Spectrum no. | Square | Layer | X | Y | Upper Z | Lower Z | Description |
|---|---|---|---|---|---|---|---|---|
| PDA-4-48 | *NTO-010; NTO-011* | E11 | 4A | 999.833 | 1010.35 | −4.683 | −4.779 | Slightly lower and more stony than sample 4-47. Below layer of angular stones |
| PDA-4-49 | *LSI-46; LSI-47* | F11 | 4B | 999.891 | 1009.76 | −4.816 | −4.961 | Fine, cm-sized limestone fragments in a sandy matrix. Large rock fall to the S. Looks like cryoturbation zone |
| PDA-4-50 | *LSI-48; LSI-49* | F11 | 4A | 999.449 | 1009.02 | −4.614 | −4.81 | Large cemented block of breccia with bone, many pores, calcite hypocoatings that appear partially replaced by apatite. Pebbles about 1–2 cm in diameter. Looks like phosphatic vegetated spring deposit |
| PDA-4-51 | *LSI-50; LSI-51; LSI-52; LSI-53* | E13 | 8 | 1001.92 | 1010.3 | −6.603 | −6.938 | Massive to coarse bedded, greasy organic sand. Some cm-thick reddish levels. Between is ~3–7 cm sediments that are concreted with very finely crystalline, shiny cement. Very hard. Loose material is called "4-51 base." Possible thin mm stringers of ash and darker bands of diffuse charcoal. Essentially similar to W part of entrance trench |
| PDA-4-52 | *MNK-28* | E13 | 8 | 1001.8 | 1011.06 | −6.54 | −6.673 | Organic-rich sand with weak brown/red bands. Much burnt flint and bone. Rests on hard, cemented band about 3 cm thick. About 60 cm N of sample 4-51. Appears less bioturbated than originally thought |
| PDA-4-53 | *LSI-54* | E13 | 8 | 1001.77 | 1010.96 | −6.73 | −6.682 | Just below sample IV -52. Cemented layer (calcite?) roughly equivalent to hard layer at base of -sample 4-51. Porous and very open work. Possible clay coatings. Some cm-sized limestone fragments tightly cemented |
| PDA-4-54 | *MNJ-25* | Sq. E13 | 8 | 1001.81 | 1010.94 | −6.784 | −6.828 | Just below sample 4-53 but distinct fine conglomerate layer. Very hard and porous |
| PDA-4-55 | *MNK-29* | Sq. E13 | 6B | 1000.24 | 1010.62 | −6.165 | −6.248 | Well-cemented massive sand with abundant flint chips oriented at a variety of orientations. Some charcoal, bone splinters. Large block. Sediments less cemented away from the rock in this square |
| PDA-4-56 | *MNJ-23* | G14 | 8 | 1002.65 | 1008.05 | −7.1 | −7.185 | From basal 7 cm: bands and red and charcoal-rich sediment similar to that of sample -51 |
| PDA-4-57 | *LSI-55; LSI-56; LSI-57; LSI-58* | G14 | 8 | 1002.63 | 1008.15 | −6.767 | −7.093 | 32 cm-long section of partly cemented, cm-thick charcoal-rich bands, rubefied. Many bones, burnt flint. Many white spots with white porous material that resembles small worm tubes. Calcined bone. Dark areas (dusky red—2.5YR3/2), "greasy" organic sand. Quite damp. Possible rubefied aggregates (1–2 cm) or pieces or weathered bedrock |

(continued)

**Table 2.3** (continued)

| Sample no. | Spectrum no. | Square | Layer | X | Y | Upper Z | Lower Z | Description |
|---|---|---|---|---|---|---|---|---|
| 2002 | | | | | | | | |
| PDA-101 | *NTO-012; NTO-001??* | E13 | 8 | 1001.7 | 1010.99 | −6.838 | | Very hard, cemented angular pebbles/scree. Flint, burnt bone and some charcoal. Many pores and intergranular voids. Quite crystalline and fresh |
| PDA-104 | *NTO-013; NTO-014; NTO-015* | D11 | 5A | | | | | Very loose mixture of burned bone, sand and stone next to large burrow sampled last year. Layer goes around to N wall where it is more compact and locally cemented. Overall, bones are quite large in this layer, which dips to SE according to lithics and bone plots; in Layer 3A, they tend to be horizontal |
| PDA-105 | *NTO-016* | F11 | 5A | 999.715 | 1009.98 | −5.481 | | Massive to cemented strong brown, silty sand with bone and flint. Laterally with angular stone fragments. Many roots and some rhizoliths. Locally cemented (1 m back, the same layer is overlain by large angular blocks of roof fall.) |
| PDA-106 | *NTO-017* | H18 | 5B | 1006.36 | 1007.46 | −5.75 | 5.794 | Fine gravelly sand with sand-sized grains of quartz and limestone. Rounded and angular, decimeter-sized blocks of rock fall. Southernmost part on E wall here. Just above Bordes' wall. Some roots |
| PDA-107 | *NTO-018; NTO-019* | D13 | 5B | 1001.53 | 1011.53 | −6.663 | | Moist pebbly sand, moderately concreted and rests above very cemented gravel unit. Some limestone but seems to be overall noncalcareous. Apparent clay films. Some bright red (heated?) rocks, burnt flakes. Definitely seems clayier than others. Generally horizontal and rests on concreted layer that slopes upward along the bedrock floor toward the back of the cave. Laterally equivalent to Layer 6 in central part of the cave to the S |
| PDA-108 | *NTO-020; NTO-021* | C13 | 6B | 1001.93 | 1012.04 | −6.343 | −6.435 | Two lithologies: Basal—organic-rich, clayey (?) sand with many burned flints. Seems to rest on bedrock, some of which seems fire reddened. Overall seems noncalcareous. Bone fragments. Some rounded cm-sized limestone grains that appear weathered and decalcifying. Lighter, upper part is sandy and seems to slope up against the back wall |
| PDA-109 | *NTO-022* | C13 | 6A | 1001.64 | 1012.12 | −6.107 | | Soft with abundant flint and bone fragments, mostly burned. No orientation or bedding, and artifacts are at different orientations (cryoturbation?) |
| PDA-110 | *NTO-023* | E13 | 8 | 1001.54 | 1010.1 | −6.877 | | Hard, calcite-cemented "breccia", consisting of burnt bone fragments (1–2 cm), <5 cm above bedrock and likely cemented because of its position above bedrock. Overlain and surrounded by reddish sediment in this area |

(continued)

**Table 2.3** (continued)

| Sample no. | Spectrum no. | Square | Layer | X | Y | Upper Z | Lower Z | Description |
|---|---|---|---|---|---|---|---|---|
| PDA-111 | *NTO-024; NTO-025; NTO-026* | F13 | 8 | 1001.62 | 1009.96 | −6.745 | −6.905 | Ca. 15 cm block that includes BF-1, BF-3 and possibly BF-5 at base. Difficult to observe in field because of moisture content and poor lighting, but overall consists of dark layer at top (BF-1; 0–6 cm) with abundant burnt bone and flint, over ~4–5 cm of lighter, reddish band in middle overlying darker unit at bottom with many burnt bones (BF-3). Sample does not include cemented unit at bottom, nor Layer 7 |
| PDA-113 | *NTO-028* | plateau | | | | | | On plateau just N of site, ca. 25 m from where path reaches plateau, steep slope down to valley, just N of cave. Sample from base of tree throw: Clayey fine sand with locally clay coats around roots. Looks like Bt horizon |
| PDA-115 | *NTO-029; NTO-030* | E13 | 8 | 1001.26 | 1010.55 | −6.862 | −6.763 | Concretion with vertical tube-like hole in the center. Appears to have vughy, geode like crystals of calcite. Could represent old drip line or stal or water emanating from phreatic water trapped at bedrock sediment contact. Some of artifacts and bones nearby are vertical and limestone fragments have different angles of dip. Only ca. 50 cm from cemented gravel in Sq. E13 |
| PDA-116 | *NTO-031* | F11 | 5A | 999.717 | 1009.26 | −5.61 | | Loose to compact, chaotic mixture of sand, flint and bone. Sand seems to vary from well sorted fine quartz to poorly sorted quartz and calcite sand. Pebbles are angular, flat to rounded and equant. Should be similar to sample -33 from D11 but is from different layer ~20 cm lower |
| PDA-117 | *NTO-032* | D12 | 7 | 1000.71 | 1011.81 | −6.476 | | Cemented sand with moderate porosity and zones of washed sand. Also looks like clay coatings locally. Seems to be above large roof fall above Layer 7 |
| PDA-118 | *NTO-033* | E11 | 5B | 999.593 | 1010.74 | −5.966 | | Rounded limestone gravel with cemented sandy matrix. Perhaps some pendants on stones |
| PDA-119 | *NTO-001* | E11 | 5B | 999.825 | 1010.44 | −5.954 | | Similar to sample -118 but more matrix. Sand, mostly noncalcified, with clay and iron concentrations. Curious that both fresh artifacts and rounded, worn chunky types of flints occur together. Many pebbles seem to exhibit Mn staining on them (possibly inherited) |
| PDA-120 | *NTO-035; NTO-036; NTO-042* | D11 | 5B | 999.769 | 1011.93 | −5.822 | −5.996 | Moist, soft, massive sand with scattered rock fragments and very abundant bone and flint at top of the layer; this is true elsewhere as well. Artifacts are fresh and angular. Layer 5B with rounded limestone [also 1 flat basalt pebble]. Lower part in Layer 6 should be roughly equivalent to PDA4-8 and is slightly above PDA4-55 |

(continued)

**Table 2.3** (continued)

| Sample no. | Spectrum no. | Square | Layer | X | Y | Upper Z | Lower Z | Description |
|---|---|---|---|---|---|---|---|---|
| PDA-121 | *NTO-037* | D12 | 7 | 1000.07 | 1011.43 | −6.592 | | Lower part: Layer 7 consists of fine, angular gravel composed of quartz, limestone, and rolled flint. Quite porous, with mostly packing voids between clasts; fines look winnowed. Cemented with microspar as in Layer 8 and sample rests on bedrock. Some secondary clays in voids. Upper part: Layer 6 is composed of poorly sorted, partly calcareous sand |
| PDA-122 | *NTO-038* | H13 | 5B | 1001.3 | 1007.39 | −5.994 | | Cemented clayey sand with rhizoliths |
| PDA-123 | *NTO-039* | H13 | 5B | 1001.28 | 1007.28 | −5.495 | | Similar to -122 but somewhat more cemented. Finer rhizoliths and no stones |
| PDA-124 | *NTO-040* | H11 | 4C | 999.843 | 1007.86 | −5.387 | | Cemented clayey sand with some roots and possibly fine rhizoliths. Looks like cemented area situated below former drip |
| PDA-125 | *NTO-041* | C16 | 8 | 1004.65 | 1012.19 | −6.212 | −6.293 | Sample has 2 parts. Lower is composed of silty sand with small domains of silt mixed with fine organic matter; possible siltans within fissures. Some limestone granules; crumbly and looks bioturbated with possible worm casts or insect burrows. Lower part, ~7 cm thick, rests on bedrock and seems to pinch out in Sq. 17. Locally cemented; modern roots. Upper part: only a small part of which is preserved here and most in sample 125 bis. Sits on the lower part with a sharp contact. Compact organic-rich silty sand with many burn bones and flints. Fairly massive but weakly bedded. Very similar to the base of Layer 8 |
| 2003 | | | | | | | | |
| PDA-4-201 | *OYF-001* | D12 | 6B | 999.979 | 1011.51 | −6.165 | −6.29 | Massive clayey sand, with angular chips of burnt bone, 1–2 cm in diameter; likely anthropogenic. Abuts against roof fall in E12. Fairly horizontal bedding, and with slight dip |
| PDA-4-202 | *OYF-002; 003* | DE12 | 6B | 999.973 | 1011.02 | −6.218 | −6.293 | Laterally equivalent to -201 |
| PDA-4-203 | *OYF-004* | F14 | 7 | 1002.54 | 1008.97 | −6.742 | 6.815 | Washed sand with well-rounded, edge-damaged flint artifacts. Some limestone chips are fresh and angular |
| PDA-4-204 | *OYF-005* | F14 | 7 | 1002.05 | 1009.13 | −6.75 | | Seems washed. With large weathered limestone block and marbled lighter and darker matrix. Partly cemented although friable. Variable thickness. Bedrock, locally rotten |
| PDA-4-205 | *OYF-006* | G13 | 7, 8 | 1001.81 | 1008.81 | | −6.799 | Cemented sandy silt. Looks washed as in sample −204. Thin tubes of calcite (hypocoatings) but could be in process of dissolving. Not many angular stones |
| PDA-4-206 | *OYF-007* | F14 | 7, 8 | 1002.47 | 1009.08 | −6.724 | −6.839 | Upper part cemented with some angular stones |

(continued)

**Table 2.3** (continued)

| Sample no. | Spectrum no. | Square | Layer | X | Y | Upper Z | Lower Z | Description |
|---|---|---|---|---|---|---|---|---|
| PDA-4-207 | *OYF-008* | F14 | 8BB1, 8BB2 | 1002.56 | 1009.08 | −6.796 | −6.83 | Top is 1–2 cm-thick cemented ash overlying soft dark reddish brown charcoal/organic matter-rich sand. Bones are black and likely burned; some calcined |
| PDA-4-209 | *OYF-009; -010* | E11 | 8 | 999.78 | 1010.27 | −6.759 | −6.884 | At base is "nougat-like" zone of reddish sand with cm-sized pebbles of white and reddish limestone. Some charcoal. This nougat climbs up along bedrock and is about 2–3 cm thick. Sample is about 11 cm thick. Bottom 5 cm has splotches of washed sand and overall are quite sandy. Not cemented but quite compact |
| PDA-4-210 | *OZV-001; -002* | F11 | 8 | 999.779 | 1009.52 | −6.817 | −6.98 | Sublayers of charcoal and rubefaction. Bedrock stained red just under sample. Washed at base |
| PDA-4-213 | *OYF-011* | J14 | 8 | 1002.69 | 1005.07 | −7.181 | −7.29 | On west side of cave entrance, from more sterile part of Layer 8, just above bedrock. Lower part sand, locally cemented and with some carbonate hypocoatings. Lower part of sample could be part of weathered bedrock. Flint and bone chips; some brown (organic matter?) dusty matrix. Abundant cultural material. Sand could be washed |
| PDA-4-215 | *OYF-012; -013; -014* | F12 | 8 | 1000.95 | 1009.39 | −6.775 | −6.995 | Layers 8AA1, 8BB1 and 8BB2; 8CC (orange layer) and 8DD. Sample also cracked in the upper 1/3, along 8CC/8DD, but probably all right. Upper part richer in bones and lithics. Lowermost part of 8CC is washed. Middle part has ghosts of limestone |
| PDA-4-216 | *OYF-015; -016; -017* | F12 | 8 | 1000.05 | 1009.34 | −6.797 | −6.997 | Upper part is mostly charcoal rich. Layer 8AA is thin veneer at top (ca. 1–2 cm); darker 8BB (~8 cm thick); redder 8CC (~5–6 cm thick) on top of darker 8DD (~2–3 cm thick). NO Layer 8EE here. Sample broke in the middle and lowermost part is disturbed. Loose samples of 8BB [dark reddish brown; 5YR3/3] and 8CC [reddish brown; 5YR4/4]. Red layer consists of poorly sorted washed sand but with burnt bone and decayed limestone. Darker 8BB is also washed sand and some carbonate hypocoatings. To west (F11) this massive dark layer spits into 3 distinct dark zones. Here, red staining on floor also increases |

square I13 it grades into sand. The contact between Layers 7 and 8 is distinct and undulates slightly (Figs. 2.5a and 2.9e). In the northern portion of the site, Layer 7 directly overlies bedrock, whereas in the southern portion it is present intermittently as patches between blocks of roof fall (Figs. 2.6 and 2.9a, b). It is generally about 5 cm thick, but in certain areas (e.g., square G14) it reaches up to 10 cm where it has been forced up between two pieces of roof fall (Figs. 2.5d and 2.7e). The upper surface of Layer 7 dips to the west and southwest pointing to erosion and/or compression of the large blocks that fell in the early stages of Layer 6B time.

**Fig. 2.5** **a** South section at end of 2002 season from bedrock at the base up to Layers 4 and 1 at the very top. Note that Layer 8 thins from *left* to *right* (east to west). Note the thin band of Layer 7 above it, and the sandy infillings between the blocks of Layers 5 and 6. The bench in the foreground is situated mostly in squares G12-14. **b** View of western half of the site in the 2001 season, showing partially excavated Western and Southern sections. Note the flat-lying limestone slabs in the center of Layer 6 that signals the initial collapse of the cave roof from within the confines of the cave; later, the brow retreated northward beginning with the accumulation of the large blocks in Layer 4. **c** View of southern area of the site in the 2001 season, showing the entrance corridor and imbricated roof collapse spanning Layers 6, 5, and 4. **d** Section along west side of the corridor leading into the site (see Fig. 2.2), showing Layers 8 through 4. Note the imbrication of the large slabs of collapsed bedrock roof fall and patches of Layer 7. The two vertical lines in Layer 8 in square G14 indicate the limits of sample 51. Note that Layer 8 tends to be thickest in squares H and I 14, and thins away from this: in square I14, dark organic sands grade into lighter colored sand to the *left* (south) as shown by the *dashed line*. As the blocks of roof fall overlie Layer 8, it is clear that the burning activities were situated just inside cave from the brow

**Layer 6 (≡Bordes' Levels J3, J3a, J3b, and J3c).** As with many units, Layer 6 is laterally variable from north to south, being generally sandy in the north (e.g., D–E lines), but becoming stonier to the south. In addition, the lower part is characterized by large (>1 m sized) blocks of roof fall that rest on Layer 7 (Figs. 2.3, 2.4 and 2.5a). Because of these latter differences, the layer was subdivided into two subunits, Layers 6A and 6B.

The basal part of Layer 6 (6B) is characterized by large, flat-lying blocks of limestone roof fall that are in direct

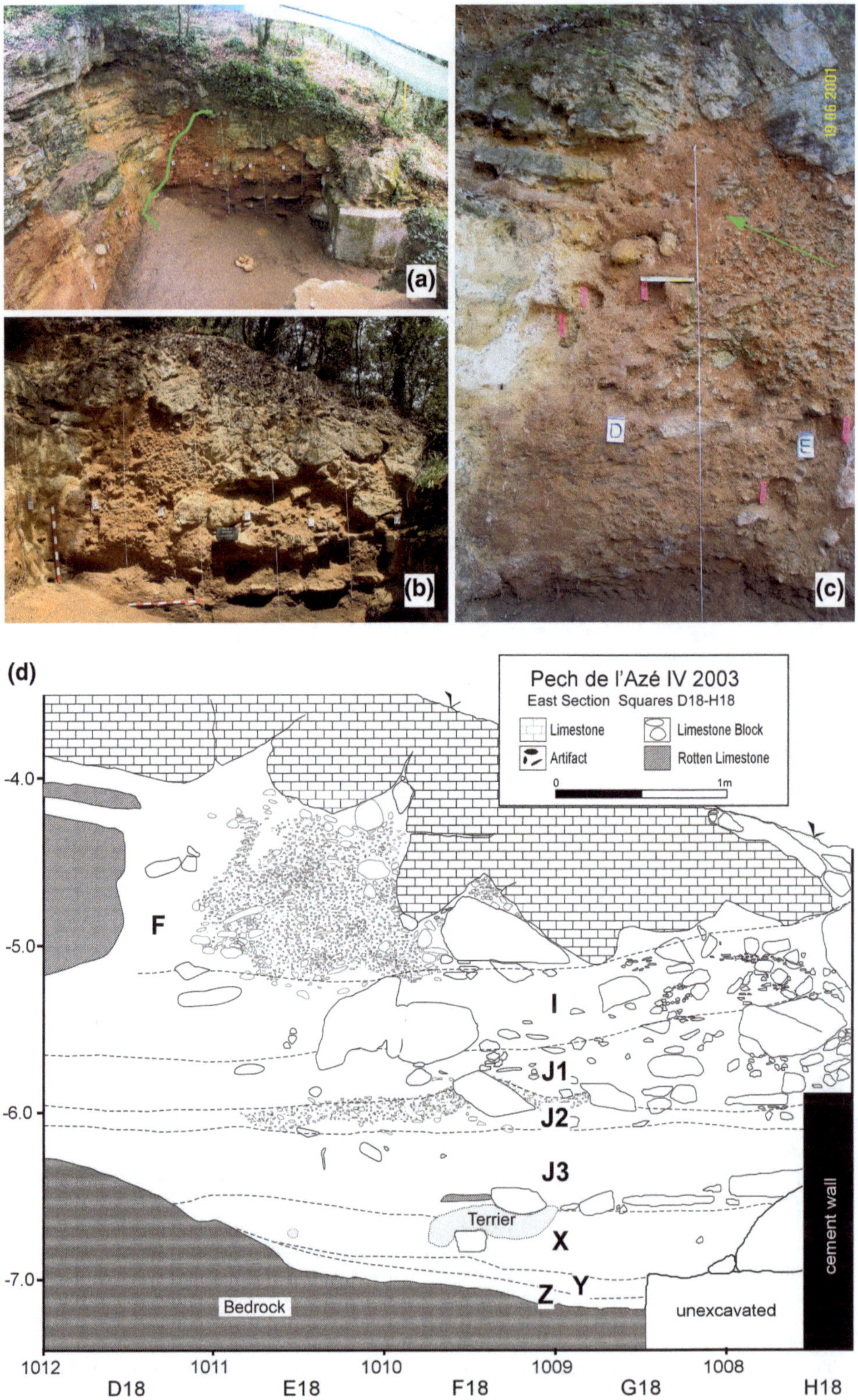


**Fig. 2.6** **a** Photograph of East section at end of 2002 season. Note how the bedrock wall on the north (*left*) side of the photo extends to the south and that the roof of the collapsed cave is more evident here in contrast to the West section, which is less close to the north wall, and nearer to the interior of the concealed entrance to the main cave. **b** East section at end of the 2003 season. Note the thickness of the rock fall at top that represents collapse of the roof that here is considerably more intact than on the West section. **c** Detail of East Section at the contact between the weathered bedrock wall at left (north) and the sedimentary fill. The arrow points to a zone of decalcified bedrock, characterized by quartz sand, some heavy minerals, and extensive red iron-rich dusty clay coatings. **d** Drawing of East Profile shown in 6B. Note that it was not possible to achieve the stratigraphic definition of the units as was the case for the West Section, so layer attributions are approximate

contact with the upper surface of Layer 7 (see Figs. 2.3c and 2.4). These blocks tend to be tabular, about 20–30 cm in average thickness, and reach up to 120 cm in length, although the larger block sizes are concentrated in the western portion of the site (squares D–E–F/12–13). Although most tend to be flat lying, many are on edge and are imbedded in the layer obliquely (Figs. 2.3, 2.4 and 2.5). Remains of some of the slabs can be still seen on the witness section of the north wall. Layer 6B also contains smaller limestone blocks, cobbles, and gravel, as well as

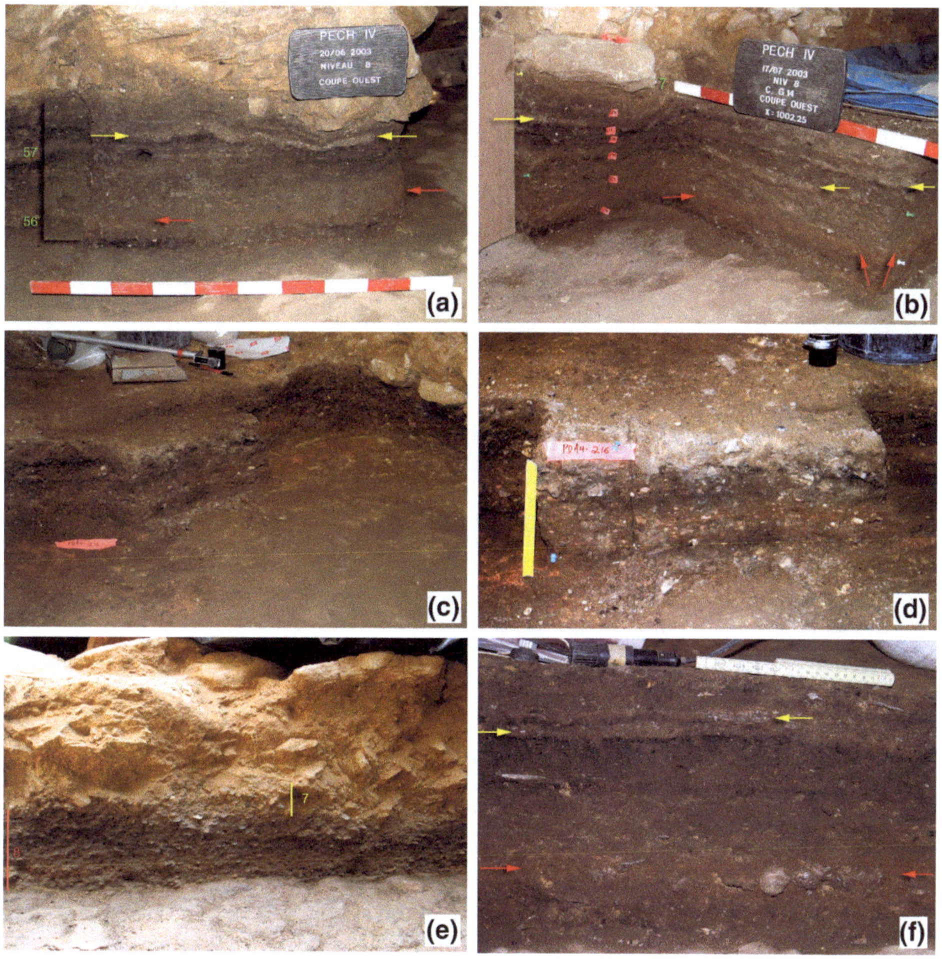


**Fig. 2.7** **a** Layer 8 looking west along the entrance trench in square J14. The *red arrows* point to reddened zones thought by Bordes to represent burning associated with hearths; rather they are apparently diagenetically formed. The *yellow arrows* indicate layers of intact calcareous ashes that overlie organic-rich zones and together represent intact hearths. Elsewhere, the dark bands are more diffuse and a result of hearth rake out and spreading out of previously combusted materials. Samples 56 and 57 come from the column in the *left hand* part of the photo. **b** Another view of Layer 8 looking southwest in sq. G14. *Yellow arrows* are ash layers whereas *red* ones point to reddish bands previously thought to be produced by heating. Layer 7 is the thin pale yellow deformed zone just underlying the piece of roof fall at the left. **c** Sample 216 from Layer 8. Note the red staining of the bedrock floor to the right. **d** Detail of sample 216 from Layer 8, showing the overall homogeneity of the deposits other than the diffuse black band at the *top*. Note the red staining on the bedrock floor and the *reddish band* in the *middle* of the photograph that stains limestone clasts. Sample 215 was removed just to the left of 216. Scale bar is 18 cm. **e** South face of Sq. F14 showing Layer 8 (*red line* is 20 cm thick), overlain by irregular cryoturbated lenses of worn flints and stones of Layer 7. **f** Layer 8, South section. Note the diffuse darker band ~10 cm below the surface, which represents hearth rake out

anthropogenic components (bone; especially in the East Section). In addition, the finer fraction is comprised of massive, soft silty sand with granule- to cm-sized clasts of limestone (Figs. 2.13, 2.14 and 2.15).

The upper part of Layer 6 (6A) differs overall from 6B in that it contains many fewer large blocks, although some tilted tabular blocks of roof fall still occur in squares E11 and E12. Otherwise, the sublayer consists of soft, yellowish red to strong brown (5YR to 7.5YR 5/6) massive sand with granule-sized limestone inclusions (Fig. 2.13). Some of the flint artifacts are vertically oriented (likely due to cryoturbation, or frost/soil creep—see below).

On the East Section, Layers 6A and 6B are less differentiated (Fig. 2.7), although the lower part (Layer 6B) also exhibits tabular slabs of bedrock; these clasts tend to be smaller than those on the West Section. In addition, the finer fraction of Layer 6 on the East Section appears to be more organic rich, and slightly darker brown (7.5YR4/4; dark yellowish brown), with possibly greater abundances of flint and bone. This latter observation suggests that burning activity took place on the eastern part of the site as well, extending further east beyond the East Section.

Layer 6 continues south of square F where in square G and H, the distinction between Layers 6A and 6B becomes

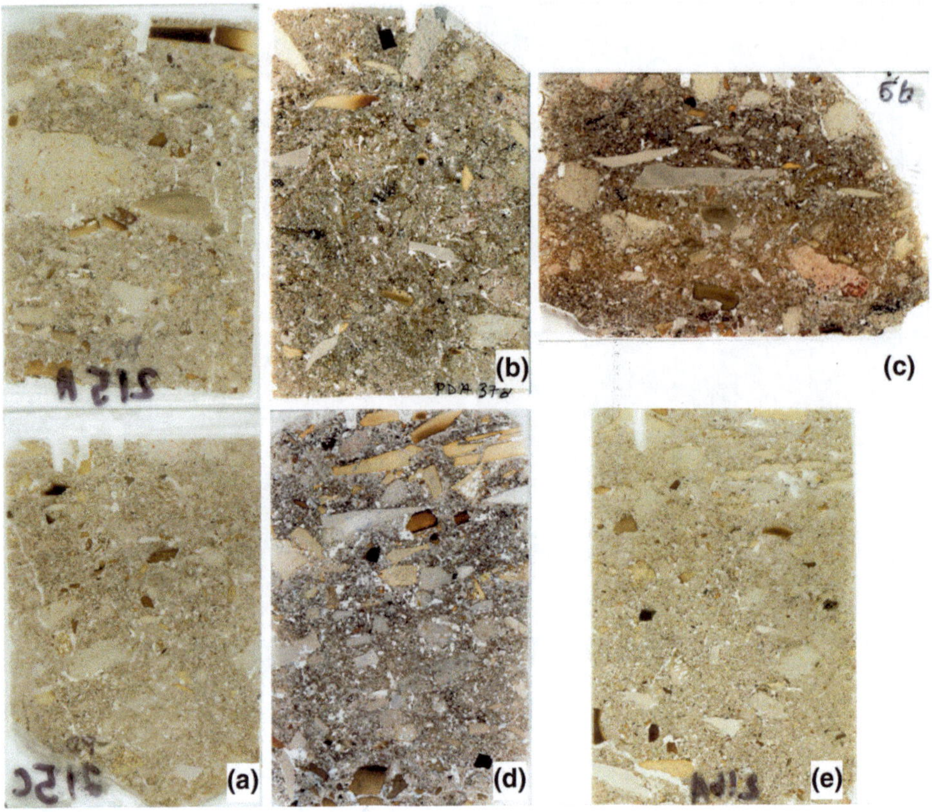


**Fig. 2.8** **a** Sample 215 (Layer 8; Sq. F12) scans of thin sections from upper (215A) and lower (215C) parts of the impregnated block. Note the distribution of brown burned bones, including a broken (trampled) one at the top of 215A. All thin sections are 50 × 75 mm and are in plane-polarized light (PPL) unless noted otherwise. **b** Sample 37A (Layer 8; Sq. D18), Layer 8, macro scan showing dense matrix containing a mixture of burnt bones, char (diffuse black domains), and angular chert clasts. **c** Sample 56 (Layer 8; Sq. G14), base of Layer 8. Visible here is red staining of the matrix (including limestone clast in the lower right of the sample. **d** Sample 57B (Layer 8; Sq. G14), from above sample 56 in Layer 8. Visible here are burned and trampled bones at the top, overlying a darker layer containing char. **e** Sample 216A (Layer 8; Sq. F12). This sample comes from the outer extent of Layer 8, which is evident by the overall light color and lower quantity of bone

somewhat less clear due to the appearance of large tabular blocks as in squares D–E–F (Fig. 2.4a). In this southern part, the finer, sandy sediments of 6A contain large tabular blocks of roof fall that are similar to those in squares D–E–F, indicating that Layer 6 represents a major period of roof fall accumulation. We note, however, that the first phase of this collapse was initially centered in squares D–E–F and with time successively migrated southward as shown for Layers 5 and 4 (Figs. 2.3 and 2.5d).

### Layers 5A and 5B (Roughly ≡ Bordes' Levels J1, J2, and J3)

This stony sandy complex does not correspond directly to Bordes' terminology, and we have regrouped some of his layers into our Layers 5A and 5B. Layer 5 rests on Layer 6A with a sharp and irregular contact (Figs. 2.3b and 2.4). Moreover, the contact between Layer 5B and the underlying 6A is generally inclined to the west and southwest. In square G11, however, Layer 5B appears to climb up over a large block of roof fall in squares G and H 11 (Fig. 2.4a). Moreover, Layer 5B thickens to the south and southwest, being about 15 cm thick in square C12, and thickening to about 40 cm in square G12. Bordes' somewhat schematic section drawing (see below) shows slight thickening of his Layer $J_2$ although it is overall thinner than our Layer 5B.

Layer 5B consists of numerous—typically rounded lithoclasts within a yellowish red (5YR5/8) silty sandy matrix (Figs. 2.16 and 2.17). Interestingly, some of the cm-sized limestone clasts are both well rounded and very fresh (Figs. 2.16 and 2.17), and have equant shapes that become more platy and angular from square E11 into D11 and C12—i.e., in the direction of the cave wall. Some of the ~4–5 cm diameter clasts are very smooth and rounded and originally interpreted as evidence only of cryoturbation (similar clasts can be seen in Pech II (Bordes 1972; Goldberg 1979); they may also be derived from older sediments within the now-concealed cave interior and represent relicts of the former phreatic system (Braillard 2000).

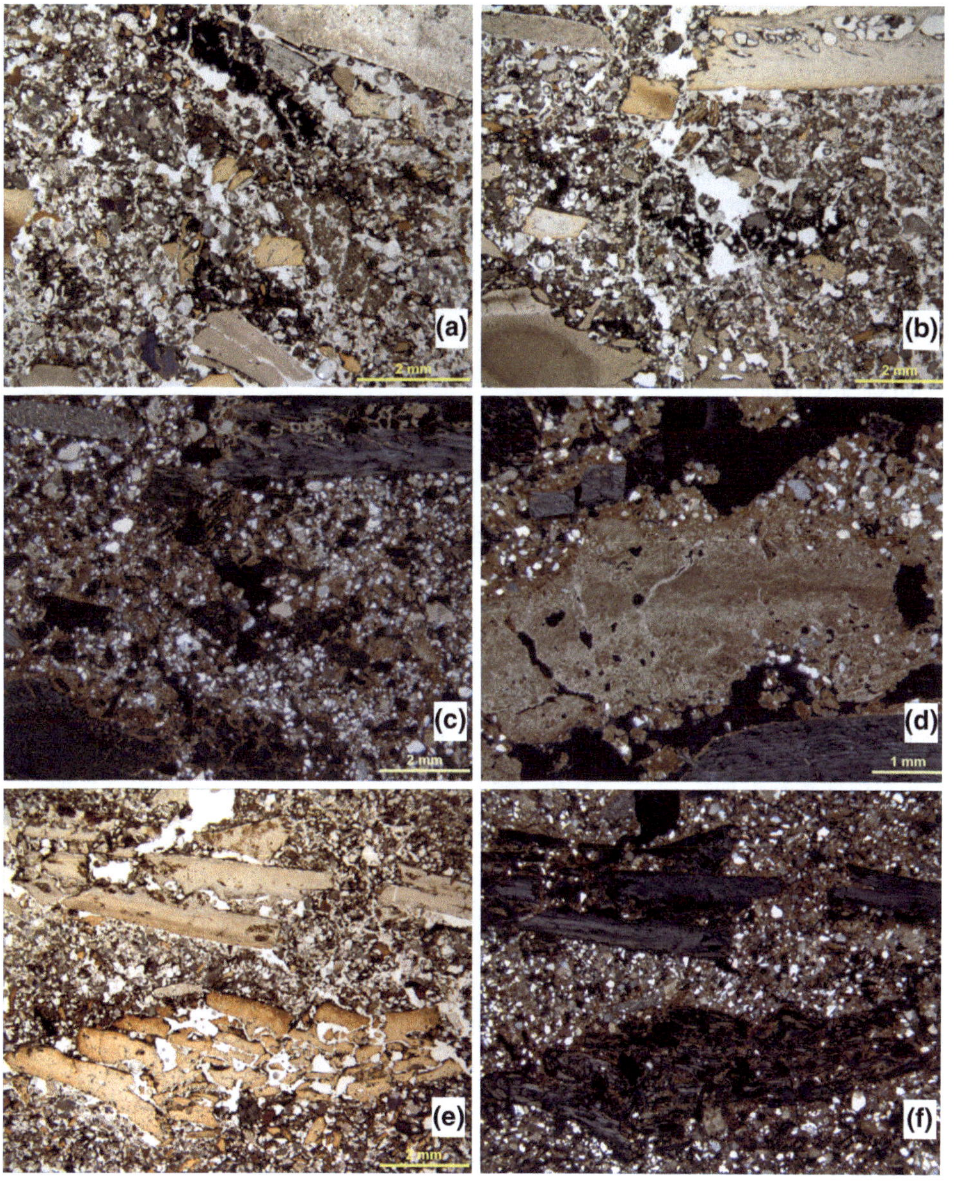


**Fig. 2.9** **a** Sample 51A (Layer 8; Sq. E13). Photomicrograph showing burned and calcined bone and darker areas of char. Plane-polarized light (*PPL*). **b** Sample 51B, (Layer 8; Sq. E13). Burned and calcined bone with zone of black char in the middle. PPL. **c** Sample 51B (Layer 8; Sq. E13). Same as b but in XPL. Note some reddish iron-rich dusty clay coatings in the irregular voids. **d** Sample 51A (Layer 8; Sq. E13). Layer of cemented, fractured, and bedded calcareous ashes from the top of thin section 51. Cross-polarized light (XPL). **e** Sample 52 (Layer 8; Sq. E13), Layer 8. Trampled and crushed bone with "scissor" fractures. PPL. **f** Sample 52. Same as in **e** but in XPL. Note the red, iron-rich coatings around the bones

Layer 5A is somewhat similar to Layer 5B but is characterized by flatter, well-bedded limestone clasts in a sandy matrix, and contains numerous generally angular decimeter-sized blocks of limestone roof fall. These are scattered throughout the layer, as well as in the East Section where larger blocks (20–40 cm across) occur at this stratigraphic level. Layer 5A thins slightly from north to south (Figs. 2.3b, c and 2.4). In thin section, both bones and lithics were relatively common (Figs. 2.16, 2.17 and 2.18), and many of the former are burned, some calcined (Fig. 2.18). In addition, in many samples (e.g., sample 32B from Layer 6A; Fig. 2.15b), remains of calcareous roots are present, indicative of plants growing on the surface beneath the dripline; as discussed below they appear to demarcate former positions of the dripline as it retreated (Figs. 2.3b and 2.4a). In thin section (Fig. 2.16b) a diffuse band of burned bones $\sim 1$ cm in diameter can be observed in Layer 5A; interestingly no combustion features were noted. Finally, sediments closer to the rear wall of the cave (e.g., sample 29; square D11) exhibit reddish grain coatings of iron-rich dusty clay (Fig. 2.18d, e). Such coatings appear to be penecontemporaneous with the calcification of roots, because in thin section, the iron/clay can be seen to impregnate some of the calcified roots but, in turn, they are pierced by fresher, unstained calcareous roots (Fig. 2.18a). In general, however, for most of the sediments observed in thin section, calcification of roots and coeval formation of hypocoatings appear to be the last post-depositional events, and for the most part,

**Fig. 2.10** **a** Sample 56 (Layer 8; Sq. G14), Layer 8. A large flint flake at top rests on cemented ash-rich bed. PPL. **b** Same as in **a** but in XPL. **c** Sample 215C (Layer 8; Sq. F12) from cemented lower part of the sample in Layer 8. Note the compact nature of the sediment, with coarser quartz and chert grains in a finer matrix composed of dusty fine calcareous sand and silt. PPL. **d** Sample 215C. Same as in (**c**) but in XPL. **e** Hand sample from square G13 of well-indurated, microsparite-cemented sediment from the base of Layer 8 that rests directly on bedrock. Cementation here appears to have been fostered by the collection of water at the sediment–bedrock interface (see also Fig. 2.11c–f). **f** Thin section scan of sample 54 (Layer 8; Sq. E13) from the base of Layer 8 in contact with the bedrock. Note the rounded cm-sized limestone grains, a few yellow bones, and the numerous vesicles and channels that perforate a compact matrix. Width is 46 mm

postdate the iron/clay coatings; some calcification could be of relatively recent age.

The lateral continuation of Layers 5A and 5B south of the G line is not very clear. However, pockets of sandy material containing well-rounded cm-sized limestone clasts are found in spaces between large blocks of fallen roof (Fig. 2.5a). These "pocket-fills" occur at elevations comparable to those where Layers 5A and 5B occur in squares D, E, and F (Fig. 2.3b, c).

The occurrence of the finer sediments between the individual limestone blocks indicates that these blocks fell prior to or are penecontemporaneous with this finer material. As in the case of Layer 6, we note that the blocks are inclined southward (Figs. 2.3c and 2.5) so that successively younger blocks are found further to the south. They do not represent one roof fall event, but rather a series of small collapse episodes. Furthermore, this imbrication denotes that the roof fall "peeled off" first from the *underside* of the roof within the cave environment. In any case, the blocks do *not* represent a total collapse or retreat of the cave brow from the outward edge toward the cliff face.

The lithologies of Layers 5A and 5B on the East Section (Figs. 2.6 and 2.7) are not as striking as those on the West Section, and correlations between them essentially follow Bordes' designations; we did not excavate this side of the site. In the East Section, Layer 5B appears to be represented by poorly sorted sandy gravel, consisting of rounded and spherical limestone gravel 1 cm in diameter, and ca. 2 cm subangular to sub-rounded gravel in a reddish yellow sandy matrix. The sediments become increasingly finer grained and redder toward the North wall (Fig. 2.6), mirroring the same phenomenon seen on the West Section. In any case, it is clear that the sediments of what appear to be Layer 5 in the East Section are eroded into the underlying softer sand, as is the case for the Layer 5/6 contact on the West Section.

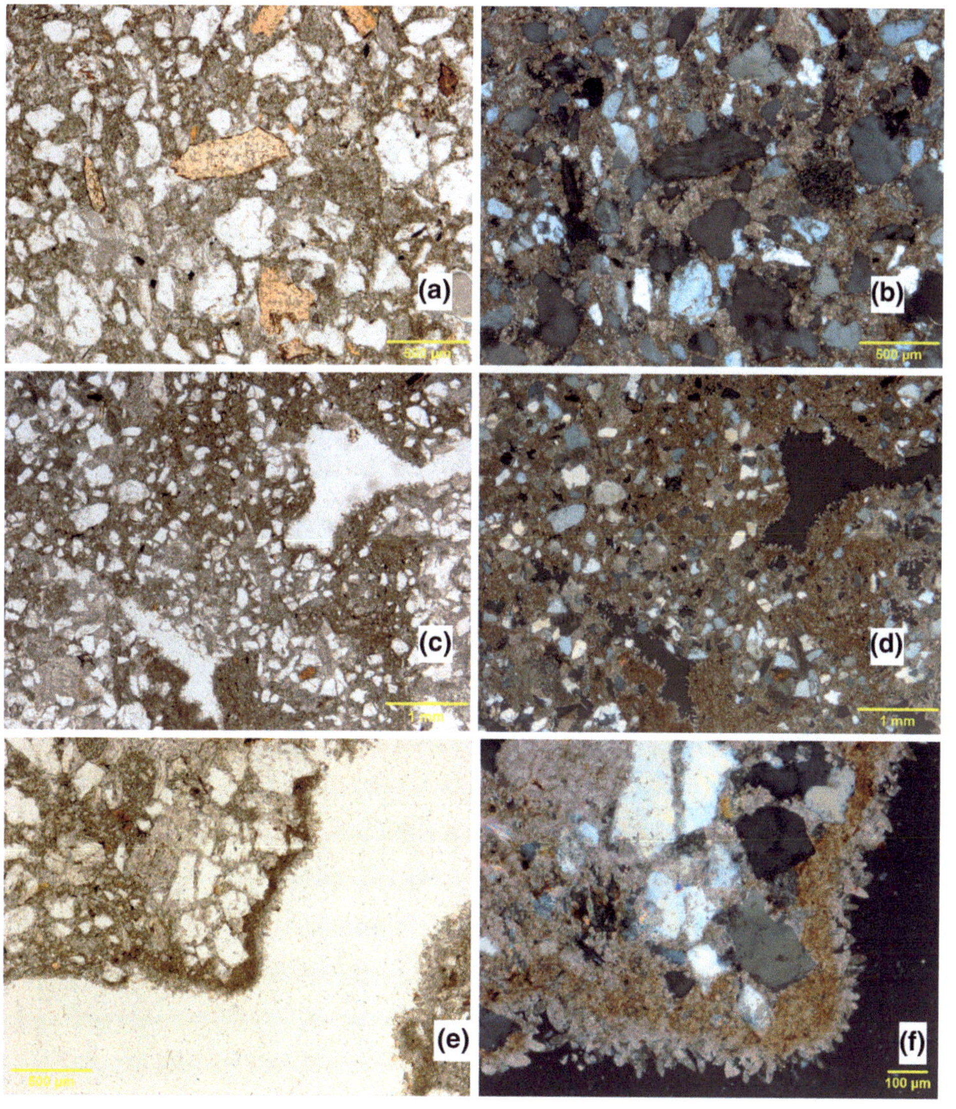


**Fig. 2.11** **a** Sample 54 (Layer 8; Sq. E13) showing angular bone fragments in a compact, calcite-cemented matrix. PPL. **b** Sample 54, XPL view of **a** at *left* with surrounding micritic cement. **c** Sample 54, showing two large voids with iron/clay coatings with subsequent phase of acicular (needles) calcite. Note the compactness of the cemented matrix. **d** Similar view as in **c** but in XPL. **e** Detailed view of **c** and **d** showing Fe/clay coating and acicular calcite. **f** Detailed view of e highlighting the acicular microsparite crystals lining the void. XPL

**Fig. 2.12** Rounded, cryoturbated pebbles from Layer 7

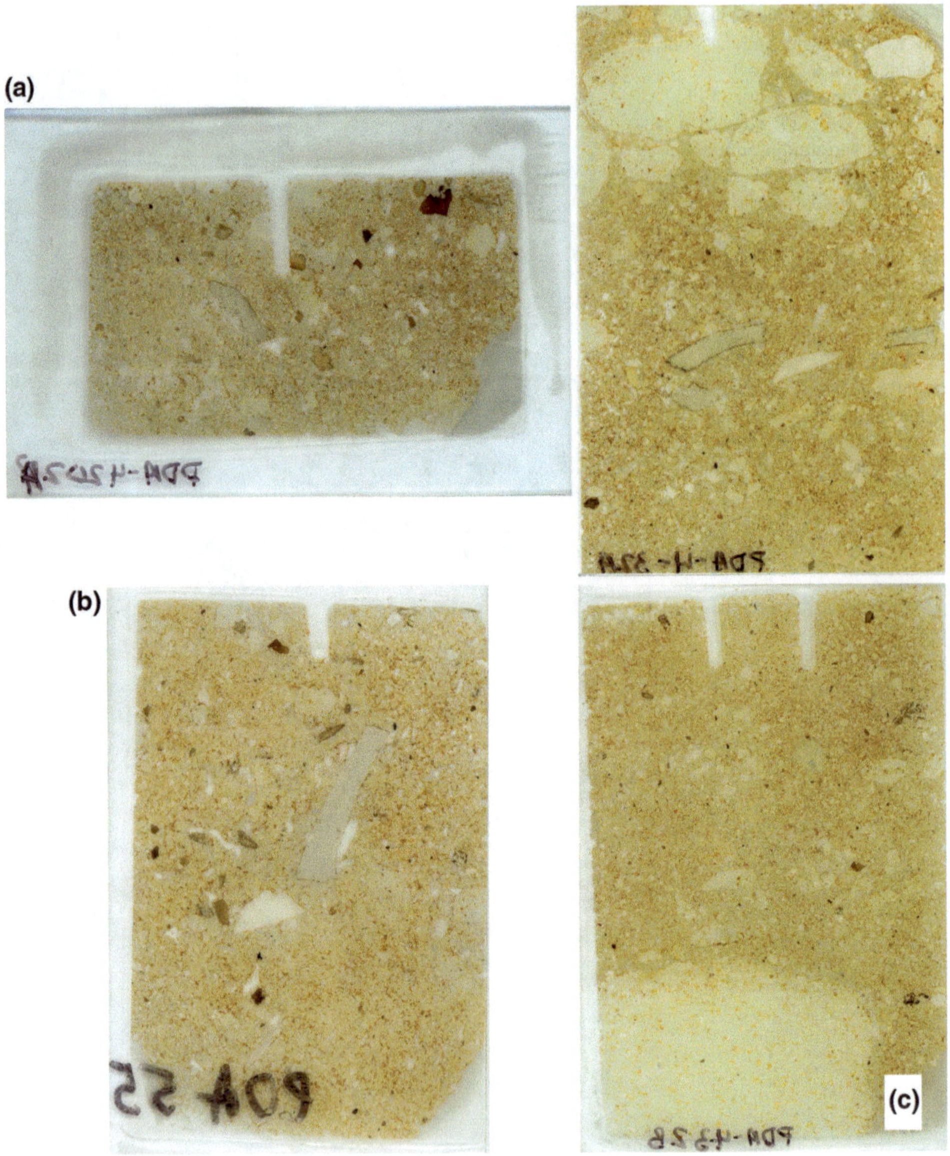

**Fig. 2.13** **a** Scan of sample 202A (Layer 6B; Sq. D/E 12) showing fine limestone sand and granules, with a piece of burned bone at the *top*. **b** Scan of sample 55 (Layer 6B; Sq. E13)—burned bone fragments (*brown specks*) and unburned bones in sandy matrix with iron-rich clay between grains. **c** Scans of samples 32A and 32B (Layer 6A; Sq. F11). Generally similar to sample 55 above but note the rounded limestone clasts and layer of angular bone and chert fragments in sample 32A

## Layers 4A, 4B, and 4C (≡Bordes' Levels G, $H_1$, $H_2$, $I_1$, and $I_2$)

This relatively thick (95–130 cm) stratigraphic unit is set off from both the overlying and underlying angular gravelly deposits by being not only generally fine grained but also by containing numerous blocks of roof fall whose abundance increases to the south; the contact with the underlying Layer 5A is sharp and erosional (Figs. 2.3 and 2.4). Layer 4 is subdivided into Layers 4C, 4B, and 4A (from bottom to top), with roof blocks being more common in Layer 4B. Stratigraphic subdivisions are clearest closer to the bedrock walls, as in square D11, and progressively to the south, there is a greater proportion of blocks of roof fall.

The shape of the roof fall is variable, as for example in squares D11, E11, and F11, where they range from tabular blocks to subangular, flattish pieces. Similar shapes are reflected in the smaller (e.g., cm-sized) clasts as seen in thin section (Figs. 2.19, 2.20, 2.21, 2.22 and 2.23). In any case, the lowermost occurrence of these blocks is in Layer 4C but they dramatically and progressively increase in Layer 4B, where concentrations of gelifracts were at the base of Layer 4B in square E11. Layer 4A also contains some roof fall but in general is finer grained, with cm-sized angular fragments of limestone in a sandy matrix (Figs. 2.19c, d, 2.20b and 2.23).

Layer 4 and its subdivisions are locally cemented (Figs. 2.3b–c and 2.4a), and Layer 4A is more calcareous in square E11 than in square D11, where it is sandier and markedly less calcareous. Interestingly, in squares F11 and G11, much of both Layers 4A and 4B is secondarily

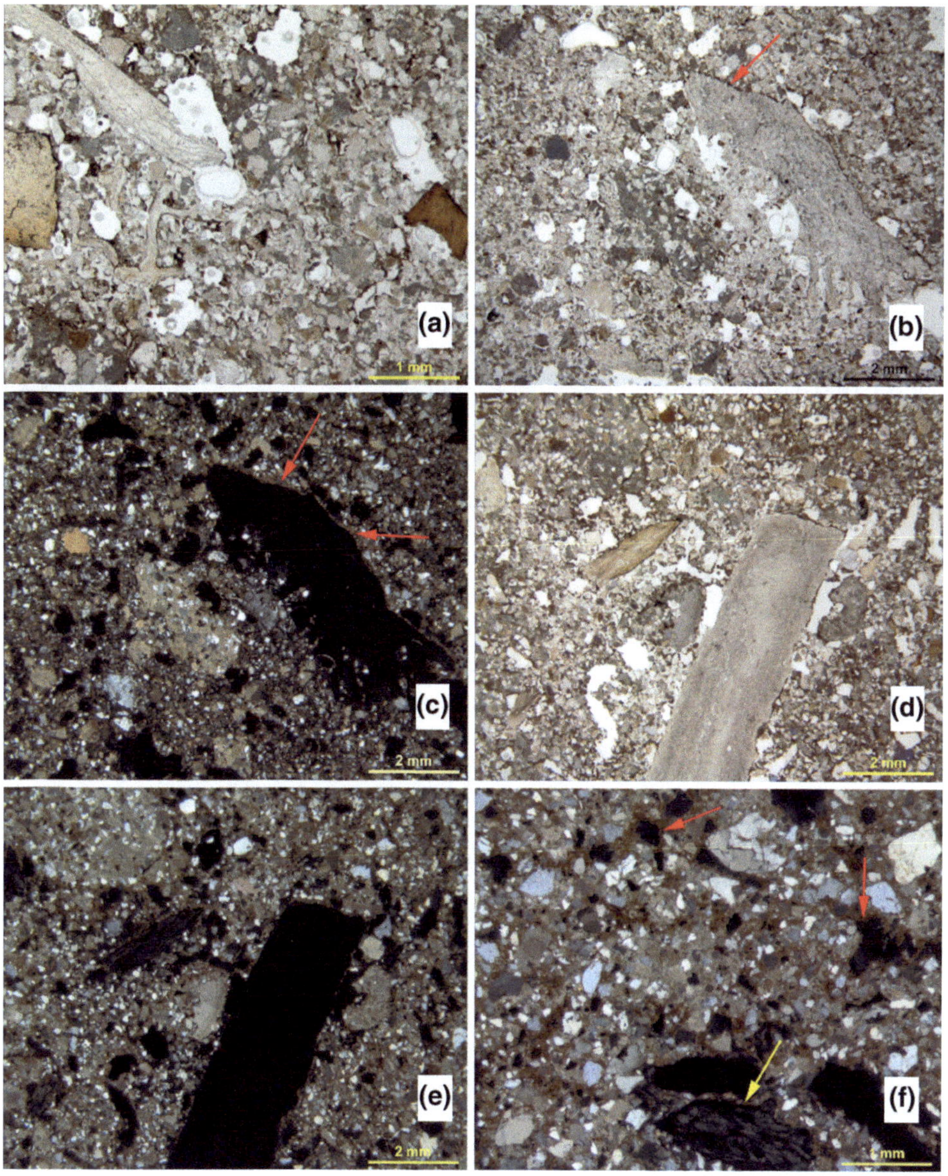


**Fig. 2.14** **a** Sample 202A (Layer 6B; Sq. D/E 12) showing yellow brown burned bones and pale yellow bone fragments in a quartzitic and calcareous sand matrix. PPL. **b** Sample 202A (Layer 6B; Sq. D/E 12) with quartzitic and calcareous sand matrix and thin calcareous silty clay coating (*arrow*) on chert grain. PPL. **c** Sample 202 A: same as **b** but in XPL. Note the relative paucity of fresh carbonate grains and relative high amounts quartz grains reflecting dissolution of the carbonate in place. This sample is close to the walls (Sq. D12) where carbonate dissolution is common. **d** Sample 55 (Layer 6B; Sq. E13). Burned bone fragments with reddish iron-rich dusty clay coatings around them PPL. **e** Sample 55 from Layer 6B, same as in **d** but in XPL. Note the presence of fresh calcareous sand and granules, which contrasts with sample 202A that is from square D12, close to the wall; this sample from square E13 is farther from the wall. **f** Detail of sample 55 illustrating its fresh calcitic nature, as well as brown, iron-rich dusty clay coatings around voids (*red arrows*) and coating grains, such as the bone fragment at the base of the photo (*yellow arrow*)

cemented as shown by interstitial micritic cements and calcified rootlets and hypocoatings (Figs. 2.21c, 2.22e–f and 2.23a, b); as mentioned above, the latter are a result of water dripping from the former brow of the cave. As is true for many of the sediments at Pech IV, iron-rich dusty clay coatings occur, and they are more abundant and pronounced near the back wall of the excavations (particularly squares C and D; Figs. 2.22c and 2.23b–d) and less so in square F for example (e.g., Fig. 2.21f).

The lateral stratigraphic correlation of Layer 4 with the large blocks occurring south of square H11 is not clear, owing to the erosional channel associated with the deposition of historic (Medieval?) deposits of Layers 1A/B/C (Figs. 2.3b and 2.4a). It appears, however, that by the time of deposition of Layer 4B, the bulk of the large blocks of roof fall (in squares I/J/K; Figs. 2.5 and 2.6) had already fallen. The roof at the time of accumulation of Layer 4B was probably situated somewhere near squares G and H (Fig. 2.3b). Thus by this time, the cave was considerably more open. Support for this inference comes from the lithology of Layer 4A, which reveals some large pieces of rock fall in squares G and H; immediately below these blocks, Layer 4B sediments are very well indurated (Figs. 2.3b and 2.4a). On the East Section, the stratigraphic correlation of Layer 4 is also not well defined, particularly for the upper part, which contains mostly large blocks of roof slump, as the cave roof was more intact at this time (Figs. 2.7 and 2.8).

The average thickness of Layer 4C is about 20–30 cm and possibly thins to 20 cm towards the south. Layer 4B is about 30–40 cm thick, although the contact with Layer 4A is limited and not very clear in squares F/G/H. Layer 4A is

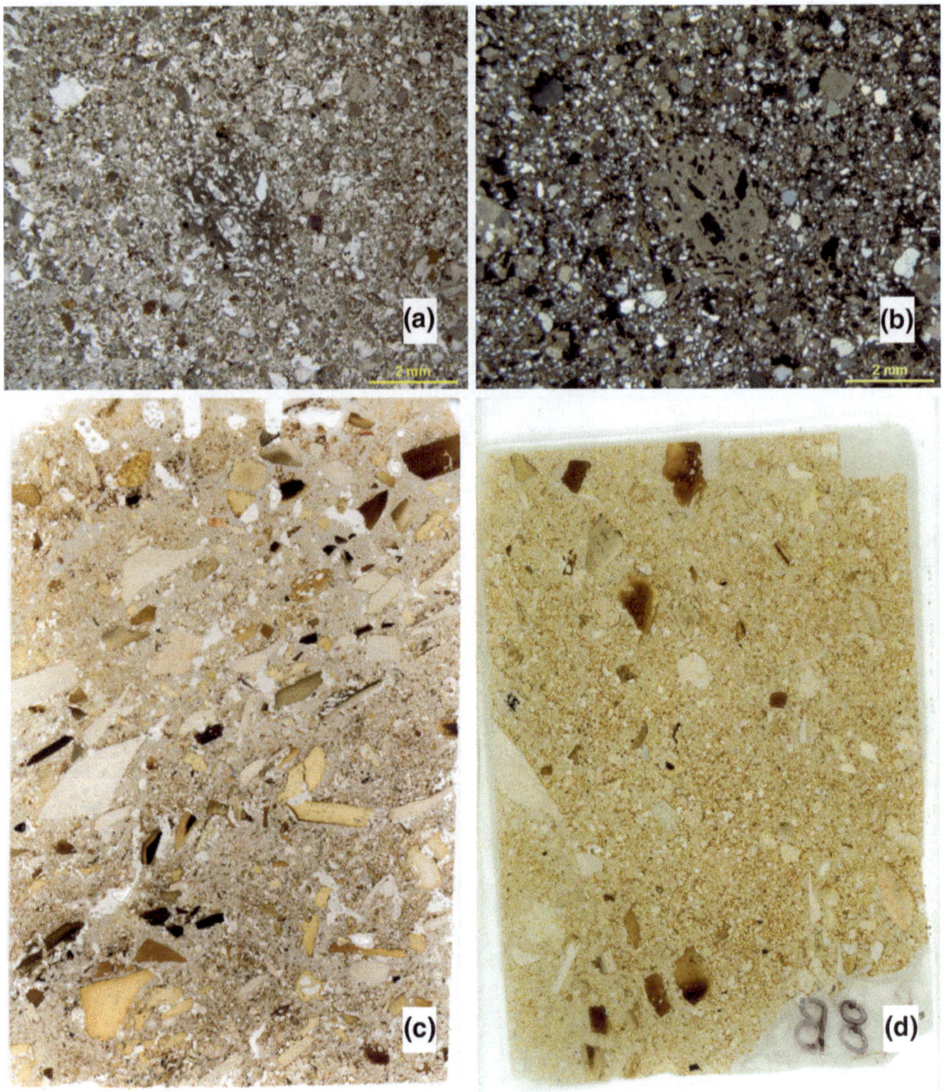


**Fig. 2.15** **a** Sample 32B (Layer 6A; Sq. F11), consisting of quartz sand and some sand-sized limestone grains. PPL. **b** Same as **a**, but in XPL. The porous area in the center is fine calcified root fillings associated with vegetation growing on the former surface at the time. **c** Scan of sample 4B (Layer 6B; Sq. E19) showing bedded bones (mostly burnt) and lithics, with a finer, sandier matrix in the upper half of the slide. **d** Scan of sample 8B from Layer 6B (Sq. D13) showing the scatter of burnt bones, although the matrix is yellow brown and not rich in organic matter

about 45–60 cm thick, but its upper part is truncated in squares G and H by the Layer 1 complex.

### Layers 3A and 3B (≡Bordes' Levels $F_1$ to $F_4$)

Layer 3, which is volumetrically much less widespread than the other layers, occurs mostly in squares D, E, and F. It is lithologically quite distinct from the Layer 4 complex, being more compact and cemented than Layer 4 and containing fewer limestone fragments (Figs. 2.24, 2.25 and 2.26). It rests on Layer 4 with a sharp, irregular contact that is inclined toward the south-southwest (Figs. 2.3a, b and 2.4a). Layer 3 is subdivided into two sublayers, which overall are lithologically similar, although Layer 3B is somewhat coarser and richer in flints and exhibits clearer bedding.

Overall, Layer 3 consists of gritty, compact to cemented quartz and limestone sand (Figs. 2.24, 2.25 and 2.26). Layer 3B (Figs. 2.24b and 2.25) is composed of somewhat coarser sand with fewer limestone clasts, but it contains a greater proportion of lithic and bone materials, which are bedded. The limestone fragments that do exist are rounded and appear to be somewhat dissolved (Fig. 2.24b); the finer matrix material is reddish and sandy, and the lower part is bedded. Many mm-sized calcareous rootlets occur throughout, and the deposits of Layer 3B are particularly well indurated toward the south (e.g., square G and part of square F; Fig. 2.25a–c) because of drip from the brow of the cave that occupied this position at this time.

The contact between Layers 3A and 3B is clear but gradational, particularly in square D, but becomes less clear toward the south. Layer 3A tends to be finer grained but contains many mm- to cm-sized rounded clasts of limestone, many of which appear to be chemically attacked (Fig. 2.24 d). In addition, Layer 3A on the whole contains a higher proportion of coarser material to the south than does Layer 3B (Figs. 2.3 and 2.4a), so that in square E11 some angular clasts of roof fall (up to 10 cm) can be observed; there is also

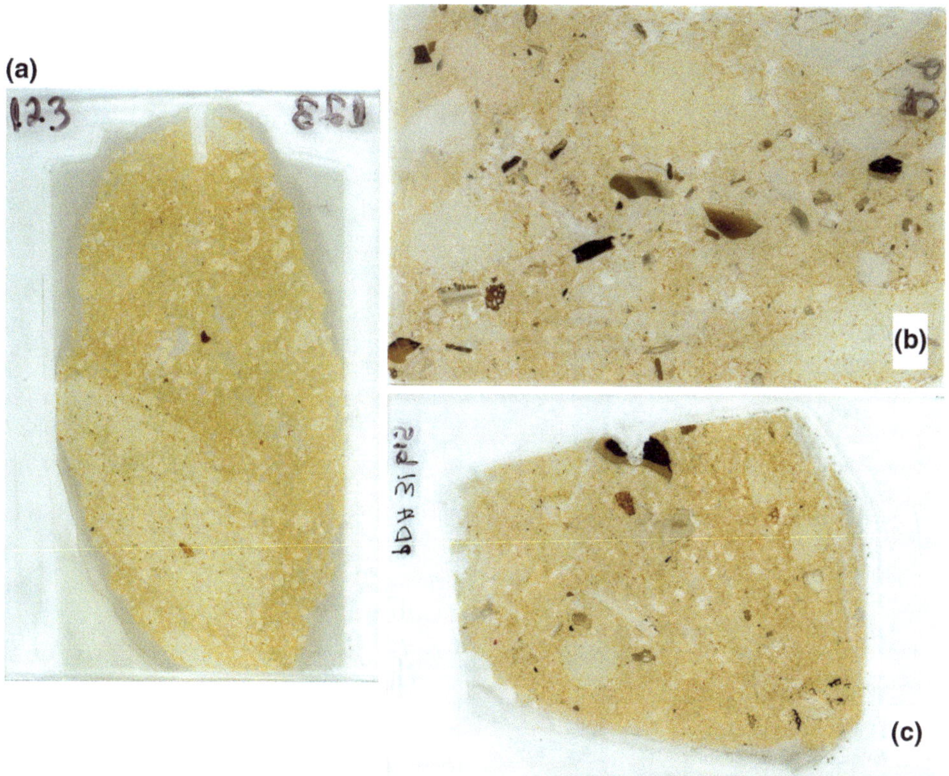

**Fig. 2.16** **a** Scan of sample 123 (Layer 5B; Sq. H13). Note the large limestone clast at the bottom and smaller, more rounded limestone grains scattered throughout the matrix. **b** Scan of sample 29 (Layer 5A; Sq. D11), showing numerous angular fragments of burned bone and well-rounded limestone clasts with reddish iron-rich dusty clay coatings. **c** Scan of sample 31bis (Layer 5A; Sq. F11). Note rounding of many of the larger, cm-sized limestone grains by cryoturbation

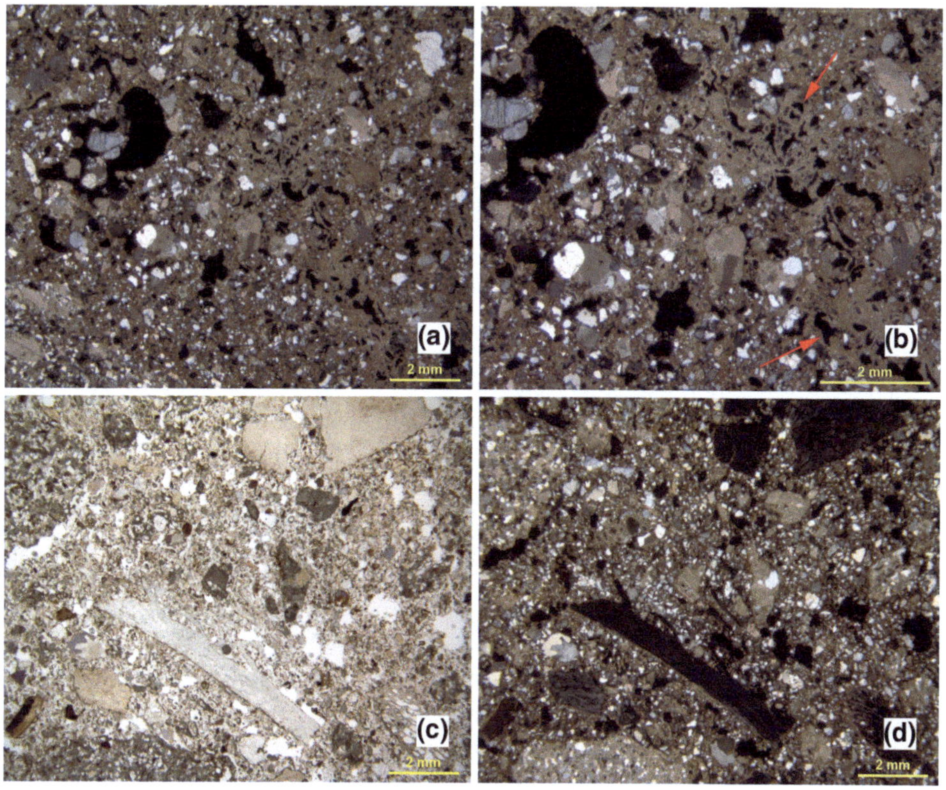

**Fig. 2.17** **a** Sample 123 (Layer 5B), which is characterized by relatively abundant calcified roots that are shown in greater detail in **b**. XPL. **b** Detail of sample 123 shown in a. The hypocoatings and porous calcified fine roots (*arrows*) are more apparent in this view. XPL. **c** Sample 31bis (Layer 5A). Angular bone and flint fragments in a sandy matrix of quartz and clasts of calcareous bedrock. **d** Sample 31bis (Layer 5A). Same as in **c** but in XPL. Note the rounding of some of the coarser bedrock sand, likely associated with cryoturbation and solifluction

a greater abundance of cm-sized rounded clasts in Layer 3A. On the other hand, towards the wall at the north end of the excavations (square D/C), the sediments become increasingly sandy, as was the case for Layer 4A. In squares C12 and 13, for example, the sediment consists of massive, firm reddish sand (Figs. 2.3b and 2.4).

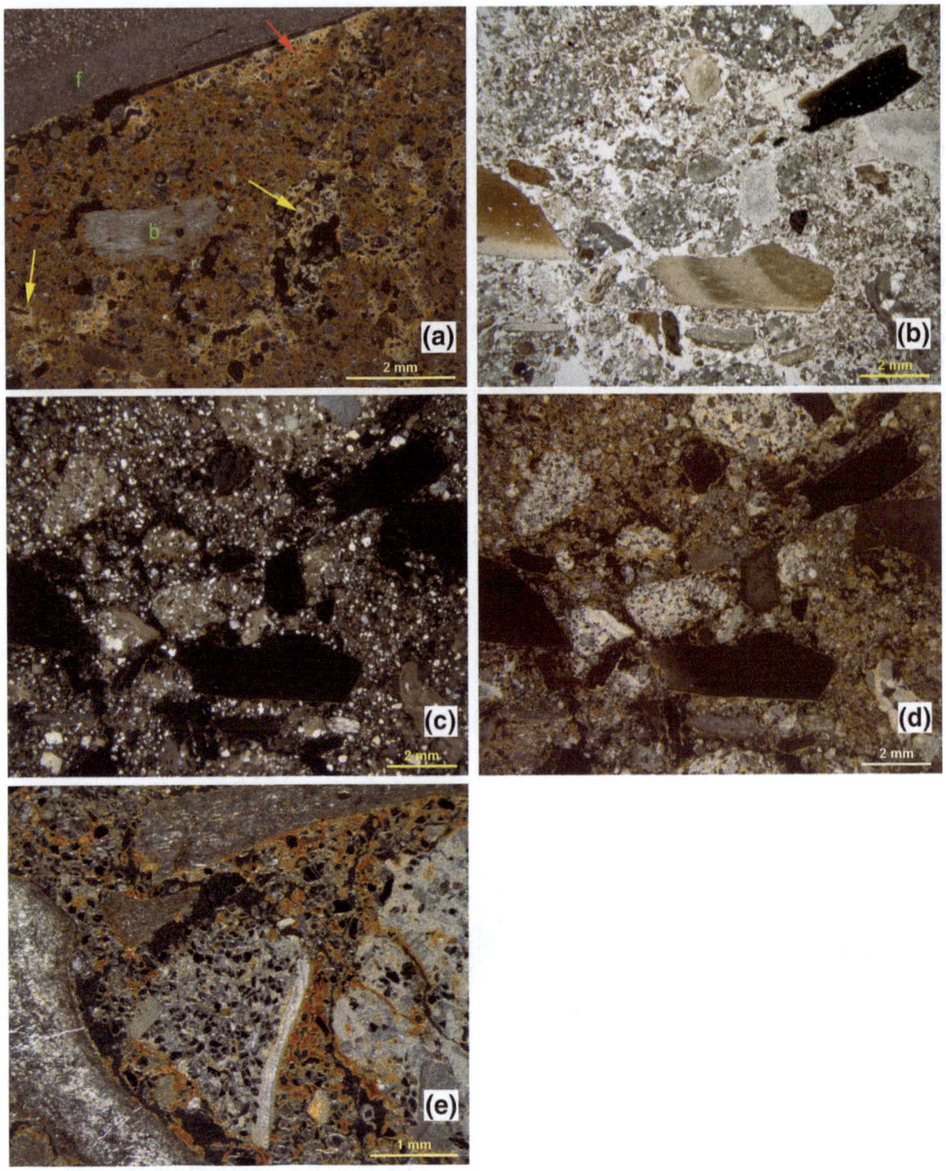


**Fig. 2.18** **a** Sample 31bis (Layer 5A). Dark field photomicrograph of flint flake (*f*) and bone fragment (*b*). Note the red staining of iron-rich dusty clay, which formed coevally with calcite around many of the roots. The latter is shown by the iron staining of calcified roots (*red arrows*), as well as a late phase of calcification that is not iron stained (*yellow arrow*). **b** Sample 29 (Layer 5A; Sq. D11) with many burned bone fragments, the lowermost of which appears calcined. PPL. **c** Same as **b** but in XPL. Note the overall calcareous nature of the deposit, with little, if any, dissolution of carbonates. **d** Sample 29 (Layer 5A; Sq. D11) as **b** and **c** but with dark field illumination, showing the well-rounded nature of the limestone clasts, and thin coatings of reddish iron-rich clay. **e** Sample 29 (Layer 5A; Sq. D11), a detailed view with dark field illumination. Note the thin iron-rich staining and coatings of many of the clasts, which are composed of limestone and chert

Layer 3B is about 15 cm thick whereas Layer 3A is about 70 cm in thickness. The correlations with the East Section are not definitive (Figs. 2.7 and 2.8), as the connecting section between them was removed during Bordes' excavations.

## Layer 2 (Possibly Bordes' Level D?)

This layer occurs only in a limited part of the West Section (Figs. 2.3b and 2.4a), partially as a result of truncation by Layer 1 in historical times. Limited exposures occur in squares F and E11 where the sediments appear as a jumble of mostly limestone cobbles of various sizes (a few mm up to 25 cm diameter) and shapes (angular to sub-rounded). In between the stones is a matrix composed of silty, mostly calcareous sand, which appears to represent local breakdown of the bedrock clasts. The matrix is partially cemented with numerous very fine, mm-sized rhizoliths.

The overlying contact with Layer 1C, which truncates it (Figs. 2.3b and 2.4a) is quite sharp, whereas the contact with Layer 3A and 3B to the north is much more subtle. Nevertheless, Layer 2 can be distinguished from Layer 3 by the fact that the matrix in Layer 2 is sandier and contains fewer granule-sized fragments of limestone than does Layer 3. The Layer 2/3 contact is somewhat vertical and almost basin shaped. The upper part is washed somewhat by movement of water associated with what probably represents historical disturbance.

Layer 2 appears to represent a small depression (either basin or trench) probably produced by water cascading from the former brow of the cave. This depression was filled as a

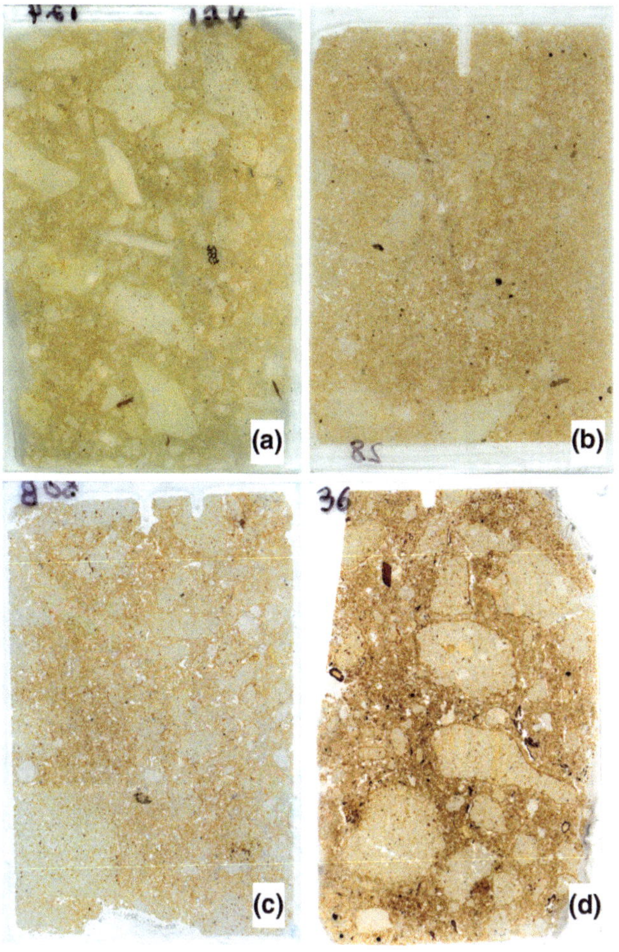


**Fig. 2.19** **a** Scan of sample 124 (Layer 4C; Sq. H11). Rounded and angular limestone clasts with isolated fragments of burned bone. **b** Scan of sample 28 (Layer 4B; Sq. D11). This sample, close to the rear wall of the excavations, is finer than equivalent samples in the same layer farther to the south, away from the wall, reflecting partial dissolution of the limestone clasts. **c** Scan of sample 50B (Layer 4A; Sq. F11). Porous mixture of rounded limestone clasts in a sandy matrix with some interstitial reddish brown iron-rich dusty clay. **d** Scan of sample 36 (Layer 4A; Sq. G10). Rounded limestone clasts with fine mm-sized bone fragments in a dusty iron-rich sandy matrix

combined result of large blocks rolling off the brow, as well as by sediment slumping into the depression from Layer 3. The latter would explain the scatter of lithics that were found here in Layer 2. The width of the layer is ~1 m and the height ~75 cm.

### Layers 1A, 1B, 1C, (Possibly Bordes' Levels $A_1$-D)

This sedimentary complex is a slope and roof fall deposit that formed during historical times, and it truncates many of the underlying deposits (Figs. 2.3b and 2.4a). It encompasses a poorly sorted mixture of angular and rounded limestone boulders, as well as smaller clasts in organic-rich silty sand. Several subdivisions were made based on texture and color:

Layer 1C contains relatively high proportions of cm-sized rock clasts derived from the underlying sediments and has a very sharp contact with these latter deposits. It generally follows the modern slope (Figs. 2.3b and 2.4a). Layer 1B, on the other hand, truncates Layer 1C and consists of sandy clay with abundant worm casts and modern roots. Layer 1A is somewhat finer grained and darker brown and contains many aggregates, roots, and tabular rock fragments. The thickness of the sublayers is generally similar: Layer 1C = ~20–35 cm; Layer 1B = ~40 cm; Layer 1A = ~35 cm. Layer 1 does not occur on the east side of the cave, possibly because the original roof is largely more intact there and there is less of a slope (Figs. 2.7a, b and 2.8).

## Specific Micromorphological Aspects of the Deposits

Micromorphological analysis of the Pech IV deposits revealed both geogenic and anthropogenic constituents. The majority of the geogenic coarse fraction in all layers is composed of mm- to cm-sized limestone bedrock fragments

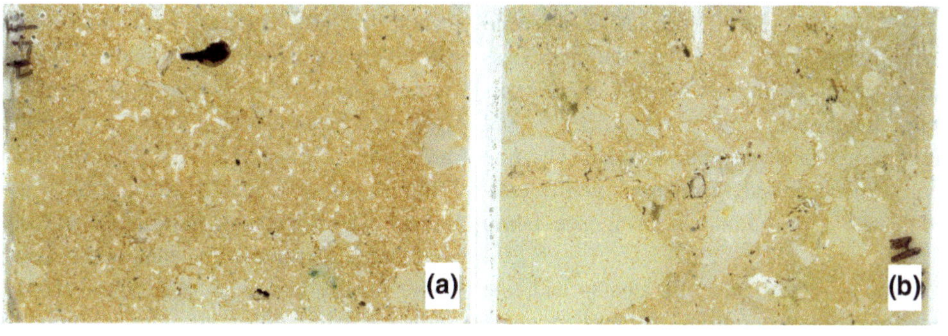


**Fig. 2.20** **a** Scan of sample 47B (Layer 4A; Sq. E11). Generally similar to sample 50B but with some burned bones. **b** Scan of samples 49B (Layer 4B; Sq. F11). This overall similar to sample 124 from Layer 4C but slightly more porous. Note also the rounding of the limestone grains

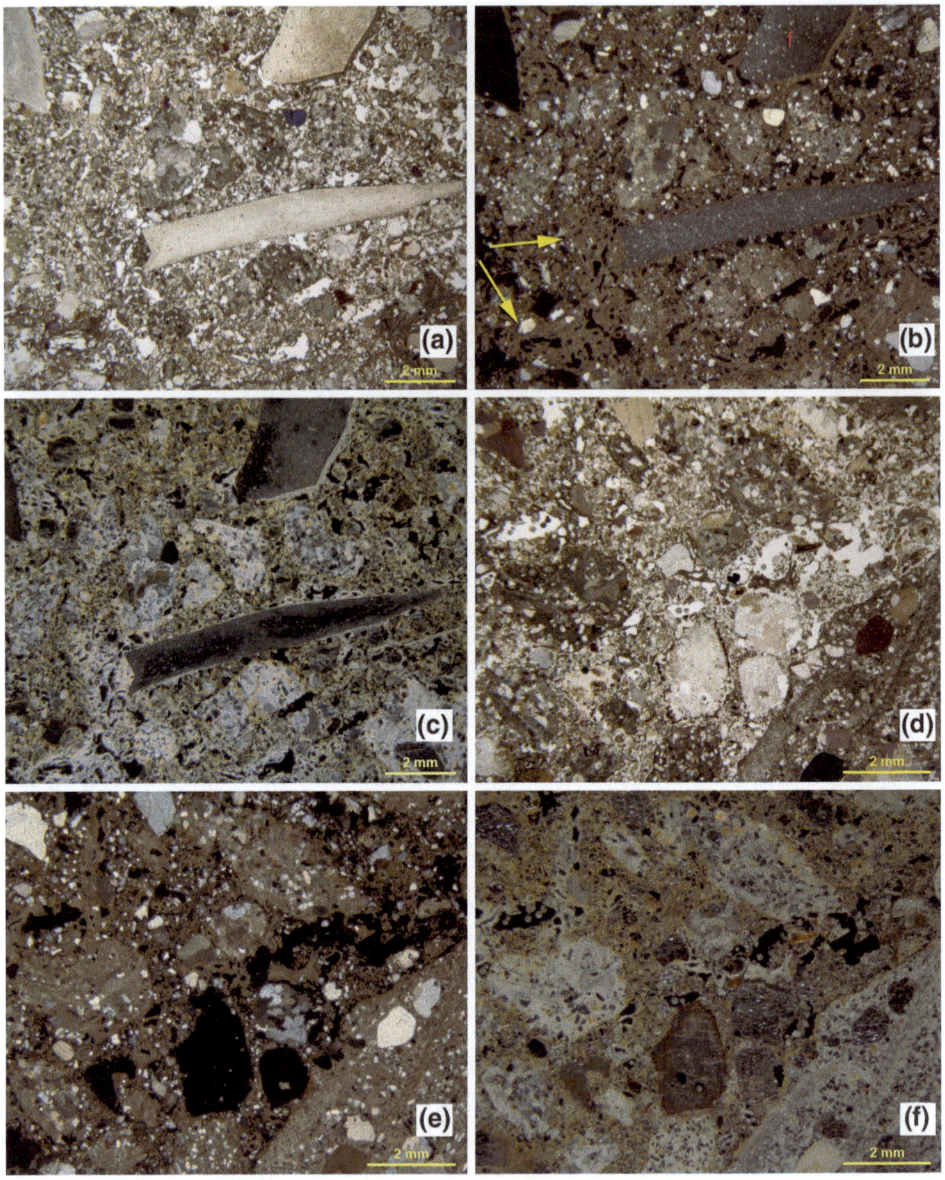


**Fig. 2.21** **a** Sample 201 (Layer 4C; Sq. H11) showing flint flakes and rounded limestone clasts. PPL. **b** Sample 124. Same as in **a**, but in XPL. Note the number of flint pieces (*f*) and overall calcareous nature of the sediment, including fresh limestone clasts and calcified roots (*yellow arrows*). **c** Sample 124 as in **a** and **b**, but with dark field illumination, where secondary calcification of roots is more evident. **d** Sample 49B (Layer 4B; Sq. F11). Note the similarity to sample 124, including the presence of limestone grains, which are mostly rounded. PPL. **e** Sample 49B as in **d** but in XPL. **f** Sample 49B with dark field illumination, which shows the orange red iron-rich dusty clay coatings and interstitial infillings

and loose quartz grains ranging in size from silt to coarse sand. Limestone fragments in thin section are commonly rounded (e.g., Figs. 2.10f, 2.13b, 2.16c and 2.19), and boundaries of grains range from sharp to gradational. Less common are natural grains of muscovite, glauconite, carbonate sands, fragments of travertine, and nodules/concretions of iron and manganese oxide. The fine fraction is composed of clays and secondary micritic (calcite) accumulations with a crystallitic birefringence fabric (b-fabric).

In Layers 8, 7, and 6 in particular, the coarse and fine fractions also contain high proportions of anthropogenic material, including bone and tooth fragments, fat-derived char and traces of charcoal, flint fragments, and ash (Layer 8 only); bones (burned and unburned) and flint artifacts occur in other layers (e.g., Layer 5; Fig. 2.16b–c), but in much lower numbers. Overall, however, fragments of bone and teeth make up the largest proportion of the anthropogenic coarse fraction. They are angular to rounded and are comparatively small, both in the field and in thin section; average fragment sizes range from ~0.25–2 cm. The bones are both fresh and burned, with a range in color from yellow to dark brown (e.g., Fig. 2.11a, b). Anthropogenic materials from samples above Layer 6 are generally isolated bone fragments, which are often sand-sized and subangular to rounded, but (e.g., sample 29, square D11, Layer 5A) they can be more locally abundant (e.g., Fig. 2.16b). Above Layer 6, biogenic materials are virtually absent, with the exception of a single coprolite fragment from sample 5 (square F19, Layer 5B).

The anthropogenic components of Layers 8, 7, and 6 deserve separate mention, particularly in the lowermost

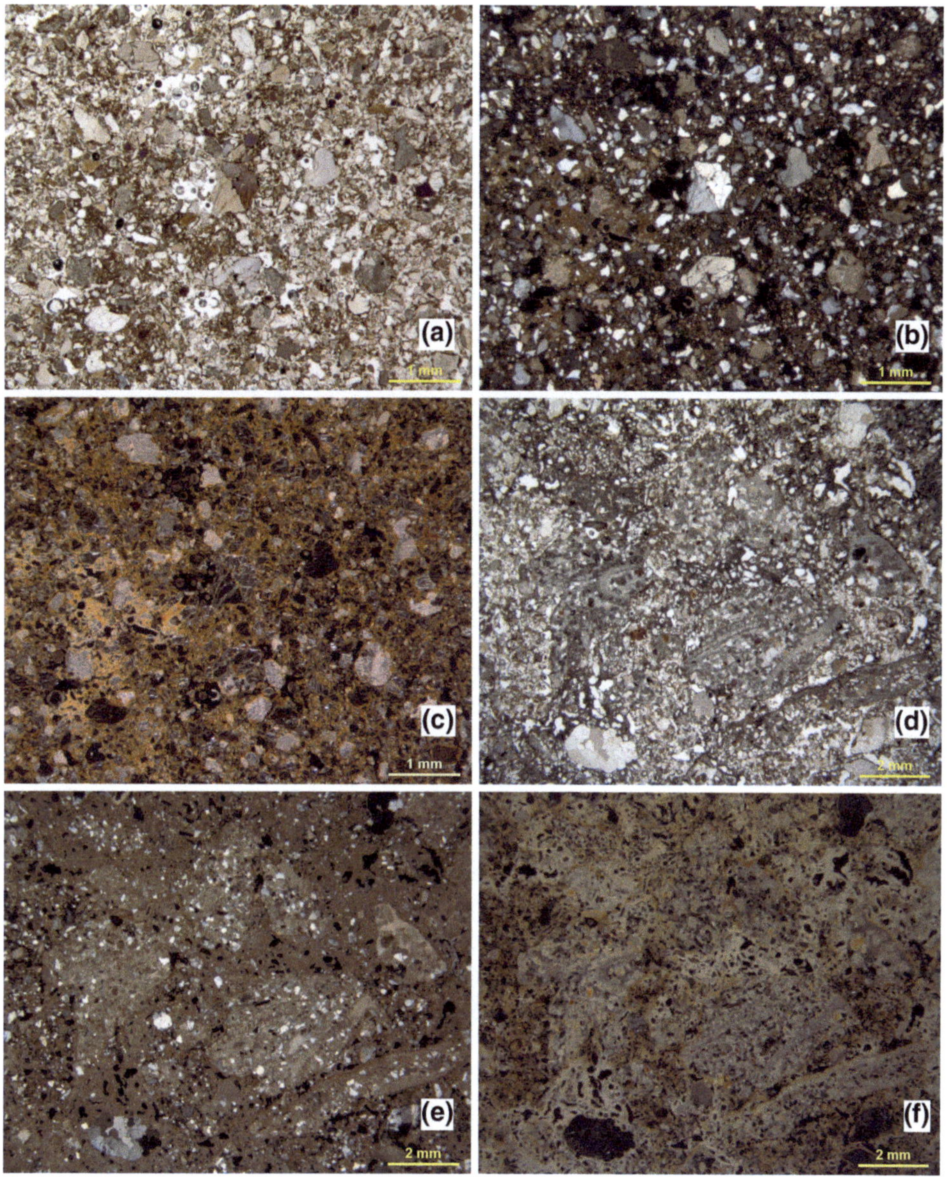


**Fig. 2.22** **a** Sample 28 (Layer 4B; Sq. D11). This sample, which is close to the north wall of the excavations, consists of quartz sand and some rounded, sand-sized grains of limestone. PPL. **b** Sample 28: as in **a** but in XPL. The greater proportion of siliclastics in this wall sediment is evident as can be seen in Fig. 2.3b. **c** Sample 28 with dark field illumination. Not the abundance of reddish iron-rich dusty clay, which is reflected in the field as well (Fig. 2.3b). **d** Sample 50B (Layer 4A; Sq. F11), overall similar to sample 124 in Sq. H11 (Layer 4C) in their degree of secondary calcification, reflecting former positions of the dripline. PPL. **e** Sample 50B but in XPL and showing the rounded clasts of limestone and abundance of secondary calcite. **f** Sample 50B same view as in (p) and (q) but with dark field illumination, showing numerous calcified rootles throughout

layer. Burned bone is present in nearly every sample, and in most slides, it is homogeneously mixed with unburned bone, and commonly exhibits layering. On the other hand, both in the field and in macro scans of the thin sections, bones can be seen concentrated into cm-thick zones, some of which correspond to large, roughly circular burned features that were excavated in 2002 (Fig. 2.9).

Several bone fragments in Layer 8 exhibit in situ breakage—scissor fractures—in which angular pieces of the bone refit and are only barely separated (Figs. 2.8a, d and 2.9e–f); these types of fracture are produced by trampling (Miller et al. 2010). Bone fragments also exhibit some post-depositional chemical features. Many are partially or fully stained with iron or coated with iron-rich dusty clay (e.g., Figs. 2.9e–f and 2.14b–c). Interestingly, such chemical changes do not appear to have affected bone preservation, as the bones are quite compact during excavation and distinct in thin section. Charcoal is present in Layer 8, but is relatively rare. Rather, the dark material that is responsible for the striking color of Layer 8 is fat-derived char resulting from burning of bones and commonly characterized by a dense vesicular black mass that is commonly fractured (Figs. 2.8b, d and 2.9a, b) (Ligouis 2006).

Wood ashes in Layer 8 (Fig. 2.7a, c, d, f) occur as three main types. The first is well preserved, nearly pure layers (~5 mm thick) containing intact rhomb-shaped calcite grains. These layers are compact, massive, and laminated and are cemented by secondary micrite (microcrystalline calcite) (Figs. 2.9d and 2.10a, b). The second type of ash occurs as isolated pieces of cemented ash layers reworked by trampling or possibly cryoturbation. The third type of ash occurs as distributed grains of particles in the groundmass

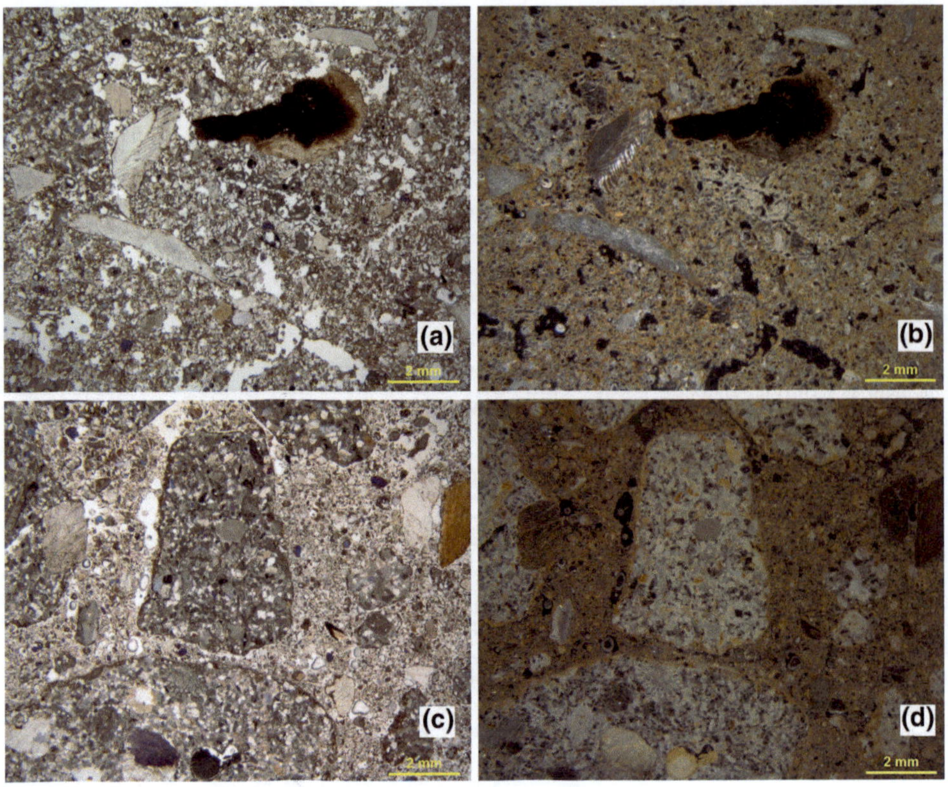


**Fig. 2.23** **a** Sample 47B (Layer 4A; Sq. E11) with burned bone fragment and rounded and platy limestone grains; an angular splinter of chert is visible at the left. PPL. **b** Sample 47B as in **a**, but with dark field illumination. More visible in this view is the presence of calcified roots just below the burned bone, as well as the reddish brown interstitial iron-rich dusty clay. **c** Sample 36 (Layer 4A; Sq. G10) with rounded limestone clasts and a burned bone. PPL. **d** Sample same as in **c** but with dark field illumination. Note the relative abundance of reddish iron-rich dusty clay as coatings around the limestone grains and as in interstitial positions between the quartz sand

and these are commonly re-cemented by calcite (Fig. 2.10a, b). This loose ash also contains rhomb-shaped grains but is dustier in appearance due to the incorporation of other materials, such as clay-sized grains and charcoal fragments. Ash mixed with groundmass often caps large fragments of limestone, flint, and bone or appears directly beneath such objects (Fig. 2.10a, b).

The composition, internal organization, and external geometry of the anthropogenic components provide clear evidence that burning events occurred during the deposition of Layer 8 and possibly Layer 6. These events, exemplified by the intact stringers of ash overlying carbon-rich layers, likely correspond spatially to the round combustion structures excavated in 2002.

The fissures present in some of the cemented layers indicate that the cementation likely occurred relatively soon after deposition and not much later. On the other hand, Layer 8 also exhibits features associated with carbonate dissolution: as mentioned above, many limestone fragments appear to be actively dissolving and many cemented ash fragments often contain thin channel voids, which are characteristic of calcite depletion (Bullock et al. 1985). Compared to Layer 5 and above, however, calcite in the lowest layers is relatively well preserved. This preservation may result from the physical protection and groundwater buffering capacity of the limestone roof blocks of Layer 6B. In any case, the ashes are primarily composed of calcite and do not show any signs of secondary phosphatic diagenesis, as is the case at Roc de Marsal (Goldberg et al. 2012).

The upper layers of the site are markedly different from Layer 8 in color, anthropogenic content, percentage of coarse material, and abundance of limestone fragments. There are also differences among these layers and lateral variations within them.

A characteristic of many of the deposits—particularly those of Layer 6 (sample 32/6A; sample 201/6B), Layer 5B (e.g., samples 119, 120) and slightly less so for Layer 5A (sample 12), Layer 4 (sample 124/4C; sample 49B; samples 20, 36, 48/4A), and Layer 3 (e.g., sample 17, 26B/3B; samples 27, 45/3A)—is the presence of cm-sized rounded clasts of limestone. Most of them are relatively fresh (e.g., sample 124—Layer 4C; Fig. 2.19a), but in some cases, where sediments have undergone some decalcification (e.g., square D11), they are somewhat etched and their surfaces are indented and not smooth (e.g., sample 50B—Layer 4A, square F11; Fig. 2.19c). In addition, many exhibit thin ($\sim 10$ µm thick) discontinuous coatings of reddish brown iron-rich dusty clay (e.g., sample 36, Layer 4A, square G10; Figs. 2.19d and 2.23c, d). The fact that these coatings usually do not coat grains uniformly or completely and are not specifically oriented on a specific part (e.g., upper surface) indicate that these grains have been moved and rotated, which suggests solifluction, cryoturbation, and/or frost creep. Such processes match the overall depositional style of

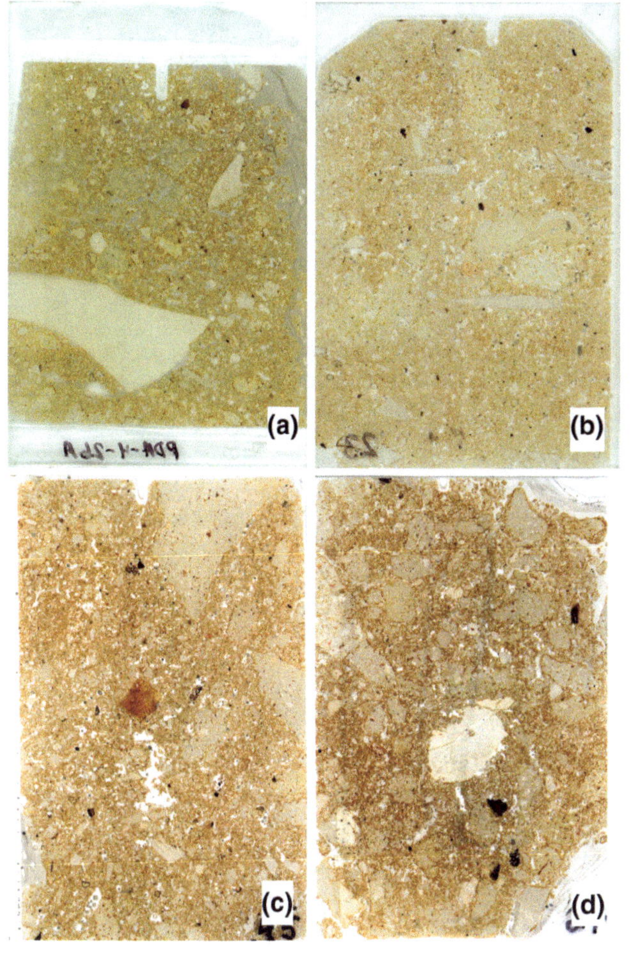


**Fig. 2.24** **a** Scan of sample 26A (Layer 3B; Sq. E11). **b** Scan of sample 23 (Layer 3B; Sq. D11). **c** Scan of sample 45 (Layer 3A; Sq. E11). **d** Scan of sample 27 (Layer 3A; Sq. D11)

the sediments as viewed in the field, particularly the sharp erosional contacts at the base of Layers 5B, 4C, and 3B (Figs. 2.3b and 2.4a).

As stated above, the sediments in Layer 8 are the most striking in the field by their color, and the presence of intact hearths and redistributed hearth materials. These aspects are complemented in the thin sections by the relative abundance of burned bone, fat-derived char, some charcoal, and various forms of calcareous ashes. Although none of the overlying layers is as striking as those in Layer 8, some do contain remains of these anthropogenic elements. For example, Layer 6—particularly in the East section (e.g., Sq. E19, sample 4B; Fig. 2.15c)—exhibits bedded burned and unburned bones, flint debris, and a dark brown, organic-rich matrix. On the West Section, accumulations of burned bone in Layer 6 are much less abundant (e.g., sample 8—Sq. D13; Fig. 2.15d) and the matrix is more sandy and less organic. Above Layer 6, the amount of burned bone visible in the field and in thin section drops off, and only a few samples exhibit them, mostly in the form of generally isolated mm- to cm-sized angular bits of burned bone (e.g., Figs. 2.8b, c and 2.9a, b, e). Interestingly, none of the deposits above Layer 6 displays organic-rich matrices.

Although some samples above Layer 8 contain burned bone, none includes charcoal, ash fragments, or organic groundmass. Layer 6 on the East Section is most similar to Layers 7 and 8 in color and content, as it also contains burned bones and occasionally, small fragments of charcoal. Layers 5 through 1 contain less groundmass and fewer limestone fragments than Layers 7 and 8.

## Artifact Orientations

The methodology used to record artifact orientations with a total station is described elsewhere (Chap. 1). Most of the techniques applied here for analyzing and representing these data have been previously described (McPherron 2005); however, also presented here are new techniques for intra-site spatial analysis to understand better variability and the effects of sample size, and to compare levels statistically.

### Methods

Because each elongated artifact is measured with only two points, its orientation can be described using only its bearing and plunge. The mean bearing angle and circular variance for a given level are calculated using circular statistics. The uniformity of the bearing angle distribution is assessed using Rayleigh's test. Significance ($p < 0.05$) indicates a nonuniform distribution of bearing angles the structure of which can be viewed using Rose diagrams. Because plunge values vary between 0 and 90° and do not cycle through this range (i.e., 90° and 0° are not the same thing and angles greater than 91° cannot exist), in contrast to previous publications (e.g., McPherron and Dibble 2007), the mean plunge angle and its variance are calculated using standard, rather than circular, statistics. The uniformity of plunge angles is assessed using a Kolmogorov–Smirnov (K–S) test with significance ($p < 0.05$) indicating a nonuniform distribution. In addition, the strength of the artifact orientations is reported here as the sum of the normalized orientation vectors (L), which includes both bearing and plunge information. Values approaching 1 indicate preferred orientations; zero indicates no preferred orientation.

To assess artifact orientations further, eigenvalues are computed on the normalized orientation vectors and transformed into shape ratios which are plotted on Benn diagrams (see Benn 1994; Lenoble and Bertran; McPherron 2005 for more details). The set of normalized three-dimensional vectors from a layer is reduced to three eigenvalues ranging from zero to one, descending in value, and totaling to one.

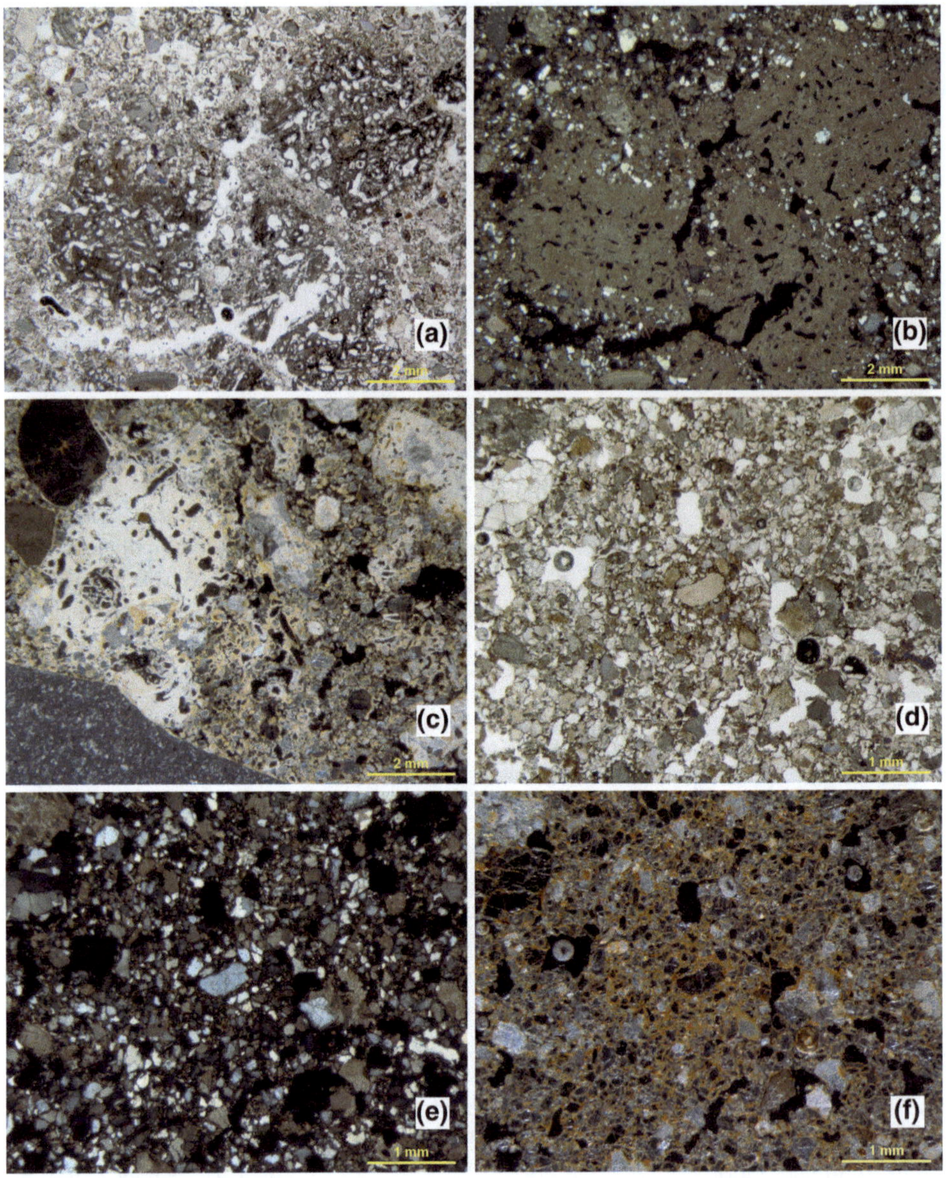


**Fig. 2.25** **a** Sample 26A (Layer 3B; Sq. E11) showing reworked clasts of reworked calcified roots. PPL. **b** XPL view of a, showing the sharp borders of the calcified root fragment. **c** Sample 26A with different phases of calcification of roots, the earlier ones with reddish iron staining. Dark field illumination. **d** Sample 23 (Layer 3B; Sq. D11) composed predominantly of sand-sized components, mostly quartz with smaller amounts of limestone grains. PPL. **e** Sample 23. Same view as d but in XPL and showing the quartz-rich nature of the sample and relatively small proportion of calcitic grains. **f** Sample 23. Dark field illumination of d and e, with reddish iron-rich dusty clay coatings and interstitial fillings and overall lack of carbonate grains or components

When the first eigenvalue is high (i.e., approaches 1) the orientations are highly linear. When the first two eigenvalues are roughly equal, the orientations are randomly distributed on a plane (though not necessarily on a horizontal plane). And when all three eigenvalues are roughly equal, the orientations are isotropic. These three patterns can be visualized using two shape ratios: elongation and isotropy. Elongation is calculated as 1—E2/E1 where E2 and E1 are the second and first eigenvalues, respectively. Isotropy is calculated as E3/E1 where E3 is the third eigenvalue.

The elongation and isotropy ratios are then plotted on a modified ternary diagram (a Benn diagram) in which the three poles correspond to linear, planar, and isotropic orientations. A ternary diagram with only two variables is preferred over a standard Cartesian *XY* plot because some regions of the *XY* space will never have points (because of the mathematical constraints on eigenvalues listed above). In a Benn diagram, the planar region is the expected outcome for artifacts dropped on a flat (though not necessarily level) surface and left unmodified by subsequent depositional processes. Site formation processes will normally alter this state and introduce some combination of linearity or isotropy into the artifact orientations. As such, these processes will pull the plotted point away from the planar region of the Benn diagram. For instance, if the surface is irregular, variance in plunge angles will increase, and this will increase the isotropy (pulling the point closer to the isotropic pole).

There are no statistical tests to show whether a layer differs significantly from an ideal planar orientation, and it is difficult to know how many artifacts are necessary to have a robust pattern. Here, these issues are addressed through resampling. The Benn diagrams for each layer are displayed

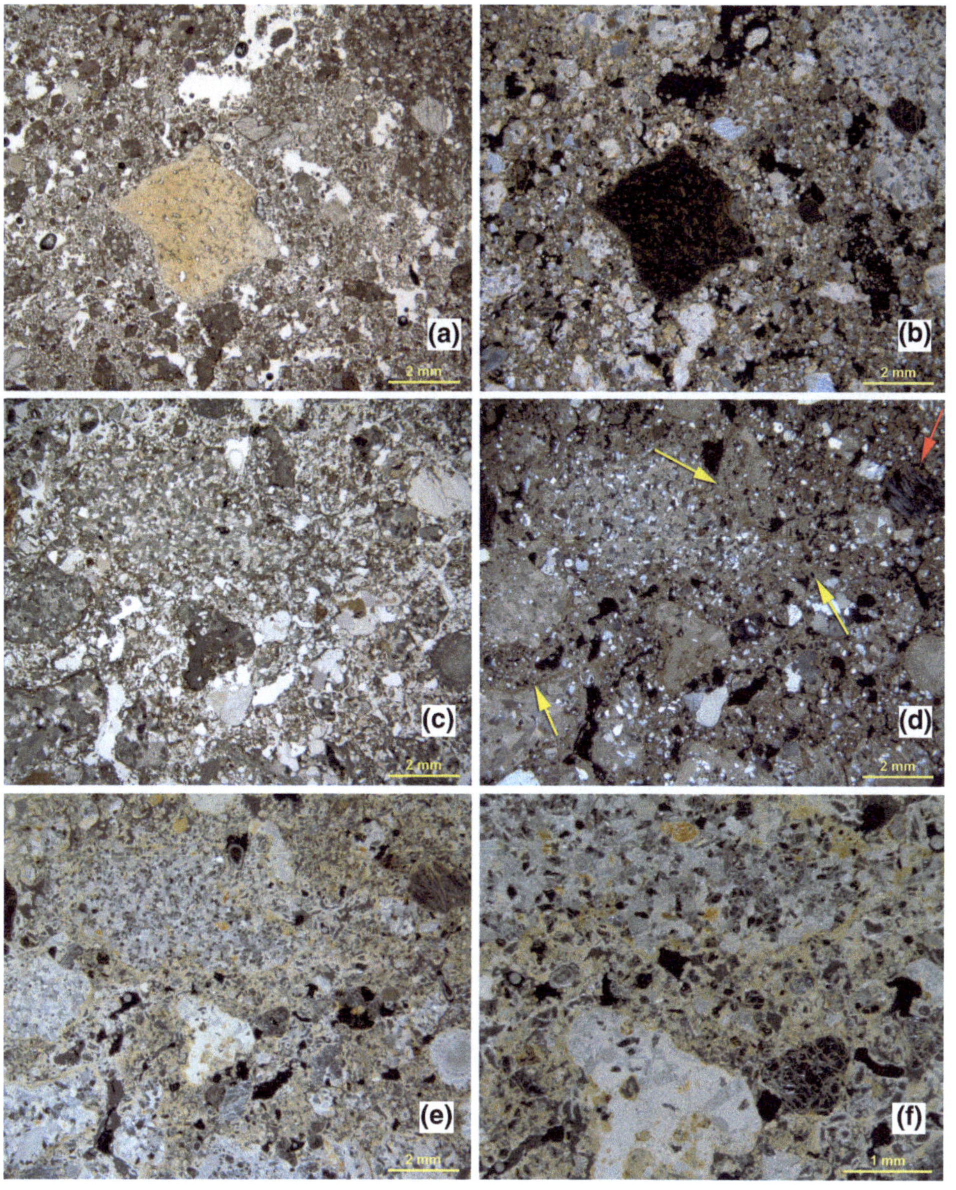


**Fig. 2.26** **a** Sample 45 (Layer 3A). Bone fragment in loose calcareous sand. PPL. **b** Sample 45. Dark field view of a showing granular nature of matrix composed of rounded coarse calcareous sand and granules. Some reddish iron-rich dusty clay is present in the matrix and coats some of the grains. **c** Sample 27 (layer 3A; Sq. E11), which is overall similar to sample 45 but somewhat more cemented by calcite. PPL. **d** Sample 27, XPL view of c showing cementation of limestone grains as well as abundant calcified roots (*yellow arrows*). A bone fragment (*red arrow*) can be seen in the upper right. **e** Sample 27 as in **c** and with dark field illumination. The widespread cementation of calcite is visible here. **f** Sample 27. Detailed view of e with secondary calcite cementing the grains, dark field

with 95% confidence intervals derived by drawing a random sample, with replacement, from the artifacts of that layer with a sample size equal to that layer. Each layer was resampled 10,000 times. This value was selected because it produced a relatively smooth and stable probability distribution. Similarly, there are no statistical tests of significance for differences in Benn ratios. Thus, to assess whether layers differed from one another, Fisher's exact test was estimated using resampling. This involved combining the two assemblages being compared and then drawing two random samples, without replacement, with sizes equal to the original two layers under comparison. Technically this involves randomly selecting a sample equal in size to one of the assemblages and then assigning the remaining artifacts to a second sample. The distribution of Euclidean distances is then derived using the elongation and isotropy ratios, and the distance of the original two layers under consideration is compared to the resampling distribution to assign a probability to that distance. A one-tail probability of $p < 0.05$ was used to assess significance because in this case, we are only interested in one portion of the distance distributions, namely those that are likely dissimilar.

In addition to these statistical procedures and the Benn diagrams, the orientation distributions are displayed in Fig. 2.27 to help with their interpretation. Thus, for each layer, a plot is made showing the spatial distribution of the artifacts used in the analysis. In addition, a Schmidt lower hemisphere diagram shows the distribution of bearing and plunge values. Schmidt diagrams place one end of the vector representing the artifact orientation at the center of a sphere and then plot where the other end of the artifact exits the sphere. Here, the lower hemisphere is plotted, meaning that

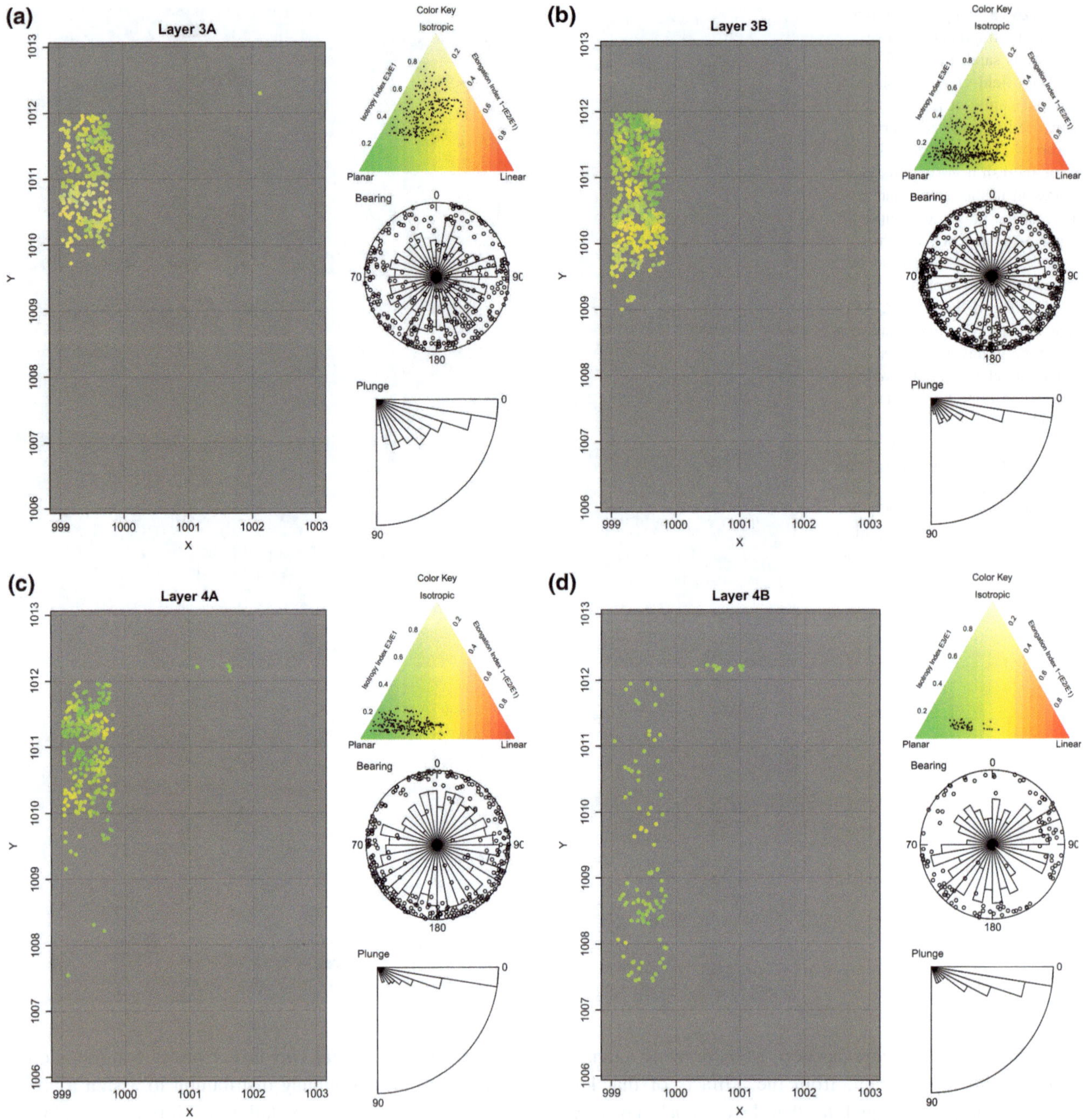


**Fig. 2.27 a–k** Pech IV artifact orientations. On the *left* is the distribution of finds used in the orientation analysis. The *color codes* represent where in the Benn diagram (on the *upper right*) an assemblage made of that artifact and its nearest 40 artifacts would fall. *Middle right* is a Schmidt lower hemisphere diagram with a superimposed Rose diagram with ten degree intervals. *Bottom right* is a modified Rose diagram showing the distribution of plunge angles

the high point on each artifact is placed at the center of the sphere and as a result, the orientation vector must exit the lower hemisphere. To view this three-dimensional object in two dimensions, the exit point is projected onto the central plane of the hemisphere (a two-dimensional circle). Artifacts that are relatively flat will plot towards the margins of the circle. Artifacts that plunge steeply will plot towards the center. The place where an artifact plots along the circle (0°–360°) indicates the direction in which the artifact is sloping. At Pech IV, angles of 0° point directly into the cliff face, 180° angles point into the valley, and 90° angles point downstream. To better appreciate potential patterns in these dot plots, a Rose diagram (a circular histogram) is superimposed on the Schmidt diagram to summarize the number

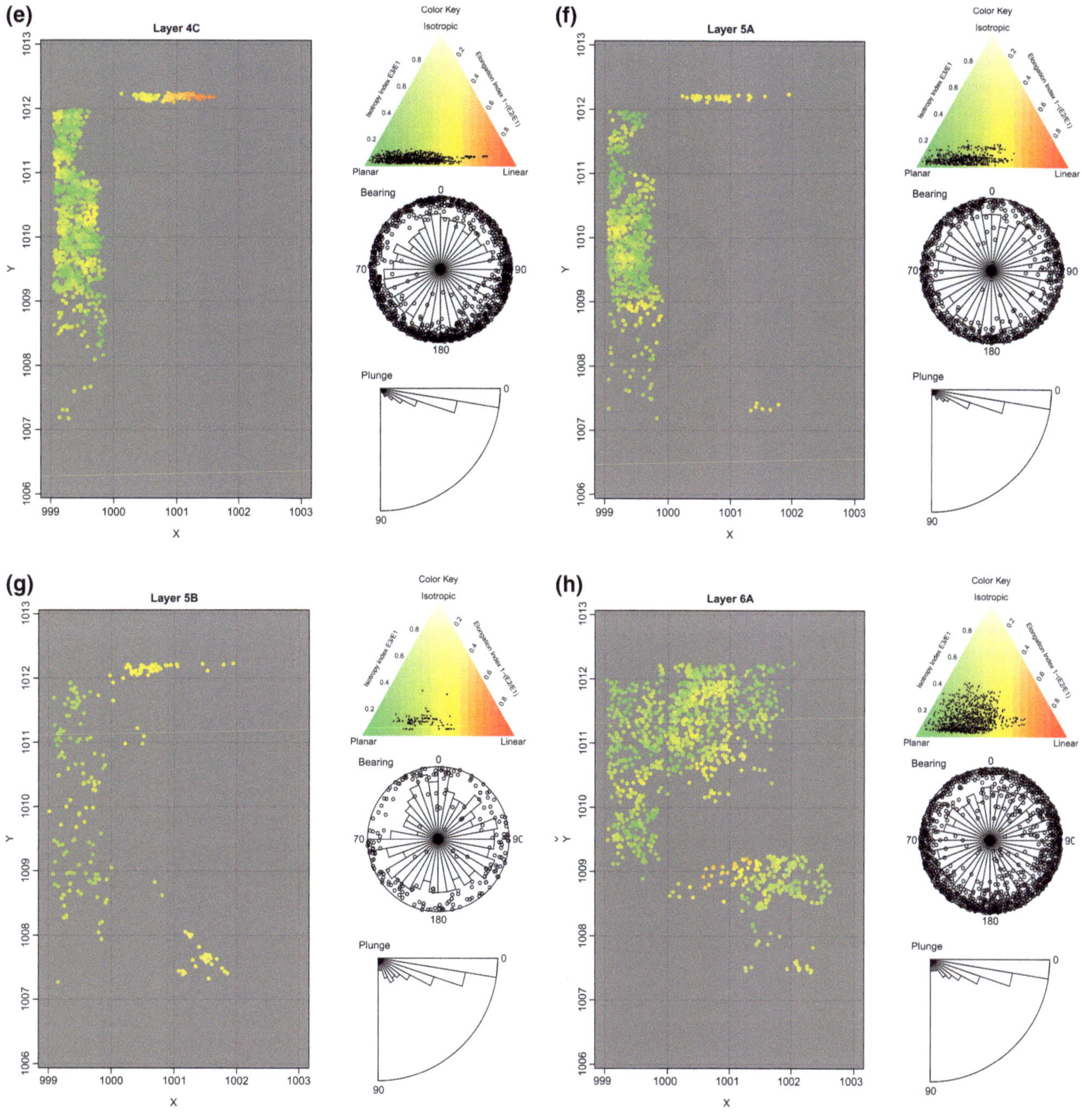


**Fig. 2.27** (continued)

of bearing angles that fall within each Rose diagram interval (here 10°). Plunge angles are summarized separately using a modified Rose diagram that shows angles from 0° to 90° (again, plunge values greater than 90° are impossible).

In addition to this figure, spatial variability within a layer is assessed using Benn values calculated based on neighboring artifacts. Here, the nearest 40 artifacts to each artifact in a layer are used to calculate Benn statistics (elongation and isotropy). A color space is then mapped onto the Benn diagram so that a plan view of each layer can be plotted with the color of each artifact representing the position in Benn space of the ratios computed from the neighboring 40 artifacts. This number of artifacts was selected arbitrarily to balance having enough artifacts to compute meaningful Benn statistics and at the same time to remain sensitive to spatial variations in these values. Next to each of these plots, a Benn diagram key is plotted with the color space indicated and with each artifact's associated Benn values plotted.

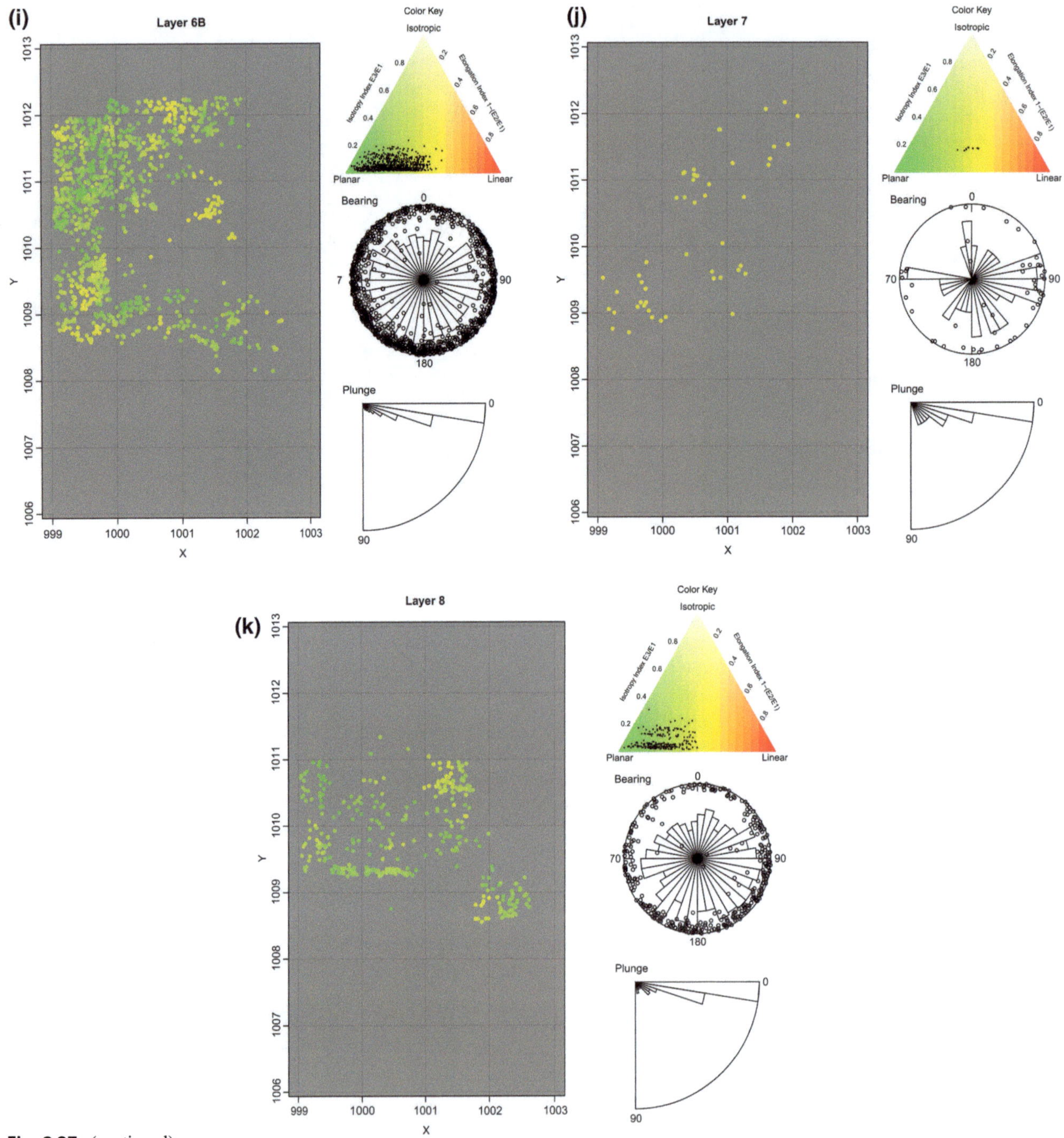


**Fig. 2.27** (continued)

These diagrams give some indication of the variability within each layer that is not apparent in the 95% confidence intervals and at the same time give some indication as to whether this variability has a spatial component.

All of the statistics and plots presented here were generated in R (R Core Team 2014). Bearing angles were assessed using the CircStats package (Lund & R port by Claudio Agostinelli 2012; CircStats: Circular Statistics 2012). The Rose diagrams were plotted with a modified version of code from this same package. Benn diagrams were modified from the ternary plots in the VCD package (Meyer et al. 2014). All of the remaining code was written by McPherron.

**Table 2.4** Bearing and plunge statistics for the Pech IV layers

| | | Length | | Bearing | | | | Plunge | | | |
|---|---|---|---|---|---|---|---|---|---|---|---|
| Layer | N | Mean | s.d. | L | Mean | Var | P | L | Mean | Var | P |
| 3A | 318 | 0.052 | 0.023 | 0.17 | 170.3 | 47.4 | 0 | 0.92 | 24.6 | 23.1 | 0 |
| 3B | 586 | 0.054 | 0.019 | 0.11 | 197.3 | 51 | 0 | 0.96 | 14 | 17.2 | 0 |
| 4A | 301 | 0.066 | 0.027 | 0.17 | 172 | 47.5 | 0 | 0.98 | 10 | 11.9 | 0 |
| 4B | 110 | 0.061 | 0.021 | 0.13 | 164.5 | 49.6 | 0.14 | 0.98 | 12.1 | 11 | 0 |
| 4C | 1143 | 0.062 | 0.021 | 0.17 | 156 | 47.4 | 0 | 0.99 | 8.6 | 8.6 | 0 |
| 5A | 764 | 0.061 | 0.022 | 0.05 | 175.7 | 54.5 | 0.16 | 0.99 | 8.9 | 9.4 | 0 |
| 5B | 210 | 0.07 | 0.026 | 0.04 | 108.3 | 55 | 0.72 | 0.96 | 14 | 15.6 | 0 |
| 6A | 1237 | 0.06 | 0.021 | 0.15 | 153 | 48.7 | 0 | 0.97 | 12.1 | 13.2 | 0 |
| 6B | 1101 | 0.063 | 0.021 | 0.18 | 198.4 | 46.7 | 0 | 0.99 | 8.2 | 8.6 | 0 |
| 7 | 50 | 0.058 | 0.012 | 0.36 | 114.9 | 36.6 | 0 | 0.97 | 12.4 | 15.3 | 0 |
| 8 | 363 | 0.053 | 0.019 | 0.23 | 151.5 | 44 | 0 | 0.99 | 7.6 | 9.5 | 0 |

The first column is the sample size. The next two columns are the mean artifact length and standard deviation based on the distance between the two points used to record its orientation. Next are the bearing statistics with the vector magnitude statistic (l), the mean bearing, the circular variance, and the Rayleigh test for uniformity. The same are presented for the plunge angles

**Table 2.5** Eigenvalues and Benn ratios (Isotropy and Elongation) for the Pech IV artifact orientations

| Layer | E1 | E2 | E3 | IS | EL |
|---|---|---|---|---|---|
| 3A | 0.397 | 0.386 | 0.217 | 0.546 | 0.029 |
| 3B | 0.476 | 0.415 | 0.107 | 0.225 | 0.129 |
| 4A | 0.491 | 0.45 | 0.059 | 0.12 | 0.082 |
| 4B | 0.55 | 0.384 | 0.066 | 0.119 | 0.301 |
| 4C | 0.499 | 0.462 | 0.039 | 0.078 | 0.073 |
| 5A | 0.484 | 0.472 | 0.044 | 0.09 | 0.025 |
| 5B | 0.456 | 0.444 | 0.1 | 0.22 | 0.027 |
| 6A | 0.48 | 0.445 | 0.075 | 0.157 | 0.074 |
| 6B | 0.507 | 0.457 | 0.036 | 0.071 | 0.1 |
| 7 | 0.582 | 0.337 | 0.081 | 0.14 | 0.42 |
| 8 | 0.498 | 0.468 | 0.033 | 0.067 | 0.06 |

## Results

Summaries of the artifact orientations results are presented in Table 2.4 and Fig. 2.27. With the exception of Layers 5 and 4B, the layers show a nonuniform bearing angle. In nearly all of these cases, the mean angle is to the south or south-southeast. Inspection of the Schmidt-Rose diagrams confirms this general pattern. The one exception is Layer 7, which has a very small sample size. As for plunge angles, all showed a nonuniform distribution, which is normal for archaeological data. In these data, Layer 3A stands out with a relatively high average plunge angle of 23.1°. This is the highest angle we have seen in our combined archaeological data sets. The rest of the angles vary between 8° and 17°.

When the bearing and plunge angles are converted to eigenvalues (Table 2.5) and plotted on the Benn diagram (Figs. 2.28 and 2.29), there is some variability between the layers. The elevated plunge angles of Layer 3A are obvious in this diagram and pull the point toward the isotropic pole. Layer 3B has high values as well, though not to the extent of Layer 3A. A permutations test (Table 2.6 and Fig. 2.29) indicates that Layers 3A and 3B are different. Additionally, when the 95% confidence interval is considered, it is clear that most of the variability within this layer relates to linearity and not isotropy. The same is true for Layer 4B. There is enough variability and a relatively low sample size to make for a broad 95% confidence interval that includes relatively undisturbed context close to the planar pole to more clearly disturbed context half way to the linear pole. These latter contexts are typically indicative of some downslope movement, due either to solifluction or water transport. As for the latter, the Schmidt-Rose diagram (Fig. 2.27) does not display the typical modal or weakly bimodal pattern typical of water transport. Layers 4C, 5A,

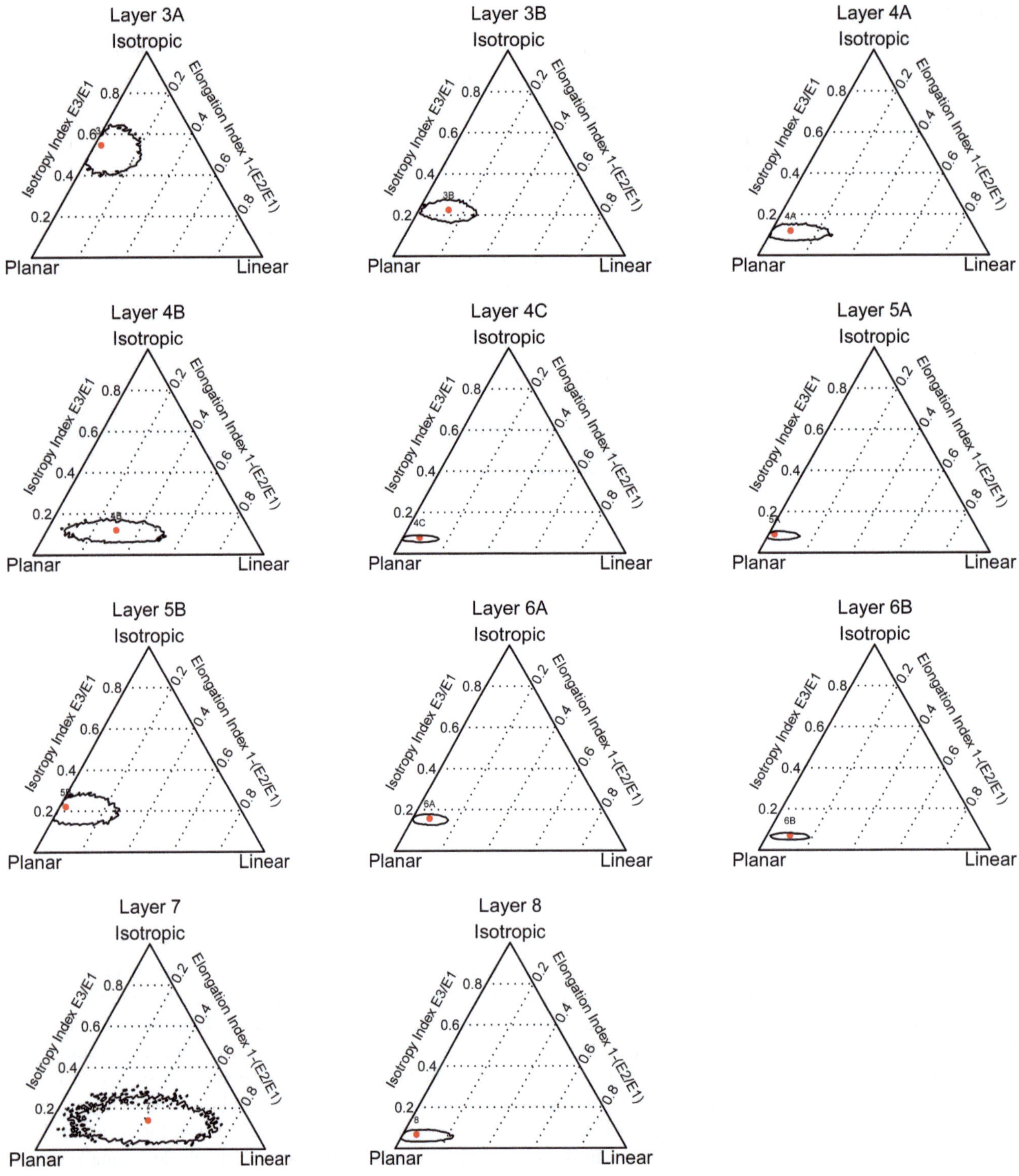


**Fig. 2.28** Benn diagrams for each of the Pech IV layers with 95% confidence intervals calculated through resampling with 10,000 iterations

6A, 6B, and 8 are clearly quite close to the planar pole even considering their 95% probability distributions, and the permutations test shows that these layers are statistically indistinguishable from one another. Layer 5B is similar to Layers 5A, 6A and 4C, but it has a bit more isotropy and will be discussed further at the end of this chapter. Layer 7 has a very large 95% confidence interval reflective of its low sample size and high internal variability. As a result, little can be said of its depositional processes based on the artifact orientations.

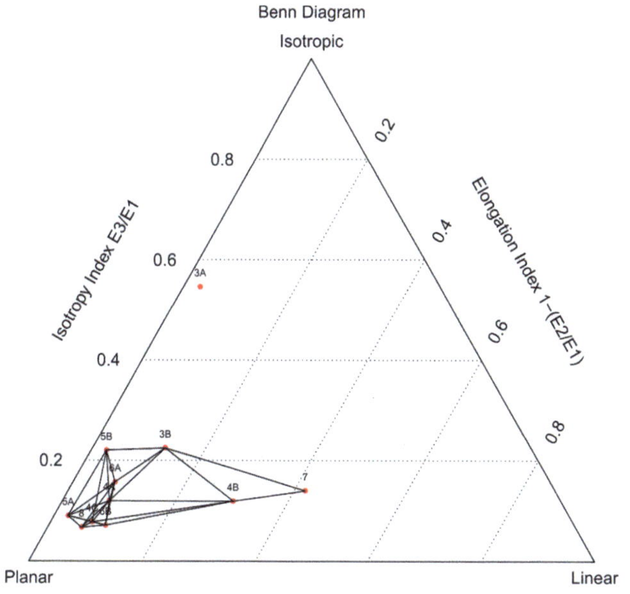


**Fig. 2.29** Benn diagram of the Pech IV layers with lines indicating which layers are statistically similar (see Table 2.6)

Some of the above-described variability is spatially patterned (Fig. 2.27). For Layers 3A and 3B, for instance, higher levels of isotropy and linearity are in the southern portion of the excavation. This is where these layers come into contact with Layer 2, which is thought to be reworked Pleistocene material. Additionally, in the southern portion of these surfaces, the deposits are not under the collapse of the Pech IV cave. Perhaps this location resulted in some reworking of these deposits after deposition or perhaps it was the collapse itself that played some role in lifting these sediments. The spatial plots also show that the portion of Layer 4C against the north limit of the excavation, meaning against the cave wall, shows high degrees of linearity. This is likely an example of the so-called "wall effect" common in cave systems where there is water percolation and movement against the cave wall. A comparison of the 95% confidence interval for Layer 4C and the spatial plot shows that in fact there is a great deal of highly localized variability within the deposit. However, these local pockets of more linear orientations are not systematically aligned enough in one direction to pull the assemblage average away from the planar pole. Layer 5A and 5B show the same wall effect pattern, though to a smaller degree than Layer 4C but with a similar level of internal variability.

Layers 8 to 6A were excavated across a larger surface, and Layer 6 was greatly impacted by large blocks of roof fall, particularly in the south and southeast portions of the excavation. In Layer 6A, these areas are where some linear and more isotropic pockets of artifacts can be found. Layer 6B, however, shows a more limited distribution, with most values having very low levels of isotropy with some pockets of higher levels of isotropy. The Layer 7 sample size is really too small to analyze spatially considering that 40 artifacts are grouped to make an analytical unit, and the layer consists of only 50 artifacts. Layer 8 shows a pattern not unlike Layer 6B.

## Edge Damage

### Methods

Edge damage is recorded with four intervals: none, 1-surface, 2-surface, and rolled. This approach differs from our study of Combe-Capelle Bas where the 1-surface interval was divided into interior versus exterior damage (Dibble et al. 1995). Assessing whether a piece is edge damaged or not (i.e., "none" vs. the other three categories) has a subjective component, whereas identifying where the damage is located and whether the damage is sufficient to classify the piece as rolled is less subjective. As for the latter, a piece is considered rolled if the edges and flake scar ridges are rounded due to abrasion. One-surface and 2-surface damage refer to whether there is damage on either the interior or exterior surface (1-surface) or on both surfaces (2-surface).

**Table 2.6** Pair-wise permutations test Benn statistics for each of the Pech IV layers. Lower values indicate more distance (less probability of two samples coming from the same population)

| Layer | 3b | 4a | 4b | 4c | 5a | 5b | 6a | 6b | 7 | 8 |
|---|---|---|---|---|---|---|---|---|---|---|
| 3A | 0 | 0 | 0 | 0 | 0 | 0 | 0 | 0 | 0 | 0 |
| 3B | – | 0.17 | 0.07 | 0 | 0 | 0.28 | 0.15 | 0 | 0.05 | 0.01 |
| 4A | – | – | 0.05 | 0.51 | 0.29 | 0.12 | 0.64 | 0.5 | 0.04 | 0.43 |
| 4B | – | – | – | 0.06 | 0.01 | 0.01 | 0.04 | 0.05 | 0.51 | 0.03 |
| 4C | – | – | – | – | 0.34 | 0.08 | 0 | 0.54 | 0.04 | 0.81 |
| 5A | – | – | – | – | – | 0.09 | 0.06 | 0.11 | 0.01 | 0.42 |
| 5B | – | – | – | – | – | – | 0.36 | 0.04 | 0.01 | 0 |
| 6A | – | – | – | – | – | – | – | 0.07 | 0.03 | 0.16 |
| 6B | – | – | – | – | – | – | – | – | 0.03 | 0.57 |
| 7 | – | – | – | – | – | – | – | – | – | 0.02 |
| 8 | – | – | – | – | – | – | – | – | – | – |

A piece is considered damaged when there are multiple edge removals typically irregular in shape and size, typically abrupt, and typically noncontiguous (McBrearty et al. 1998).

An effort is made to exclude excavation and curation damage. The latter is likely not a large issue with the assemblages presented here as they were analyzed soon after excavation, and the artifacts were individually bagged. At one extreme, a single ding on an edge or even 2 or 3 very small and dispersed removals (cf. McBrearty et al. 1998) are not enough to be considered edge damage and at the other extreme, it can be difficult to separate edge damage from light scraper retouch. We consider these two cases to be the main area of subjectivity in this classification system. Note that the scale does not vary based on the geological context, that is, these observations are made independent of an object's geological context. Note too that this system does not distinguish pre-depositional use damage from post-depositional damage. An experimental trampling study designed to address this issue (McPherron et al. 2014) demonstrated (a) that edge damage is correlated with edge angles and (b) that all other variables being equal, exterior surfaces show more damage. Unfortunately, this study also demonstrated that size measures are only a weak predictor of edge angles and thus a poor predictor of edge damage. Likewise, as mentioned above, which surface was damaged (interior versus exterior) was not recorded for the Pech IV assemblage. Finally, the damage rates reported here are based on complete and proximal flakes only. Tools are excluded because they are presumably more likely to have use damage in addition to damage induced by trampling and other site formation processes.

### Results

A summary of edge damage percentages is presented in Table 2.7. A few aspects are worth noting. First, Layers 8, 6A, 5A, 4C, 4B, 4A, 3B, and 3A are similar with high levels of undamaged pieces (between 66 and 77%). Second, Layer 7 stands in contrast with the lowest percentage of undamaged pieces and a very high percentage of two-surface damage and rolled pieces. Layers 6B and 5B fall in between these two groups, with Layer 5B showing higher levels of damage than in Layer 6B. The pattern in Layer 5B will be discussed below. While there are many possibilities, the pattern in Layer 6B could reflect some mixing of material at the contact between Layers 6B and 7, which was clear but not sharp, it could be the result of the roof fall in this layer, or it could represent increased trampling. As for the former, section plots of the rolled material from Layer 6 do not show that more pieces come from the bottom of the layer, as one would expect if mixing were the cause.

## Breakage and Heated Flints

### Methods

Lithic breakage rates are easily quantified as the percentage of proximal blanks (retouched and unretouched flakes retaining at least 50% of the platform) in the total sample of blanks (retouched and unretouched proximal and complete flakes):

**Table 2.7** Edge damage on complete and proximal flakes. Layer 5B is reported for the whole layer and divided into samples labeled 5B-1 and 5B-2 (see below for an explanation)

| Layer | No damage | 1 surface | 2 surfaces | Rolled | N |
|---|---|---|---|---|---|
| 3A | 77.3 | 16.7 | 5.9 | 0.1 | 982 |
| 3B | 70.3 | 21.7 | 8.1 | 0 | 1685 |
| 4A | 69.4 | 18 | 12.6 | 0 | 111 |
| 4B | 70.4 | 22.2 | 7.4 | 0 | 27 |
| 4C | 69.3 | 23.9 | 6.9 | 0 | 348 |
| 5A | 66.9 | 25.6 | 7.4 | 0.1 | 844 |
| 5B | 38.6 | 30.7 | 23.8 | 6.9 | 319 |
| 5B-1 | 26 | 27.3 | 31.3 | 15.3 | 150 |
| 5B-2 | 47.2 | 33.3 | 19.4 | 0 | 180 |
| 6A | 66.7 | 23.2 | 9.9 | 0.2 | 1179 |
| 6B | 48.2 | 30.9 | 19.4 | 1.5 | 1111 |
| 7 | 16.4 | 21.7 | 46 | 15.9 | 1719 |
| 8 | 72 | 20.1 | 6.6 | 1.3 | 1168 |

**Table 2.8** The percentage of heated flints through the Pech IV sequence

| Layer | N unburned | N burned | % Heated |
|---|---|---|---|
| 3A | 1089 | 12 | 1.1 |
| 3B | 1849 | 11 | 0.6 |
| 4A | 131 | 1 | 0.8 |
| 4B | 31 | 0 | 0 |
| 4C | 404 | 3 | 0.7 |
| 5A | 911 | 11 | 1.2 |
| 5B | 361 | 3 | 0.8 |
| 6A | 1311 | 61 | 4.5 |
| 6B | 1177 | 72 | 5.8 |
| 7 | 1813 | 165 | 8.3 |
| 8 | 1096 | 286 | 20.7 |

**Table 2.9** A comparison of breakage rates in heated and unheated flakes

| Layer | Complete blanks (*N*) | Broken blanks (*N*) | Ratio | Complete heated blanks (*N*) | Broken heated blanks (*N*) | Ratio | Chi-square | *P* |
|---|---|---|---|---|---|---|---|---|
| 6A | 861 | 450 | 1.91 | 15 | 46 | 0.33 | 42.63 | <0.0001 |
| 6B | 703 | 474 | 1.48 | 18 | 54 | 0.33 | 33.53 | <0.0001 |
| 7 | 791 | 1022 | 0.77 | 50 | 115 | 0.43 | 10.99 | <0.0001 |
| 8 | 680 | 416 | 1.63 | 124 | 162 | 0.77 | 32.55 | <0.0001 |

$$\text{Breakage Index} = \text{Proximal Blanks} / (\text{Proximal Blanks} + \text{Complete Blanks}) * 100$$

However, as is demonstrated below, it is clear that heating either breaks pieces or makes them more susceptible to breakage (or both). Heated lithics are a useful proxy for the use of fire since they are resistant to the kinds of taphonomic processes that might remove other lines of evidence, such as charcoal or heated sediment [e.g., Sandgathe et al. (2011), though it is important to bear in mind that the heat from a fire can alter objects located several centimeters below the fire itself (Aldeias et al. 2016)]. Identification of heating is reasonably straightforward: a lithic is considered heated if pot lids or crazing is present on its surface or on its edges. Color, such as reddening, is a less reliable measure of heating since different materials react differently to heating. Evidence for heating was assessed on all piece-provenienced lithics. Thus, here the frequency of heated artifacts and breakage in heated artifacts is considered as well, and finally the rate of artifact breakage on only the unheated lithics is presented.

## Results

A summary of the percentage of heated artifacts is presented in Table 2.8. The frequency of heated artifacts is greatest in Layer 8 and decreases gradually upward. In Layers 5, 4, and 3, heating stays below 1.2%; Layer 6, which is dark colored and contains burned bones on the East Section contains approximately 5% heated lithics. These data also show that heated artifacts are more likely to be broken (Table 2.9). Therefore, to control for this factor, the breakage patterns are presented here (Table 2.10) with heated artifacts removed from the calculation.

Even with heated artifacts removed, breakage rates are highest in the lower portion of the section. Layer 7 again clearly stands apart from the rest with elevated levels of broken artifacts. Above and below Layer 7, the rate drops below 50%, and in the rest of the sequence, breakage rates are between roughly 25 and 35%.

# Small Finds

## Methods

As discussed in Chap. 3, all sediments from the layers reported here were water screened through two size fractions: 2 and 6 mm. Each water-screened sample comes from one bucket of sediment excavated typically over an area equivalent to one-quarter meter square (though not necessarily aligned with square boundaries). Buckets are filled to a standardized volume of seven liters unless a layer change is encountered in which case they are closed prior to being full. Stones larger than 10 cm in maximum dimension are not included in the bucket volume. In our experience, depending

**Table 2.10** The percentage of complete flake blanks along with their size measurements and two shape ratios

| Layer | Complete blanks | Proximal blanks | % complete | Length (mean) | Width (mean) | Thickness (mean) | Refinement (mean) | Elongation (mean) |
|---|---|---|---|---|---|---|---|---|
| 3A | 828 | 273 | 75.2 | 33 | 23.7 | 8.1 | 2.9 | 1.4 |
| 3B | 1344 | 516 | 72.3 | 33.1 | 23.6 | 7.1 | 3.3 | 1.4 |
| 4A | 98 | 34 | 74.2 | 36.8 | 27.3 | 7.7 | 3.6 | 1.4 |
| 4B | 26 | 5 | 83.9 | 39 | 27.6 | 8.7 | 3.2 | 1.4 |
| 4C | 282 | 125 | 69.3 | 36.6 | 26.4 | 6.9 | 3.8 | 1.4 |
| 5A | 654 | 268 | 70.9 | 37.8 | 24.2 | 6.4 | 3.8 | 1.6 |
| 5B | 228 | 136 | 62.6 | 39.1 | 25.8 | 7.7 | 3.3 | 1.5 |
| 6A | 876 | 496 | 63.9 | 39.8 | 26.8 | 7.2 | 3.7 | 1.5 |
| 6B | 721 | 528 | 57.7 | 33.8 | 24.6 | 7.2 | 3.4 | 1.4 |
| 7 | 841 | 1137 | 42.5 | 34.2 | 24.3 | 8.5 | 2.9 | 1.4 |
| 8 | 804 | 578 | 58.2 | 36.5 | 24 | 6.1 | 3.9 | 1.5 |

on the number of stones larger than 10 cm and on sediment compaction, a 5 cm spit excavated over a quarter meter square corresponds to roughly 2.5 buckets. Buckets are provenienced by recording a point at the center of the excavated area once the bucket is full.

To assess Bordes' excavation techniques, we excavated, screened, and analyzed in exactly the same manner 87 buckets of sediment from his backdirt. An analysis of these data is presented here.

For Combe-Capelle Bas, the small fraction (2–5 mm) was analyzed (Kluskens 1995), but at subsequent sites that we excavated this fraction was not analyzed (but all of these samples are now archived in the MNP). These data are not discussed further here. For the large fraction (6–25 mm), the material was sorted into bones versus stones and treated separately. For the stone artifacts, we passed the large fraction through a set of stacked sieves with sizes of 20, 16, and 10 mm resulting in four size classes (20–25, 16–20, 10–16, and 6–10 mm), which were counted and weighed.

The analyses presented here are based solely on the stone data. In preparing this data set, artifacts that should have been piece provenienced (i.e., artifacts larger than 25 mm), have been removed from the small finds and included in the large find counts. Similarly, provenienced artifacts smaller than 25 mm were removed from the lithic analyses by giving them a special code ("toosmall"), and in the analyses presented here these pieces have been reintegrated into the small finds data set when layer totals are provided but not integrated when considering small finds on a bucket by bucket basis. The reason for this is that in our excavation system at that time piece-provenienced lithics were not easily associated with a particular bucket. Note that generally, there are relatively few such finds, particularly in relation to the overall sample size. In Layer 8, for instance, 84 small flakes were incorrectly provenienced (due to their small size) from a total of 27,736 small flakes contained in the 6–25 mm fraction.

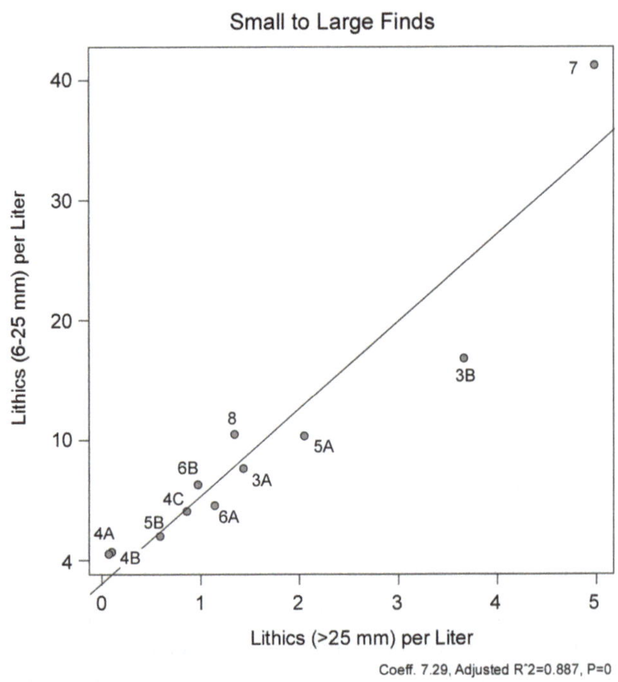


**Fig. 2.30** A plot of the density of large lithic finds (>25 mm) per liter versus small lithics finds (6–25 mm) per liter by layer

## Results

Variability in the small finds from Pech IV can be analyzed in comparison to the large finds and internally using the data from the stacked screens. With regard to the former, it has been shown experimentally that the distribution of size

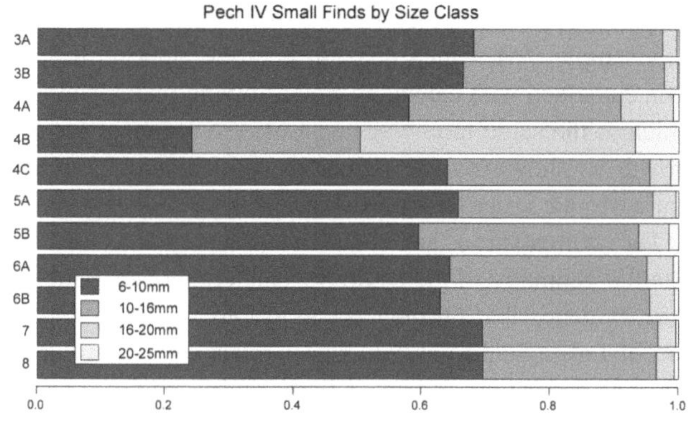


**Fig. 2.31** The proportion of various size classes with the 6–25 mm lithics fraction recovered from the screens

classes is not effected by reduction intensity or raw material size (Lin et al. 2015). It is clear, however, that various site formation processes do affect size distributions. At Pech IV the ratio of large finds to small finds per liter show a highly linear relationship (Fig. 2.30). Layer 3B has proportionally more large flakes and Layer 7 has proportionally more small ones, and both layers have relatively high artifact densities in comparison with the other layers. This was already clear from excavation of these units, especially Layer 7, which other measures and the geological observations presented here have shown underwent severe post-depositional disturbance. With regard to internal variability in size classes (i.e. within the 6–25 mm fraction), with the exception of Layer 4B, the data show very similar distributions throughout the sequence (Fig. 2.31).

Setting aside Layer 4B for the moment, what is interesting about these data is perhaps *not* what they tell us of site formation processes, but how little they vary considering the amount of variability in lithic technology that exists in the sequence. For instance, Layers 3A and 3B, which show similar distributions and display the presence of biface production, have relatively low retouched tool counts, and have a blank technology based on Levallois and discoidal approaches. The Layer 3 distribution, however, is very similar to that of Layer 8, which has a high percentage of retouched tools and a high incidence of Levallois, and to Layer 7, which is difficult to describe technologically because it is so broken and damaged due to cryoturbation. The industries of Layer 6 are characterized by low tool production but an increased production of small flakes using a variety of techniques including Levallois (Dibble and McPherron 2006). These small flakes drop below the 25 mm size cut-off for proveniencing artifacts, and it perhaps shows in the size class distributions with the larger flakes being slightly more numerous than in, for instance, Layer 8. However, Layers 5B through 4C, which plot similarly to Layer 6, do not show an emphasis on small flake production, but they do show evidence of Levallois technology and a return to retouched tool production. Layer 4A represents an entirely different technology of blank production (Quina) with very high retouched tool frequencies and very little evidence of on-site blank production (see Chap. 6). Thus, as was shown experimentally already (Lin et al. 2015), we are not able to see a relationship between technology and small flake distributions in our data.

The pattern in the one outlier, Layer 4B, is interesting. The technology of this layer is intermediate between that of Layers 4C and 4A, but the size class distribution is not suggestive of a mix of the two layers. The smallest size class (6–10 mm) shows a deficit, which is consistent with the selective removal of small finds by water. The artifact orientations for this layer show a rather high level of linearity and especially a high contrast to Layers 4A and 4C, which have very low linearity. Linearity (or bimodality) is also consistent with water flow. However, there is no apparent evidence for running water that is observable in the thin sections (e.g., sample 28, Fig. 2.19b; sample 49B, Fig. 2.20b), so perhaps the size distribution may be a result of artifact import, rather than removal of material.

Note that in terms of large (>25 mm) artifact densities and small (6–25 mm) find densities, Layer 7 shows several patterns distinct from the rest of the sequence. Both the number of large lithic artifacts per liter (5.0) and small lithic artifacts per liter (50.2) are far greater than any other layer at the site. This is not surprising given the high level of breakage. However, when breakage—including heat-related breakage—is corrected for in the large artifact data by considering only platform flakes and tools, the artifact density (2.3 per liter) is still at the high end of the range for the site as a whole, but below that of Layer 3B (2.3). Additionally, the ratio of small to large lithics (10.08), while at the high end, is within the range of variation seen within the sequence.

## Discussion

The stratigraphy, general nature of the deposits and site formation based on artifact analyses are discussed above. Here, we turn to a discussion of the field and microscopic observations that contribute to understanding the entire site history, including geogenic and anthropogenic aspects of site formation.

The thin sections from Pech IV show that the quartz and carbonate sand, as well as larger limestone fragments are derived from breakdown of the quartz-rich limestone bedrock. This conclusion is based on the clear similarity in grain size, sorting, and roundness of the loose quartz in the matrix to the quartz in the limestone. In addition, most minor coarse fraction components such as glauconite and mica are also found as trace components in the limestone bedrock. Many micromorphology samples reveal fragments of limestone that are actively dissolving and contributing quartz sand to the sediment in situ (dissolving limestone is distinguished by its undulating grain edges). Some fragments also weather internally and contain vughs and infillings of fine material, or coatings of reddish brown iron-rich clay. In some samples, the carbonate component of the limestone was completely dissolved, leaving behind a residuum of quartz sand, and some muscovite and glauconite. Therefore, the origin for nearly all of the coarse sediments in the site is ultimately the cave environment.

A volumetrically minor component of the fine sediment fraction is also derived from limestone weathering and the breakdown of its insoluble components. This material is represented by weathered glauconite and muscovite grains, as well as silt-sized grains of iron (Fe) and manganese (Mn) oxides, which are partially responsible for giving many of the deposits a redder color (Fig. 2.6). In addition, some of the Fe/Mn fraction observed in the groundmass also originated from the relict Tertiary soils and deposits on the plateau (*sidérolithique*; Texier [2009]). Finally, a more significant factor in imparting the red color is the formation of reddish brown dusty clay coatings in voids and on grains. These coatings are widespread in the Pech IV deposits, but are particularly well developed in the East Section and in deposits in squares D and E in the West Section.

Many thin sections also display post-depositional calcium carbonate that occurs as localized fine micritic root coatings (Figs. 2.18a and 2.22f); they can also be found as isolated individual coatings or as aggregated ones (e.g., Figs. 2.21 c,2.22e and 2.25a–c). In the case of the latter, the surrounding matrix is well indurated, and it is possible to observe on the West Section a broad zone of cemented deposits (Fig. 2.4a). This zone includes Layers 4 and 3 from squares H through F11 and as such, they appear to indicate locations of previous driplines: they give the impression of migrating upward and northward toward the back of the cave (see Fig. 2.4a). It is worth noting that the calcification of the rootlets is generally the last diagenetic event because, for the most part, the rootlets developed after the formation of most Fe-rich dusty clay coatings. However, in some instances both seem to have precipitated alternately, as Fe-rich zones are precipitated or formed on carbonate, which are then later pierced by calcified roots (Fig. 2.18a).

Another, rather different type of secondary carbonate occurs at the base of Layer 8 (sample 54, square E13—Fig. 2.11c–f), which in the field appeared as a large well-indurated block of "breccia", with bone, many pores, and calcite hypocoatings. Visible in thin section are cm-sized clasts of limestone, quartz sand, and bone tightly cemented with microsparite (Fig. 2.11e–h). In addition, acicular microsparite needles coat voids (Fig. 2.11c–f), some of which exhibit previous coatings of iron-rich dusty clay. This sample (54), which rests on bedrock, appears to have been cemented with calcite by water flowing on the bedrock, which at the same time was favorable environment for root growth as shown by the numerous vesicles and chambers that perforate a dense matrix (Fig. 2.11c).

The composition and both internal organization and external geometry of the anthropogenic components provide clear evidence that burning events occurred during deposition of Layer 8 and possibly Layer 6. Layers 8–6 all show elevated percentage of heated flints. These burning events, exemplified by the intact stringers of ash overlying carbon-rich layers, likely correspond to the round combustion structures excavated in 2002.

The fissures present in some of the cemented layers indicate that the cementation likely occurred relatively soon after deposition and not much later. On the other hand, Layer 8 also exhibits features associated with carbonate dissolution: as mentioned above, many limestone fragments appear to be actively dissolving and many cemented ash fragments often contain thin channel voids, which are characteristic of calcite depletion (Bullock et al. 1985). Compared to Layers 5 through 1, however, carbonate in the lowest layers is relatively well preserved. This preservation may result from the physical protection and groundwater buffering capacity of the limestone roof blocks of Layer 6B. In any case, the ashes, which are primarily composed of calcite, do not show any signs of secondary phosphatic diagenesis as is the case at Roc de Marsal (Goldberg et al. 2012). In fact, secondary phosphatization was extremely rare in any of the samples.

## Trampling

Most of the bone fragments visible in thin section are typically equant to rectangular in shape. However, some of the elongated ones—particularly in Layer 8—are fractured but contiguous and conjoinable (Figs. 2.9e, f and 2.10a, d), features that point to clear evidence of trampling (Goldberg et al. 2012; Miller et al. 2010). Such trampling supports the intensive fire activity demonstrated by the superposed hearth and rake-out layers in Layer 8. This activity also seems to be localized at the cave entrance, just inside the extant dripline at the time. On the other hand, trampling certainly occurred in overlying layers that were away from the entrance, but the lack of appropriate sediments (mostly gravel, pebbles and sand with few bones) preclude their traces from being preserved, or at least evident. In terms of edge damage on lithics, the amount of damage seen in Layer 8 is typical of most of the layers at the site, with only some other layers showing higher amounts of damage (particularly Layers 7, 6B and 5B). In any case, trampling of varying intensity in Layer 8 may also contribute to the compact nature of the ash layers, as well as to mixing of burned and unburned bones, ashes, and limited charcoal within the layer. In addition to trampling, such episodes of mixing may also have been produced by hearth rake-out activities, whereby the remains of existing hearths were spread out laterally from their original location, providing a substrate for the next fire episode. In this regard, as pointed out above, we note that the thickest part of Layer 8 appears in squares H/I 14 and trends SW from there (Figs. 2.2, 2.3a, c and 2.5b,c).

## Depositional Processes

Micromorphological and field evidence indicate relatively few depositional processes operating at Pech IV, although as pointed out above, we have excavated only at what was the former entrance and its immediate surroundings. The most important source of sediment is the bedrock, which furnishes a variety of clasts, the most striking of which are roof fall in the form of large slabs that are concentrated at the entrance of the cave (Fig. 2.4a). These large spalls are dramatic, and as pointed out above, they began to accumulate in Layer 6 in the central part of the cave (Sq. F and G) with initial collapse of the overhanging ceiling. With time, successively larger decimeter-sized slabs fell from the ceiling toward the direction of the entrance where we witness an accumulation of imbricated slabs that prograded outward to the south (Figs. 2.3a–c and 2.5d): blocks in Layer 6 occur in square G14, whereas those in Layers 5 and 4 can be found in squares G14-J14 and H14-K14, respectively. Less volumetrically important are smaller, cm-sized blocks that are particularly abundant in Layer 4 and in the southern part of Layer 3 (Figs. 2.3a–c and 2.4a).

The dislodging of the limestone clasts from the roof is presumably a combined effect of chemical and physical weathering. The accumulation of the larger slabs, however, while likely fostered by weathering, more readily represent instability of the roof as it was weakened by initial collapse in Layer 6 time. However, eventually its configuration as a relatively thin, overextended roof protruding from the cliff face (see present-day Pech I for a modern analogy) produced the "internal exfoliation" of collapse from the interior outward (Fig. 2.32). It is worth stressing here that at the time of accumulation of the large slabs of Layers 5 and 4, the entrance to the cave was somewhat obstructed, and occupation took place in a restricted space between the back wall to the north and the large slabs of roof fall to the south.

The finer fraction (predominantly quartz sand and limestone pebbles) was liberated by dissolution of the bedrock, accumulated initially by gravity. However, lateral transport took place by solifluction, as can be seen in the field by sharp contacts between layers, especially between Layers 5B/6 and 3B/4 (Fig. 2.3a–c); in addition, many of the cm-sized clasts in these units are well rounded (see above; e.g., Figs. 2.12, 2.13c, 2.9b–c, 2.19a, d and 2.20b). Moreover, movement on a smaller (granular scale) was likely accomplished through the action of frost/soil creep (Bertran et al. 1997; van Vliet-Lanoë 1985; Vliet-Lanoë 1982).

Other fine materials are derived from red soils and sediments on the plateau (*sidérolithique*) that have been incorporated into the sedimentary fill via colluviation and gravity. Moreover, a significant part of the red color in the sediments from the East Section (Fig. 2.7b)—and particularly close to the bedrock walls in the West Section (the so-called "wall effect"; Fig. 2.7c)—is a result of weathering/dissolution of the bedrock (e.g., iron-rich minerals such as glauconite), as well as translocations of iron-rich dusty clay that coat voids and grains (e.g., Figs. 2.9c, f and 2.18a, d, e). The iron-rich reddish coatings are strongest against the north wall (Figs. 2.3b and 2.7a) where water would naturally concentrate, leading to preferential dissolution here. But dusty iron-rich coatings occur away from the wall so their origin is more likely associated with dripping water from the roof or perhaps some slight runoff that allows for translocation of the detrital silty clay. It is not clear whether these wall effects were of sufficient intensity to account for the linearity seen in the artifact orientations along the north wall as well.

As stated above, most of the Pech IV sediments were generated internally from dissolution of bedrock leading to the deposition of coarse (cm- to dm-sized) limestone clasts and finer cm-, granule-, and sand-sized quartz and limestone grains; during excavation several pieces of travertine were recovered (e.g., square E11). These lines of evidence support the concept that Pech IV is a collapsed cave and is analogous to the situation visible at Pech I, ~80 m to the NW (Fig. 2.1b). As the hillslope eroded slightly obliquely to the axis of the jointing and Pech de l'Azé karstic system, the overhangs of both caves were destabilized, resulting in the collapse of the cave roof as outlined above—at Pech IV first as internal collapse starting with Layer 6 and later by the retreat of the brow. This retreat is recorded in the microstratigraphy as secondary calcite accumulations that appear to migrate northward and stratigraphically upward through time. Final roof collapse resulted in the reduction of the overhang to a few meters, traces of which are visible on the West Section of the current excavations. Remains of the overhang are more prominent in the East Section where the roof collapse is thicker and appears to represent a final collapse that took place *after* many of the layers had already been deposited (see Figs. 2.7a, b and 2.8). In passing, it is interesting to note that the retreat of the cave towards grid northwest means the current excavation sampled an ever-changing location relative to the dripline.

During his excavations, Bordes observed evidence in his Level $J_1$ (our Layer 5A) of cryoturbation in the form of rounded limestone blocks and damaged flints (see also

Chap. 1), both of which appeared to be more pronounced in the front of the site. In the field and in thin section, cryoturbation is generally conveyed by rounded limestone pebbles and distinct artifact inclination and mixing (see above). Interestingly, in Pech II cryoturbation was visible as pockets of rounded/abraded pebbles in Layer 4D (also Layer 5 and "X"; (Bordes 1972; Laville et al. 1980); other micromorphological evidence at Pech II for cold climate features was illustrated by silty cappings and coatings on grains, as well as ice lensing (Goldberg 1979). Although some of the layers at Pech IV do contain rounded fragments of limestone, micromorphological observations do not reveal similar types or degrees of reorganization of the fine fraction into cappings as at Pech II; some dusty clay cappings can be seen at Pech IV but they are relatively muted in comparison to those of Pech II. This lack of microscopic expression of cold climate suggests that rounded limestone blocks from Layer 6 were possibly not rounded in situ through freeze–thaw, as is the case at Pech II. Rather, they could be reworked from older non-prehistoric, possibly phreatic gravels from inside the cave.

On the other hand, the distribution of limestone clasts in Layer 5 and the abundance of relatively fresh micro-éboulis (cm-sized limestone grains) are suggestive of solifluction (Bertran and Texier 1999). Evidence in support of this can also be found in the archaeological measures of site formation processes in Layer 5B. The artifact orientations show higher levels of isotropy than the layers immediately above and below. There is also an elevated percentage of rolled pieces (6.9%). When the distribution of rolled pieces is plotted (Fig. 2.33), it is clear that they are not distributed throughout the layer but rather form a lens within it. This lens was not noticed and isolated during excavation. Therefore, here a 2D kernel density estimate is applied to the projected Y and Z coordinates of the rolled pieces, and a 0.5 contour interval on the resulting density matrix is used to isolate the lens (Layer 5B-1) from the rest of the layer (5B-2). With these two subsamples, some archaeological site formation indicators can be compared. As shown in Table 2.11 and Fig. 2.34, the artifacts of subsample 5B-1 show a much higher level of linearity in their orientations. The sample size for subsample 5B-1 is low ($N = 43$); however, in this case permutations test show that while the 95% confidence intervals do overlap to some extent, the two samples (5B-1 and 5B-2) are statistically distinct at the $p < 0.05$ level. Next, with regard to edge damage, sample 5B-1 shows a lower percentage of undamaged pieces and a higher percentage of pieces with damage on two surfaces (Table 2.12). This is unsurprising, as generally in our data sets two-surface damage is elevated in assemblages with elevated rolled pieces, but it further confirms the integrity and the associated differences in site formation processes affecting the two subsamples. Finally, breakage levels are higher in subsample 5B-1 (Table 2.13), but the difference is not significant ($X^2 = 2.41$, $p = 0.12$).

Because Pech IV was occupied during a series of climatic fluctuations (Jacobs et al. 2016), it is interesting that the sediments do not record more evidence of past environments. Although cold climate features, such as ice lensing require both cold temperatures and a source of moisture in order to form, they also need fine material to retain water. Thus, a reasonable explanation of the differences in expression of micromorphological features between Pech II and IV may be related to differences in sediment grain size: the Pech II sediments are generally finer and contain more clay- and silt-sized particles, which enable the development and preservation of freeze–thaw textures because they retain moisture and inhibit water drainage. Hence, at Pech IV the paucity of finer groundmass and predominance of coarse sediment (sand- through pebble sizes) may, therefore, not preserve most macroscopic expressions of climate processes. Whereas roof collapse and the deposition of limestone blocks have also been linked to climatic processes—specifically cold and dry periods (Laville et al. 1980)—their specific interpretations are not clear. Thus, although the deposits at Pech IV contain both large roof blocks and smaller clastic contributions that may in a general way reflect climate, nothing more definitive can be said at this point.

## Sequence of Formation

When occupation of Pech IV began, the morphology of the site was more consistent with a cave than a rockshelter. The earliest deposition was primarily anthropogenic in nature (Table 2.14; Fig. 2.31). Activity took place directly on the exposed surface of the bedrock. This activity included the deposition of anthropogenic materials (e.g., bone, lithics, char, some charcoal, ashes) mixed with natural cave detritus (e.g., quartz sand, some silt). Most ash is present as stringers and cemented ash grain clusters, which implies an interesting sequence of events. The ash was likely deposited in several thin layers corresponding to ephemeral hearth-like features or single rake-out events. Immediately after deposition, gravitational settling and other sedimentary processes caused some unconsolidated ash to filter through the sediment profile and shelter nearby large grains. The remaining ash layers were quickly cemented by (drip or soil?) water. After cementation, the ash layers were fractured by trampling, and some of the sediments became mixed, possibly during later hearth rake-out activities. The ash fragments were then subjected to some calcite dissolution, as evidenced by the development of voids (vughs) and a few highly porous fragments.

Layer 7 marks a major cold event as shown by its cryoturbated structure. Moreover, it contains a rather rich lithic assemblage dominated by heavily damaged, rolled, and abundant broken pieces but without the differential removal of a particular size class. This latter point makes water transport unlikely. The artifact densities are comparable to those in Layer 3B, a layer that shows little post-depositional alteration. Bone densities in Layer 7 are the lowest in the Pech IV sequence and may be a result of these same post-depositional processes.

The first major roof fall event (Layer 6B) produced massive, flat-lying blocks of limestone that protected the combustion features from further trampling and acted as a chemical buffer that saturated groundwater in carbonate before it reached the lowest layers. This saturated water perched above the bedrock and locally cemented the bottom of Layer 8.

The first roof collapse event likely changed the morphology of the cave to a more open spatial configuration, as the sediments immediately above the limestone blocks contain iron and quartz concretions of subsurface pedogenic origin (Fig. 2.31). Combustion activities may have continued for a short period or shifted in location, as evidenced by the burned bones of Layer 6 on the East Section and by the continued presence of heated lithics in this layer. The resulting sediments were not protected from carbonate dissolution, however, and the only remnants of these activities are burned bones, heated lithics, and occasional small fragments of charcoal. The large roof fall concentrated in squares D–E–F–G 11–12–13–14 is also contemporaneous with noncalcareous sand accumulation from inside the cave system. A thin activity surface was deposited before a second collapse. Following this second collapse, accumulation of massive sand continued, punctuated by smaller roof falls (individual blocks), spreading to G–H 11–12–13–14. The northern squares (D–E–F 11–12–13–14) experienced marked erosion of Layer 6A followed by infilling of stony sand. Strong inclination of the surface in E–F 11 combined with chaotic orientation of stones and thickening of Layer 5B in square G11 suggest a possible solifluction lobe banked up against the roof fall. Although this feature is not visible in microscale, an analysis of artifact orientations, breakage, and edge damage shows a feature with Layer 5B that is consistent with this interpretation. This phase (Layer 5) marks the last major period of block collapse at the entrance (squares I, J, K) and the beginning of a different mode of collapse: the backward retreat of the brow towards the north. At this point, the morphology of the site becomes more like a rockshelter than a cave. Again, the above scenario shows that retreat of cave overhang towards grid northwest indicates that our excavation samples from bottom to top represent a changing location relative to the dripline.

We can surmise that once the cave began to collapse during the onset of Layer 6, the surface of the deposits was periodically covered with snow. Thus, whereas much of the sedimentary infill is composed of breakdown products of the limestone bedrock with initial sedimentation by gravity, snowmelt must have played a part in mobilizing what was lying on the ground. Initially in Layer 6 time, a sloping surface was inherited from or conditioned by the bedrock floor, which slopes up to the north (Figs. 2.3a–c and 2.4a). This slope, though slight, was enough to cause movement of the sediment by solifluction or soil/frost creep, which explains many of the features observed in thin sections (e.g., granular, loose nature of the fabrics). In addition, the presence of water and percolation beneath former surfaces in the past would also encourage weathering of the material on the surface, leading to the formation of red iron-rich dusty clay coatings and cappings seen on many of the grains.

Reddish brown silty clay accumulates in the north part of the site, while imbricated roof blocks mark the retreat of the brow in the south (Fig. 2.31). Cemented zones in the

**Table 2.11** Eigenvalues and Benn shape ratios for the sublevels of Layer 5B

| Level | E1 | E2 | E3 | IS | EL | N |
|---|---|---|---|---|---|---|
| 5B-1 | 0.555 | 0.334 | 0.111 | 0.2 | 0.397 | 43 |
| 5B-2 | 0.49 | 0.421 | 0.089 | 0.181 | 0.141 | 152 |

**Table 2.12** Edge damage for the sublevels of Layer 5B. Rolled is removed because it is the basis of separating the two sublevels

| | None | Interior | Exterior | Exterior/interior | N |
|---|---|---|---|---|---|
| 5B-1 | 30.7 | 7.9 | 25.2 | 36.2 | 27 |
| 5B-2 | 47.2 | 11.7 | 21.1 | 20 | 180 |

**Table 2.13** The percentage of complete flakes for the sublevels of Layer 5B

| | Complete | Proximal | % complete |
|---|---|---|---|
| 5B-1 | 108 | 74 | 59.34 |
| 5B-2 | 136 | 65 | 67.66 |

**Table 2.14** Summary of Pech de l'Azé IV history

| Phase | Layer | Events |
|---|---|---|
| 1 | 8 | Anthropogenic accumulation of bones, ash, char and organic matter, and inwashing of sand from interior of cave and with little rock fall. Some intact hearths, but mostly laterally displaced combustion materials as a result of hearth rake out or simply trampling. The anthropogenic deposits thin laterally toward the north and south and are thickest just inside the former position of the brow, which trended ENE/WSW. They are deposited directly on bedrock and are well indurated and calcite-cemented at the base by water trapped at bedrock/sediment interface |
| 2 | 7 | Influx of heavily abraded/rolled (cryoturbated) flints in sandy matrix. Subsequent deformation by rock fall of Layer 6 hinders our ability to determine if Layer 7 was originally of uniform or variable thickness (i.e., whether it took the form of a solifluction lobe). Although no inclination of the layer was observed, it is reasonable to assume it was derived (in part) from the direction of the interior of the cave. Flat contact at base. Upper part is irregular and eroded by Layer 6, resulting in variable thickness of layer. *Brief cold period?* |
| 3 | 6B | Accumulation of large slabs of rock fall in squares D–E–F–G 12–14, which are mostly broken along bedding planes and originate from the central part of the then-extant cave roof. Coevally, (corresponding to the base of Layer 6B) some sandy deposits began to accumulate by disaggregation of the roof and soil/frost creep, particularly near the north wall. Anthropogenic inputs of burned bones, particularly in the East Section. Rounded limestone clasts with cappings suggest cryoturbation and weak freeze–thaw. *Possible climatic significance = cold with alternating freeze–thaw?* alternating freeze-thaw, although this is not supported by the faunal analyses |
| 4 | 6A | Continued deposition of homogeneous, massive sand by soil/frost creep that is penecontemporaneous with deposition of angular slabs of rock fall in square E12 and large blocks in squares F–G 12, 13, and 14; the latter continue to fall from the interior of the cave ceiling, producing an imbricated structure, with successive blocks being deposited further out toward the entrance. Local overtopping of large blocks by sand. *Same conditions as Layer 6B* |
| 5 | 5B<br>5A | Loose blocks continue to fall in squares G–H 11–12–13–14. To north (squares D–E–F 11–12–13–14) marked erosion of Layer 6A by Layer 5B followed by deposition of calcareous disaggregated limestone sand and rounded by soil/frost creep or solifluction. Stones are fresh and tend to be well rounded but they are flatter and more angular in square C closer to the wall. Strong dip/inclination of surface in E–F 11 toward the W. Chaotic orientation of stones and thickening of Layer 5B in square G11 point to distal part of solifluction lobe banked up against the roof fall to the south, and associated with stone lifting. With time, the clasts become more angular and flatter. Dispersed mm- to cm-sized rounded and angular burned bone fragments indicate reworking of anthropogenic inputs; no intact burning features were observed |
| 7 | 3A | Accumulation of fine gravel (some angular but mostly rounded) in a sandy matrix that are eroded into Layer 3B with a sharp contact. It contains rare larger, fist-sized pieces of roof fall, and extends basically from squares D to F. Artifacts are also bedded and tend to follow the dip, although some are on edge. The roof at this time seems to have been above the F line, as the sediment in these squares is cemented. The cave was nearly full at this time, and the deposits came close to the level of the ceiling. This layer appears to represent fine-grained solifluction of the geological sediments, but the artifacts are quite sharp and fresh, so no or little transport of the flints seems to have occurred [it is also possible that because the matrix is so fine grained, that artifacts are not abraded]. Bone and limestone clasts have thin Mn coatings representing presence of water, perhaps associated with former dripline |
|  | 3B | These deposits are laterally truncated by a channel situated under the then-extant drip line that is later filled in by Layer 2 |
| 8 | 2 | Cutting and filling of channel into Layers 3A and 3B beneath the then-extant overhang, situated over squares E and F. It is filled with cm-sized pieces of bedrock consisting of both angular blocks and rounded cobbles produced by roof retreat and slumping from the adjacent hillslope, as well as also reworking of Layer 3A material that slumped in from the side of the channel |
| 9 |  | Final collapse of remaining nub of roof above square D11 as shown on the W section by the pivotal rotation of the large tilted block at the top of the section that is close to the present bedrock back ledge of the cave. On the E Section, this last phase of collapse is shown by thick accumulation of large slumped blocks of roof that also seal the deposits |
| 10 | 1A<br>1B<br>1C | Medieval (?) erosion of slope, coupled with colluviation of cm-sized rocks in dark brown organic clay sand |

sediments represent positions of former drip lines, which move northward with each collapse event. Truncation and erosion of Layer 4A by angular gravelly sand with a sharp contact indicates possible cryoclastism. These cold climate deposits are laterally truncated by a channel that formed from flowing water under the then still extant drip line. Deposition of Layer 2 fills the channel with cm-sized pieces of bedrock consisting of both angular blocks and rounded cobbles produced by roof retreat and slumping from the adjacent hill slope. This layer also contains some reworking of Layer 3A material that slumped in from the side of the channel. During historic times, final slumping of roof occurred, as shown by the rotation of the roof block above square D11. Final erosion of the slope, coupled with colluvial accumulation, buried the site by cm-sized rocks in dark brown organic clay sand.

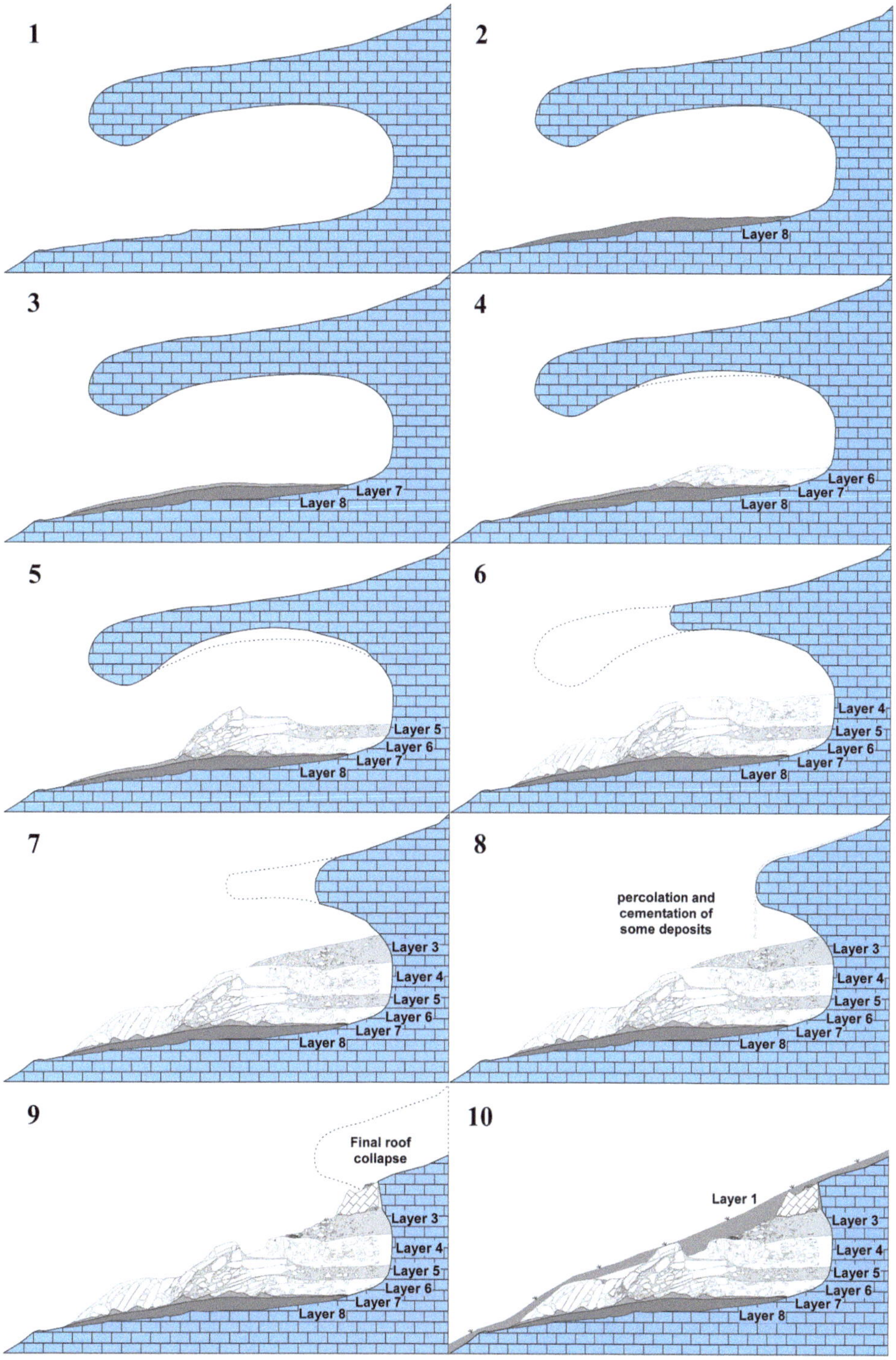


**Fig. 2.32** Evolution of Pech de l'Azé IV

## Integrity of the Archaeological Record

As noted above, the geological dynamics described above show that sediments are essentially locally derived and transported and that any transport was local, i.e., confined to the immediate environs of the site itself. The vast majority originated from the limestone as roof fall and disaggregation products; only small contributions from the plateau (e.g., *sidérolithique*) were noted. Furthermore, lateral transport was confined to creep, which mobilized sediments on the

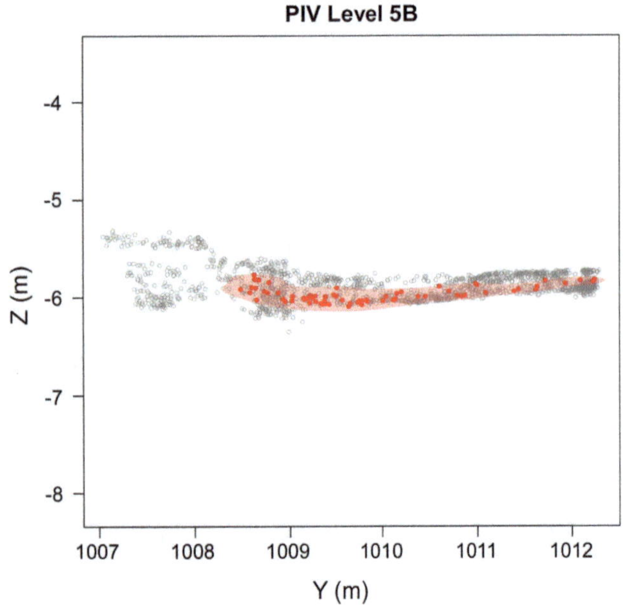


**Fig. 2.33** The distribution of rolled lithics in Layer 5B as shown from a sagittal section view. The *red line* represents a kernel density estimate of the rolled pieces. All artifacts that fall within this *red line* are labeled as 5B-1. The remaining artifacts are 5B-2

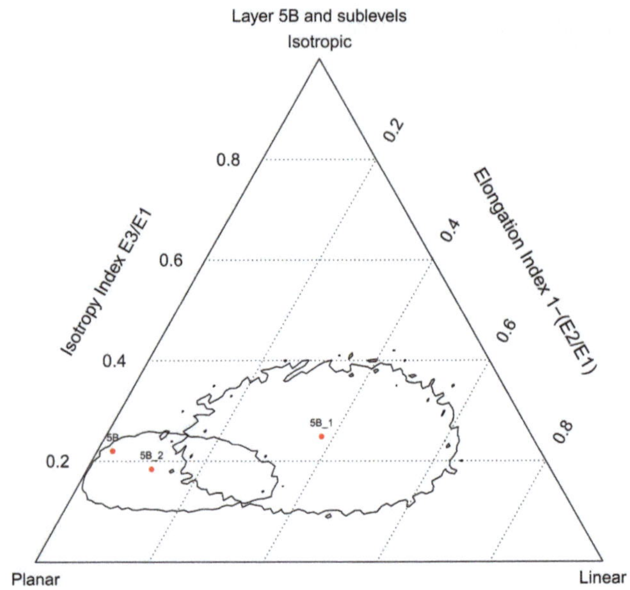


**Fig. 2.34** Benn diagram for Layer 5B complete and the sublevels of Layer 5B. The boundaries on 5B-1 and 5B-2 represent 95% confidence intervals. A permutations test shows them to be statistically distinct

granular scale. Some solifluction was noted in Layers 5B, 4C, and 3B. Although the artifact orientations suggest some re-alignment of the artifacts with the orientation of the cave system, there is no evidence from the various observations presented here of massive reworking of the archaeological material. A few layers show some substantial post-depositional disturbances (particularly Layer 7 and parts of Layer 5), but these appear to have modified the assemblages very locally and essentially in place, if at all. In other words, such processes did not likely result in substantial artifact transport: sediments moved but the artifacts did not.

The relatively widespread use of fire in Layer 8 did not change the displacement of artifacts to a major extent, as was the case at Roc de Marsal (Aldeias et al. 2012). Trampling did result in comminution of bone and perhaps mm to cm vertical displacements of artifacts, and rake out possibly moved artifacts over a surface in the order of perhaps cm or dm; probably not more than that since individual lenses in Layer 8 are at the most 30 cm across. On the other hand, in the lower layers, fire resulted in increased levels of breakage, inflated small find counts, and burned bones.

## Concluding Comments

Geoarchaeology and site formation at Pech IV were evaluated using detailed field observations, soil micromorphology, and attribute analysis of the lithic finds, including their orientations, size, density, and breakage characteristics. In addition to updating Bordes' stratigraphy, we now have a clearer understanding of the site history and formation processes.

Layer 8—and to some extent, Layer 6 (particularly in the East Section)—is characterized in the field by its dark color and bedded anthropogenic components. These include chert, burned bone, char, and ashes, which are preserved to some extent as intact lenticular combustion structures, but mostly as redistributed components through the action of hearth rake out. These combustion structures represent isolated events, as we do not tend to find superposition of intact hearths one directly on top of another.

Whereas both intact hearths and associated sediments occur throughout Layer 8, sediment thickness is arranged in a linear west-northwest trend that generally parallels the position of the former brow of the cave, and occupation appears to have occurred just inside of the dripline, where we observe in thin section numerous examples of trampling. Thus, although Layer 8 is characterized by burning activities, they are not distributed evenly throughout the cave space or specifically through time, as the top of Layer 8 appears to be richer in organic matter and ashes, although the latter occur as both intact lenses and reworked clasts of cemented ashes. Numerous attempts were made to excavate the Layer 8 sediments using *décapage* technique but, owing to the thin, lenticular nature of the combustion features and their redistribution by rake out, it was impossible to follow surfaces over more than the scale of centimeter. The intact lenticular nature of the hearths are reminiscent of some of those at Hayonim and Kebara Caves (Israel) (Goldberg and Bar-Yosef 1998) and like virtually all Middle Paleolithic hearths are not associated with rock paving or enclosures.

Observations of the site's morphology and setting, as well as the location, bedding inclinations and directionality of the lithics, data demonstrate that the site is in fact a remnant of a phreatic cave system (similar to Pech I and II): it was breached by backward erosion of the valley, which cut the phreatic tube diagonally. This type of erosion resulted in the overhang that Bordes observed and which gave the impression of the site being a rockshelter.

Overall, our stratigraphic subdivisions broadly match those of Bordes, although there is not a one-to-one correspondence. This is not surprising, as his subdivisions were based on a mix of lithological and typological attributes. Furthermore, based on lithostratigraphy and comparative lithic analysis between his collections and ours, we were able to link the East and West Sections, thereby showing that contemporaneous occupations occurred throughout the excavated space. We are not dealing with two separate "sub-sites."

The most striking evidence for Neanderthal use of fire is found within Layer 8, with its large quantities of highly burned bone and lithic contents, ash and char. The amount of burned bone and lithics decreases in Layer 6, which is still dark colored, and drops off dramatically above that. Nevertheless, micromorphological analysis showed the presence of burned bone in Layers 5A, 4A, and 3A. They occur in these upper layers without any association of combustion features, charcoal, or ashes. At the moment, they demonstrate that fire was used—even though apparently to a very low extent in these layers, but without any other associated burned materials, it is difficult to assess what their implications for fire are at the site. As the bone fragments in these upper samples are in low concentrations, it is likely that they were not heated in place but rather that they were derived from somewhere relatively close by, perhaps on the order of meters away. Nevertheless, burned bone counts from the layers, more or less mirror the observation of burned bones observed in thin sections.

The high degree of isotropy in Layer 3A is intriguing and unique. The micromorphology shows a number of the limestone clasts with a variety of orientations and a high degree of bioturbation as reflected in the porosity. Moreover, a greater degree of trampling in this part of the site does not seem likely as Layer 3 has the lowest amount of rolled pieces. The reasons for this amount of isotropy remain unclear.

Overall, there is no *striking* evidence for climate change within the deposits except for the clear cryoturbation in Layer 7 and solifluction in Layer 5B. In addition, there are some cappings in the deposits, but they are sporadic and not well developed. As mentioned, the lack of ice lensing or other indicators of cold climates are not present, likely due to the lack of fine material in the sediments.

## References


Aldeias, V., Dibble, H., Sandgathe, D., Goldberg, P., & McPherron, S. (2016). How heat alters underlying deposits and implications for archaeological fire features: A controlled experiment. *Journal of Archaeological Science, 67,* 64–79.

Aldeias, V., Goldberg, P., Sandgathe, D., Berna, F., Dibble, H., McPherron, S., et al. (2012). Evidence for Neandertal use of fire at Roc de Marsal (France). *Journal of Archaeological Science, 39*(7), 2414–2423.

Arpin, T. L., Mallol, C., & Goldberg, P. (2002). Short contribution: A new method of analyzing and documenting micromorphological thin sections using flatbed scanners: applications in geoarchaeological studies. *Geoarchaeology—An International Journal, 17*(3), 305–313.

Benn, D.I. (1994). Fabric shape and the interpretation of sedimentary fabric data. *Journal of Sedimentary Research, 64*(4), 910–915.

Berna, F., & Goldberg, P. (2008). Assessing Paleolithic pyrotechnology and associated hominin behavior in Israel. *Israel Journal of Earth Sciences, 56,* 107–121.

Bertran, P., Hétu, B., Texier, J.-P., & Van Steijn, H. (1997). Fabric characteristics of subaerial slope deposits. *Sedimentology, 44*(1), 1–16. doi:10.1111/j.1365-3091.1997.tb00421.x.

Bertran, P., & Texier, J.-P. (1999). Facies and microfacies of slope deposits. *CATENA, 35*(2–4), 99–121.

Bordes, F. (1972). *A tale of two caves*. New-York: Harper & Row Publishers.

Bordes, F. (1975). Le gisement de Pech de l'Azé IV: Note préliminaire. *Bulletin de la Soçieté Préhistorique Française, 72,* 293–308.

Braillard, L. (2000). La dynamique de mise en place et l'évolution du remplissage de l'abri sous roche Sur-les-Creux à Tanay (Vouvry VS). *Bulletin de la Murithienne, 118,* 41–58.

Bullock, P., Fedoroff, N., Jongerius, A., Stoops, G., Tursina, T., & Babel, U. (1985). *Handbook for soil thin section description*. Wolverhampton: Waine Research.

Capdeville, J. P., & Rigaud, J. P. (Cartographer). (1987). Notice Explicative de la Feuille Sarlat-la-Canéda à 1/50 000.

Courty, M.-A., Goldberg, P., & Macphail, R. I. (1989). *Soils and micromorphology in archaeology*. Cambridge: Cambridge University Press.

Dibble, H., Lenoir, M., Holdaway, S., Roth, B., & Sanders-Gray, H. (1995). Techniques of Excavation and Analysis. In H. Dibble & M. Lenoir (Eds.), *The middle paleolithic site of Combe-Capelle Bas (France)* (pp. 27–40). Philadelphia: University of Pennsylvania, University Museum Press.

Dibble, H., & McPherron, S. (2006). The missing mousterian. *Current Anthropology, 47*(5), 777–803.

Goldberg, P. (1979). Micromorphology of Pech-de-l'Azé II sediments. *Journal of Archaeological Science, 6,* 1–31.

Goldberg, P., & Bar-Yosef, O. (1998). Site formation processes in Kebara and Hayonim Caves and their significance in Levantine prehistoric caves. In T. Akazawa, K. Aoki, & O. Bar-Yosef (Eds.), *Neandertals and modern humans in Western Asia* (pp. 107–125). New York: Plenum.

Goldberg, P., & Berna, F. (2010). Micromorphology and context. *Quaternary International, 214*(1–2), 56–62.

Goldberg, P., Dibble, H., Berna, F., Sandgathe, D., McPherron, S., & Turq, A. (2012). New evidence on Neandertal use of fire: Examples from Roc de Marsal and Pech de l'Azé IV. *Quaternary International, 247*(1), 325–340.

Goldberg, P., & Macphail, R. I. (2003). Strategies and techniques in collecting micromorphology samples. *Geoarchaeology, 18*(5), 571–578.

Goldberg, P., & Macphail, R. I. (2006). *Practical and theoretical geoarchaeology*. Oxford: Blackwell Publishing.

Jacobs, Z., Jankowski, N., Dibble, H., McPherron, S., Soressi, M., Sandgathe, D., et al. (2016). The age of three Middle Palaeolithic sites: single-grain optically stimulated luminescence chronologies for Pech de l'Azé I, II and IV in France. *Journal of Human Evolution, 95,* 80–103.

Karkanas, P., Rigaud, J.-P., Simek, J. F., Albert, R. M., & Weiner, S. (2002). Ash bones and guano: A study of the minerals and phytoliths in the sediments of Grotte XVI, Dordogne, France. *Journal of Archaeological Science, 29*(7), 721–732.

Kluskens, S. L. (1995). Archaeological taphonomy of Combe-Capelle Bas from artifact orientation and density analysis. In H. Dibble & M. Lenoir (Eds.), *The Middle Paleolithic Site of Combe-Capelle Bas (France)* (pp. 199–243). Philadelphia: The University Museum Press.

Laville, H., Rigaud, J.-P., & Sackett, J. (1980). *Rock shelters of the Perigord*. New York: Academic Press.

Ligouis, B. (2006). Jais, lignite, charbon et autres matières organiques fossiles: Application de la pétrologie organique à l'étude des éléments de parure et des fragments bruts.. In J. Bullinger, D. Leesch, & N. Plumettaz (Eds.), *Le site magdalénien de Monruz, volume 1: premiers éléments pour l'analyse d'un habitat de plein air. Archéologie neuchâteloise* (Vol. 33, pp. 197–216). Neuchâtel: Service et musée cantonal d'archéologie.

Lin, S.C., McPherron, S.P., & Dibble, H.L. (2015). Establishing statistical confidence in Cortex Ratios within and among lithic assemblages: A case study of the Middle Paleolithic of southwestern France. *Journal of Archaeological Science, 59,* 89–109.

Lund, U., & R port by Claudio Agostinelli. (2012). CircStats: Circular statistics. (2012). S-plus original. Retrieved from http://CRAN.R-project.org/package=CircStats.

Madella, M., Jones, M. K., Goldberg, P., Goren, Y., & Hovers, E. (2002). Exploitation of plant resources by Neanderthals in Amud Cave (Israel): The evidence from phytolith studies. *Journal of Archaeological Science, 29*(7), 703–719.

McBrearty, S., Bishop, L., Plummer, T., Dewar, R., & Conard, N. (1998). Tools underfoot: Human trampling as an agent of lithic artifact edge modification. *American Antiquity, 63*(1), 108–129.

McPherron, S.J. (2005). Artifact orientations and site formation processes from total station proveniences. *Journal of Archaeological Science, 32*(7), 1003–1014.

McPherron, S., & Dibble, H. (2007). Artifact orientations from total station proveniences. In A. Figueiredo & G. Velho (Eds.), *The world is in your eyes*, pp. 161–66. Tomar, Portugal: CAA.

McPherron, S., Dibble, H., & Goldberg, P. (2005). Z. *Geoarchaeology, 20*(3), 243–262

McPherron, S., Braun, D. R., Dogandžić, T., Archer, W., Desta, D., & Lin, S. C. (2014). An experimental assessment of the influences on edge damage to lithic artifacts: A consideration of edge angle, substrate grain size, raw material properties, and exposed face. *Journal of Archaeological Science, 49,* 70–82.

Meyer, D., Zeileis, A., & Hornik, K. (2014). vcd: Visualizing categorical data. R package version 1.3–2.

Miller, C. E., Conard, N. J., Goldberg, P., & Berna, F. (2010). Dumping, sweeping and trampling: experimental micromorphological analysis of anthropogenically modified combustion features. *P@lethnologie: Revue francophone en Préhistoire*. http://www.palethnologie.org.

Salomon, J.-N., & Astruc, J.-G. (1992). Exemple en zone tempérée d un paléocryptokarst tropical exhumé (la cuvette du Sarladais, Dordogne). In J.-N. Salomon & R. Maire (Eds.), *Karsts et évolutions climatiques* (pp. 432–447). Bordeaux: Presses Universitaires de Bordeaux.

Sandgathe, D., Dibble, H., Goldberg, P., McPherron, S., Turq, A., Niven, L., et al. (2011). On the role of fire in Neandertal adaptations in Western Europe: Evidence from Pech de l'Azé and Roc de Marsal, France. *PaleoAnthropology*, 216–242.

Soressi, M., McPherron, S., Lenoire, M., Dogandzic, T., Goldberg, P., Jacobs, Z., et al. (2013). Neandertals made the first specialized bone tools in Europe. *PNAS, 110*(35), 14186–14190.

Stoops, G. (2003). *Guidelines for analysis and description of soil and regolith thin sections*. Madison, WI: Soil Science Society of America.

Team, R.C. (2014). R: A language and environment for statistical computing. Vienna, Austria: R Foundation for Statistical Computing.

Texier, J.-P. (2009). *Histoire géologique de sites préhistoriques classiques du Périgord une vision actualisée la Micoque, la grotte Vaufrey, le Pech de l'Azé I et II, la Ferrassie, l'abri Castanet, le Flageolet, Laugerie Haute*. Paris: Édition du Comité des travaux historiques et scientifiques.

Turq, A., Dibble, H., Goldberg, P., McPherron, S., Sandgathe, D., Jones, H., et al. (2011). Les Fouilles Récentes du Pech de l'Azé IV (Dordogne). *Gallia préhistoire, 53,* 1–58.

van Vliet-Lanoë, B. (1985). Frost effects in soils. In J. Boardman (Ed.), *Soils and quaternary landscape evolution* (pp. 117–158). Chichester: Wiley.

Villagran, X. S., Schaefer, C. E. G. R., & Ligouis, B. (2013). Living in the cold: Geoarchaeology of sealing sites from Byers Peninsula (Livingston Island, Antarctica). *Quaternary International, 315,* 184–199.

Vliet-Lanoë, B. V. (1982, 1981 March 2–6). *Structures et Microstructures Associées à la Formation de Glace de Ségrégation: Leurs Conséquences*. Paper presented at the Fourth Canadian Permafrost Conference, Calgary, Alberta.

# An Absolute Chronological Framework for Pech IV

3

Nathan R. Jankowski

## Introduction

All archaeological questions begin as an initial questioning of sample context in terms of both space and time. The ability to position archaeological traces within their correct chronological context is therefore as important to archaeological investigations as the accurate recording of physical sample location or the understanding of processes of site formation. Without such temporal constraint, the archaeological record remains dislocated from its chronological context—simply anthropogenic traces superimposed on a geological background. The need for accurate age control at Pech IV was, therefore, at the forefront of the research design for the current excavations at the site.

Four independent dating methods were employed to construct the chronological framework for Pech IV. These techniques were radiocarbon ($^{14}C$) (McPherron et al. 2012), electron spin resonance (ESR) (Turq et al. 2011), thermoluminescence (TL) (Richter et al. 2013; Richter et al., in press), and optically-stimulated luminescence (OSL) (Jacobs et al. 2016). Each of these methods (a) used sample materials collected during or subsequent to the recent excavations; (b) used a different type of material to act as a time-keeper (or chronometer); and (c) has its own particular strengths and weaknesses. While these dating techniques were still in their infancy, or were even non-existent, during the time of Bordes' original excavations at Pech IV, they have subsequently become common chronological methods in archaeological settings (Grün et al. 2010; Guérin et al. 2012; Jacobs et al. 2015; Soressi et al. 2007; Valladas et al. 2003) thanks in part to significant advances in both technologies and experimental methodologies, resulting in greater accuracy and precision of the resulting age estimates.

N.R. Jankowski (✉)
School of Earth and Environmental Sciences, Centre for Archaeological Science, University of Wollongong, Wollongong, NSW 2522, Australia
e-mail: njankows@uow.edu.au

The aim of this chapter is to provide an assessment of the Pech IV chronology as it currently stands. Initially, a short overview of the four independent dating techniques applied at Pech IV is provided, followed by a review of the previously published age estimates for the site. Finally, a chronological framework that integrates these various techniques into a cohesive whole will be presented.

## Setting the Clock Ticking: An Overview of the Absolute Dating Methods Used at Pech IV

The four techniques used at Pech IV all provide numerical estimates of age for the material being dated. These techniques can be subdivided into two categories according to the way in which they act as time-keepers: (1) radioactive decay and (2) electron capture methods. They also differ in terms of the event being dated. Here, a brief outline of each of the four techniques use provided and their respective similarities and points of difference indicated.

Radiocarbon dating is the only technique used at Pech IV that measures the time elapsed as a function of radioactive decay. This dating method is perhaps the most widely recognized beyond academic contexts given that it was developed in the late 1940s by Willard Libby and colleagues (Libby et al. 1949) and has been used widely in archaeology ever since. Radiocarbon dating is based upon the generation of the radioactive $^{14}C$ isotope in the upper atmosphere—the result of a collision between cosmic radiation from outer space and an atom of atmospheric nitrogen ($^{14}N$) (Taylor 1987, p. 1–3). The resulting $^{14}C$ atom then combines with oxygen to form a carbon dioxide ($CO_2$) molecule that is taken up by photosynthetic organisms before being passed through the various trophic levels of the food chain. As such, the $^{14}C$ concentration within an organism is in equilibrium with that in the atmosphere while the animal is living. Once the organism dies, the exchange of $^{14}C$ ceases and the


H.L. Dibble et al. (eds.), *The Middle Paleolithic Site of Pech de l'Azé IV*,
Cave and Karst Systems of the World, https://doi.org/10.1007/978-3-319-57524-7_3

concentration of $^{14}C$ atoms begins to decay as a function of its radioactive half-life (the time taken for half of the original concentration of radioactive $^{14}C$ atoms to decay away) which is ~5730 years (Godwin 1962). The $^{14}C$ chronometer, therefore, begins ticking at the time of death of the organism. By convention, all $^{14}C$ ages are reported as time before the present (BP), where the 'present' is defined at 1950 AD (Flint and Deevey 1962).

Given the large time-depth expected for Pech IV, both contamination and calibration are two important aspects to consider for $^{14}C$ dating. Beyond the ~45 ka mark, a numerical age estimate becomes increasingly difficult. Here, the addition of only 1% 'modern' carbon to a sample of 'infinite' age would produce an 'apparent' $^{14}C$ age of between 35 and 40 ka BP (c.f., Chappell 1991). This issue can be overcome with rigorous pre-treatment techniques such the ultrafiltration method (Higham et al. 2006) that was used for all the bone samples measured at Pech IV.

Reported $^{14}C$ ages, unlike those for the other dating techniques described below, require calibration from 'radiocarbon' years into 'sidereal' or calendar years (Ramsey et al. 2006). This is achieved using an internationally ratified calibration curve (currently IntCal13, Reimer et al. 2013) that allows for samples with radiocarbon ages of less than 45 ka BP to be accurately determined. Calibrated age estimates are, by convention, reported as 'years cal. BP.' Calibration is currently not possible for samples approaching this age or older, and ages may be reported simply as 'years BP' and remain several ka younger than their true age.

The three remaining chronometric techniques used at Pech IV—ESR, TL, and OSL—fall under the umbrella of 'trapped charge' dating methods. This family of absolute dating techniques are based upon the capture and storage of electrons over time as a result of natural background radiation damage (Aitken 1985). Natural materials, such as teeth in the case of ESR (Grün et al. 2008), flint in the case of TL (Mercier and Valladas 2003), and quartz grains for OSL (Jacobs 2010), act as naturally occurring dosimeters, absorbing radiation energy as a function of the radiation dose delivered to them over time from the sedimentary deposit itself; acting effectively as geological 'clocks' (Huntley et al. 1985).

The radiation dose recorded by the dosimeters is comprised of alpha and beta particles, and gamma rays (Aitken 1998). These radiation types are derived both internally and externally of the dosimeter used and have different penetrating distances in sediments: alpha particles travel 20–40 μm, beta particles up to 3 mm, whereas gamma rays can reach 30 cm. Cosmic rays from outer space also provide a contribution to the captured charge (Prescott and Hutton 1994) and can penetrate up to tens of metres into rock and sediment. Collectively, the total background radiation flux experienced by the dosimeter is termed the environmental dose rate ($D_r$ or sometimes $\dot{D}$). When these radiation particles and rays collide with the crystal structure of the dosimeters, the energy released during the damaging collision is captured by the crystal as electrons and electron vacancies, known as 'holes'. These electrons and holes can then become lodged within crystallographic defects in the dosimeters mineral lattice, called 'traps.' As the dosimeters remain buried for later periods of time, the number of electrons stored within these traps proportionally increases (Rhodes 2011). This stored energy is termed the equivalent dose ($D_e$), although some publications also refer to these stored electrons as the paleodose ($P$) or the burial dose. If both the $D_e$, measured in Gray (Gy; where 1 Gy is equal to the absorption of 1 J of energy per 1 kg of matter), and the $D_r$ (measured in Gy/ka) are known, then the age of the sediment (in thousands of years; ka) can be determined using the following (simplified) equation:

$$\mathrm{Age(ka)} = \frac{D_e(\mathrm{Gy})}{D_r(\mathrm{Gy \cdot ka^{-1}})}$$

The uncertainty term associated with these trapped charge dating techniques are often reported at $1\sigma$ and incorporates various intrinsic random and systematic errors associated with the measurement of $D_e$ and $D_r$.

While each of these three charge capture techniques are interrelated they differ substantially in the materials and the event being dated. At Pech IV, ESR dating was carried out solely on fossil animal teeth. In life, the crystalline structure of tooth enamel is effectively free from radiation damage but, once buried, radiation exposure begins knocking electrons out of their original, magnetically neutral positions resulting in a weak paramagnetic signal (Grün 1989). Thus, time zero can be considered the moment of incorporation of the tooth into the sediment. Problems with ESR, however, are often encountered when it comes to modelling the U uptake histories of the teeth being dated, which act as chemically open systems (Grün 2007).

TL dating of burnt flints provides an estimate of the time elapsed since the flint tool was last heated in antiquity (Aitken 1985). Prior to this heating event, the flint retains much of its 'geological' dose—i.e. the charge accumulated while the flint was locked within the bedrock. It is only once the flint tool is sufficiently heated to temperatures greater than ~400 °C that this 'geological' dose is reset back to zero (Valladas 1983). Thus, time zero for TL dating is this heating event, and is assumed to have took place at the time that the artefact was in use by hominins. Finally, OSL dating (Jacobs and Roberts 2007) at Pech IV was conducted on ubiquitous sand-sized grains of quartz found throughout the deposit. In the same way as TL, the quartz grains have an antecedent 'geological' dose that is required to be zeroed. However, this resetting process is achieved with a few seconds of light exposure. OSL measures the time elapsed since the quartz grains were last exposed to sunlight (Murray and Wintle 2000). Thus, OSL provides an age estimate for the depositional context, rather than a specific anthropogenic activity; effectively, OSL is dating archaeological remains by association (Aitken 1998).

## Marking Time: Published Age Estimates for Pech IV

The absolute chronology for the samples collected during the recent excavation at Pech IV has been reported in a series of four publications (Jacobs et al. 2016; McPherron et al. 2012; Richter et al. 2013; Turq et al. 2011). The TL ages of Bowman et al. (1982) are not considered here as previous authors have called into question their accuracy (c.f., Richter et al. 2013). Initially, preliminary TL ages for Layer 8 only were reported by Dibble et al. (2009). These six ages were subsequently finalized and reported by Richter et al. (2013), along with eight additional age estimates from Layers 5A, 4C, and 3B. ESR age estimates for Layers 3A and 3B, five in total, were published in Turq et al. (2011) as part of their site synthesis. McPherron et al. (2012) present $^{14}C$ age estimates, again, for Layers 3A and 3B. Finally, a series of 33 single-grain OSL age estimates were published by Jacobs et al. (2016) that sampled each of the preserved layers at Pech IV, with the exception of Layer 7. A summary of these age estimates is provided in Table 3.1 and displayed graphically in Fig. 3.1. It should be noted that the application of these four independent dating techniques at Pech IV not only makes it one of, if not the, best-studied Middle P sites in the French Perigord region in terms of its chronology, but also provides us with a means of cross-examining the accuracy of each dating method. Unless otherwise stated, all uncertainties are quoted at $\pm 1\sigma$.

The deposition of the sediments comprising Layer 8 is dated by single-grain OSL to ~95 ka. A series of six age estimates were published for this layer and range in age ($\pm 1\sigma$) from 98.6 ± 11.1 to 91.5 ± 7.6 ka (Fig. 3.1). Jacobs et al. (2016) statistically show that these samples are identical and therefore calculated a weighted mean age of 95 ± 4 ka for this layer (shown as the black-filled diamond in the darkest grey band in Fig. 3.1). In doing so, the authors assumed that the individual age estimates represent a single event or a series of events spread over time that is significantly shorter than the size of the individual age uncertainties.

Of the at least 15 individual flint samples submitted for TL dating (Dibble et al. 2009) only six were heated sufficiently during antiquity to provide accurate age estimates for the firing event. The individual TL age estimates range from 106 ± 12 to 89 ± 9 ka (Fig. 3.1). Statistical tests conducted by Richter et al. (2013) demonstrated an internal consistency and statistical normality allowing a weighted mean to be calculated for these six samples. As such, a weighted mean age of 96 ± 5 ka was obtained (shown as the black-filled circle in the darkest grey band in Fig. 3.1). This agrees well with the published single-grain OSL chronology, with both weighted mean age estimates being indistinguishable from each other at $1\sigma$. Given the consistency between both the TL and single-grain OSL weighted mean estimates, confidence in the accuracy and applicability of the procedures used in $D_e$ and $D_r$ determinations for both independent dating methods can be assured.

It should be noted that Jacobs et al. (2016) and Richter et al. (2013) calculated their weighted mean age estimates in slightly different ways. Where Jacobs et al. (2016) removed the individual systematic errors prior to calculating the weighted mean, before adding it back to the random error on the weighted mean age in quadrature, Richter et al. (2013) included the systematic errors on both occasions, before and after the weighted mean calculation. However, if the Jacobs et al. (2016) ages were calculated in a similar fashion as Richter et al. (2013), there would be neither a significant change in the calculated weighted mean age estimate nor an increase in the associated total uncertainty.

Layer 6 has only been dated using single-grain OSL. A series of five samples were taken over both subunits 6A and 6B, two from each sublayer and one on the boundary between both. The age estimates were all found to be internally consistent at $1\sigma$, ranging from 79.8 ± 6.1 to 74.1 ± 5.6 ka (Fig. 3.1), allowing for a weighted mean age of 77 ± 4 ka to be calculated (the black-filled diamond in the dark grey band in Fig. 3.1).

Layer 5 is dated by both TL and single-grain OSL age estimates. Four single-grain OSL samples span both subunits; two from Layer 5A, one from 5B and an additional sample straddling the boundary between both. Individual age estimates, ranging from 77.7 ± 5.7 to 74.3 ± 5.3 ka (Fig. 3.1), were again internally consistent, with a weighted mean single-grain OSL age of 76 ± 4 ka calculated (shown as the white-filled diamond in the dark grey band in Fig. 3.1). 18 burnt flints were recovered from the upper sublayer 5A. Of these 18 flints, only four were sufficiently heated to allow for accurate TL age estimates to be obtained. The individual TL age estimates are internally consistent, ranging from 88.3 ± 9.5 to 68.3 ± 6.4 ka (Fig. 3.1), and have a weighted mean age of 74 ± 5 ka (shown as the white-filled circle in Fig. 3.1). This weighted mean age is again indistinguishable for the corresponding single-grain OSL weighted mean age for Layer 5.

Each of the three subdivisions of Layer 4 was dated using single-grain OSL, with sublayer 4C also being dated with TL.

**Table 3.1** Summary of published age estimates for samples collected during or after the 2000–2003 excavation seasons

| Layer | OSL[a] | TL[b] | ESR[c] | $^{14}C$[d] |
|---|---|---|---|---|
| 3A | 51 ± 2 | – | 50 ± 3 (EU), 52 ± 5 (LU) | 44.1–46.4, 44.6–47.8 |
| 3B | 51 ± 3 | 46 ± 4*, 55 ± 5* | | 41.5–43.1, *49.0–52.2* |
| 4A | 57 ± 3 | – | – | – |
| 4B | 62 ± 3 | – | – | – |
| 4C | 68 ± 4 | 69 ± 7*, 72 ± 7* | – | – |
| 5A | 76 ± 4 | 74 ± 5 | – | – |
| 5B | | – | – | – |
| 6A | 77 ± 4 | – | – | – |
| 6B | | – | – | – |
| 8 | 95 ± 4 | 96 ± 5 | – | – |

All ages are reported in ka and at $\pm 1\sigma$, with the exception of the $^{14}C$ ages which are $2\sigma$ calibrated ranges BP using IntCal13 (where possible). All age estimates have been rounded to the nearest 1000 years, apart from the $^{14}C$ ages with have been rounded to the nearest 100 years

[a]Single-grain OSL ages published in Jacobs et al. (2016). All ages presented are weighted mean estimates

[b]TL ages published in Richter et al. (2013). Age estimates mark with an (*) are individual estimates, all other estimates are weighted mean ages

[c]ESR ages published in Turq et al. (2011) including all 5 individual estimates using the early- (EU) and linear-uptake (LU) model

[d]$^{14}C$ ages published in McPherron et al. (2012). Age estimates provided are the youngest and oldest individual ages from each sublayer. The range presented in *italics* has not been calibrated

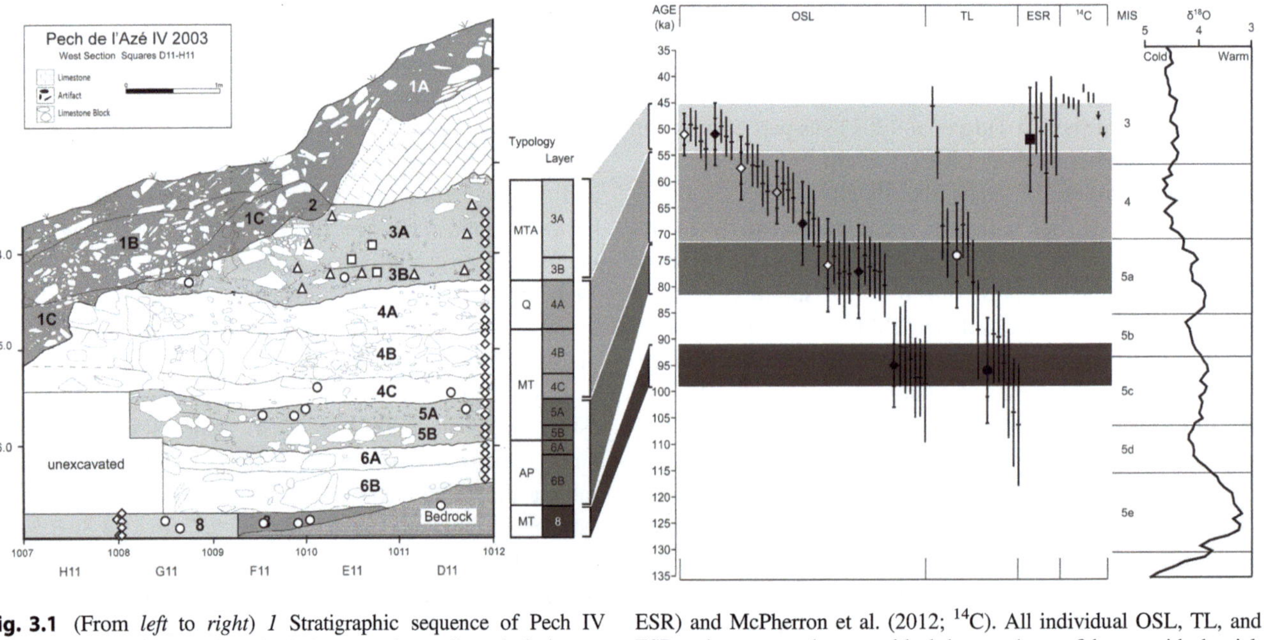


**Fig. 3.1** (From *left* to *right*) *1* Stratigraphic sequence of Pech IV showing the positions of each dating samples collected during or following the 2000–2003 field season. OSL samples are shown as diamonds, TL samples as circles, ESR samples as squares, and $^{14}C$ as triangles. *2* A stylized section showing the relationship between stone tool typology and stratigraphic level. Mousterian of Acheulian Tradition is shown as MTA, Quina as Q, Typical Mousterian as MT, and Asinipodian as AP. *3* Age estimates shown are those presented in Jacobs et al. (2016; OSL), Richter et al. (2013; TL) Turq et al. (2011, ESR) and McPherron et al. (2012; $^{14}C$). All individual OSL, TL, and ESR estimates are shown as black bars at $1\sigma$ confidence, with the tick mark representing the midpoint. Where weighted mean age estimates were calculated for sublayers/layers, the estimates are shown as filled symbols on the left-hand side of the individual estimates, with the tick marks delimiting the 1 and $2\sigma$ uncertainties. 'Best estimate' ages are delimited by the four horizontal bands. *4* The $\delta^{18}O$ record and marine isotope stages (MIS) of Lisiecki and Raymo (2005)

Eleven samples in total were collected from Layer 4 for single-grain OSL age estimation; five from sublayer 4A, three each from sublayers 4B and 4C. Age estimates for sublayer 4C range between 72.4 ± 5.6 and 65.9 ± 5.2 ka, for 4B from 63.1 ± 4.8 to 60.4 ± 4.2 ka and for 4A from 61.8 ± 4.7 to 52.9 ± 3.7 ka (Fig. 3.1). Within each sublayer, the age estimates were statistically shown to be internally consistent providing a progression of weighted mean age estimates (in stratigraphic order) of 68 ± 4 (4C), 62 ± 3 (4B) and 57 ± 3 ka (4A) (shown as black, grey and white-filled diamonds in the mid-grey band in Fig. 3.1, respectively). The single-grain OSL age estimates for sublayer 4C is consistent with that obtained using TL. Of the nine burnt flints tested, only two were usable and gave individual age estimates of 71.8 ± 6.7 and 68.5 ± 6.6 ka (Fig. 3.1).

The age of Layer 3 is the only such layer to be examined using all four absolute dating techniques. The single-grain OSL ages for sublayer 3B range from 52.5 ± 3.4 to 49.5 ± 3.2 ka (Fig. 3.1) and have a weighted mean age of 51 ± 3 ka (black-filled diamond in the light grey band in Fig. 3.1). These are indistinguishable from sublayer 3A, which range between 53.8 ± 4.0 and 49.2 ± 3.4 ka (Fig. 3.1) and have a weighted mean age of 51 ± 2 ka (white-filled diamond in the light grey band in Fig. 3.1). Two TL ages of 54.6 ± 5.0 and 45.7 ± 3.9 ka were obtained for sublayer 3B (Fig. 3.1). ESR ages were obtained for 5 samples coming from three separate teeth recovered from both sublayer 3B and 3A (Fig. 3.1). Given that both linear and early uranium uptake ages were calculated for each of the five samples, the resulting age ranges presented in Fig. 3.1 represent the entire $1\sigma$ range of both uptake model estimates, with the tick mark representing the midpoint of this range. The weighted mean early uptake model estimate dates the deposit to 50 ± 3 ka ($n = 5$), while the corresponding linear-uptake model age estimate of 52 ± 5 ka ($n = 5$) is consistent at $1\sigma$. The black-filled square in Fig. 3.1 represents the midpoint of both weighted mean age estimates with the range encompassing the corresponding $2\sigma$ uncertainty.

Ten cut-marked bone samples were selected from both sublayers 3A and 3B for $^{14}$C analysis. Of these ten samples, only one was rejected for its low collagen yield and its significantly younger $^{14}$C age. Of the remaining nine age estimates, two could not be calibrated due to their $^{14}$C ages being greater than that of the current calibration range of the IntCal13 curve; 50.6 ± 0.8 and 47.4 ± 0.7 ka BP (shown as arrows in Fig. 3.1). As such, these ages should be considered minimum age estimates only as uncalibrated $^{14}$C ages remain several thousand years younger than their calibrated counterparts. The $^{14}$C ages presented in McPherron et al. (2012) were calibrated using IntCal09, here we have recalibrated them using IntCal13 (Table 3.2). The calibrated ranges of these samples fall between 46.2 ± 1.6 and 42.3 ± 0.8 ka cal BP. Note that these values (and those shown in Fig. 3.1) are the midpoint of the $2\sigma$ calibrated range with the uncertainty giving the entire $2\sigma$ range. Overall, there is remarkable agreement among all of the age estimates obtained for Layer 3 using the four independent absolute dating techniques.

## Pech IV Chronological Framework: Age Synthesis

In this section, the published data is synthesized into a coherent chronological framework for the Pech IV site. To do so requires that the available data presented above be critically evaluated, keeping in mind the key criterion of 'what is the event being dated?' As such, there are two aspects that we are most interested in understanding: first, what is the timing of site formation (i.e. sedimentary deposition)? And, second, is this timing contemporaneous with the human occupation? The single-grain OSL age estimates were used to provide an estimate of elapsed time since the sediment grains were last exposed to sunlight (i.e. the depositional age). In contrast, those layers where either TL, ESR or $^{14}$C age estimates are available are used to determine contemporaneity with the depositional age of the sediments as each of these methods can be inextricably linked to a specific archaeological process. As such, the following timeline for the formation of Pech IV sedimentary sequence is our best estimate using the currently published chronologies.

Layer 8 is proposed to have been deposited over an ~8 ka period between ~99 and 91 ka ago based upon the $1\sigma$ total uncertainty on the weighted mean single-grain OSL age for this unit. This is shown in Fig. 3.1 as the darkest grey band. The weighted mean TL age estimate (96 ± 5 ka) is indistinguishable from this best estimate. Thus, it was during this ~8 ka period that Layer 8 was deposited and the cave occupied by the Neandertals. Presumably, this coincided with the opening up of the cave system allowing sediments and human occupants to enter (see Chap. 2). Climatically, this period coincides with the latest part of marine isotope stage (MIS) 5c and the earliest portion of the relatively cooler 5b interstadial. This climatic variability is reflected in the shift in the faunal assemblage recovered from this layer (see Chap. 5).

At present one can only speculate as to the age of Layer 7. The significant roof fall event that caps this layer is assumed, here, to have occurred during the cooler conditions of the MIS 5b interstadial, suggesting that the deposition of Layer 7 is closer in time to that of Layer 8, than Layer 6B. This supposition is corroborated by the abundance of edge damaged stone tools recovered from this layer, the preservation state of which are considered the result of cryoturbation damage (Turq et al. 2011; see also Chap. 2).

The best age estimate for the deposition of Layers 5 and 6 is between ~81 and 72 ka, shown as the dark grey band in Fig. 3.1. These two units are considered to represent a period

**Table 3.2** $^{14}C$ age estimates and IntCal09 calibrated age estimates published in McPherron et al. (2012), and IntCal13 using OxCal online

| Layer | Sample code | MPI lab code | AMS lab code | $^{14}C$ age ($\pm 1\sigma$) | IntCal09 range ($2\sigma$) | IntCal13 range ($2\sigma$) |
|---|---|---|---|---|---|---|
| 3A | E11-537 | S-EVA 3940 | OxA-V-2344-11 | 43,050 ± 400 | 46,860–45,350 | 45,240–43,500 |
| 3A | D11-162 | S-EVA 3941 | OxA-V-2344-12 | 43,910 ± 450 | 47,750–45,770 | 46,370–44,150 |
| 3A | D11-636 | S-EVA 3942 | OxA-V-2333-35 | 44,720 ± 700 | 48,310–45,860 | 47,780–44,640 |
| 3A | E11-493 | S-EVA 3943 | OxA-V-2333-36 | 37,400 ± 370 | Rejected | Rejected |
| 3A | E11-231 | S-EVA 3944 | OxA-V-2344-13 | 43,720 ± 450 | 47,580–45,680 | 46,140–43,990 |
| 3B | E11-1350 | S-EVA 3946 | OxA-V-2344-14 | 42,930 ± 450 | 46,840–45,240 | 45,230–43,350 |
| 3B | D11-2062 | S-EVA 3947 | OxA-V-2344-15 | 50,580 ± 800 | Beyond cali. curve | Beyond cali. curve |
| 3B | E11-1332 | S-EVA 3948 | OxA-V-2344-16 | 47,400 ± 650 | Beyond cali. curve | Beyond cali. curve |
| 3B | E11-1935 | S-EVA 3949 | OxA-V-2344-17 | 40,760 ± 400 | 45,700–44,180 | 43,140–41,550 |
| 3B | D11-1835 | S-EVA 3950 | OxA-V-2344-18 | 42,690 ± 500 | 46,710–45,070 | 45,080–43,090 |

of continuous deposition because the weighted mean single-grain OSL ages estimates for these layers cannot be statistically distinguished. The ~9 ka age range for this package of sediment is estimated from the $1\sigma$ uncertainties on the weighted mean single-grain OSL age estimates from these two units and encapsulates the weighted mean TL age estimate for sublayer 5A. We consider, therefore, that the deposition of the sediments and archaeological remains found within these two layers most probably occurred entirely within the MIS 5a interstadial. Again, the faunal record (Chap. 5) recovered from this layer appears to indicate similar climatic conditions as Layer 8, which is consistent with this best estimate range.

Layer 4 was deposited over a more protracted period of ~18 ka duration, from 72 to 54 ka ago. This range, shown as the mid-grey band in Fig. 3.1, is defined by the $1\sigma$ total uncertainties on the single-grain OSL weighted mean ages for the 4A and 4C sublayers and is broadly consistent with the two TL ages obtained by Richter et al. (2013) for sublayer 4C. We can, therefore, constrain the deposition of Layer 4 to within the cooler climatic conditions of MIS 4 with uppermost sublayer, 4A, probably extending beyond the MIS 3/4 boundary into MIS 3.

Layer 3, containing an industry consistent with a Mousterian of Acheulian Tradition (MTA) designation, was formed over the span of ~9 ka, between 54 and 45 ka ago (shown as the light grey band in Fig. 3.1). The upper age limit of this range equates to the $1\sigma$ uncertainty on the weighted mean single-grain OSL age estimate for Layer 3B, whereas the lower age limit is defined by the $2\sigma$ extent of the same estimate. As such, this layer was deposited within MIS 3. Our best estimate range encapsulates both TL age estimates from Layer 3B at $1\sigma$. With respect to the ESR ages, it is suggested that not as much weight be placed upon these particular age estimates, for the following reasons. First, the original publication of these ages in Turq et al. (2011) lacks crucial information regarding the methods of both the $D_e$ and $D_r$ determinations, e.g., what procedure was used to determine the $D_e$, what equipment was this conducted on, were there any quality control tests undertaken to validate these procedures, what method/s were used in calculating the $D_r$ and how was the uncertainties on the $D_e$, $D_r$, and resulting age estimates propagated? Second, this information has subsequently been lost with the retirement of Prof. Rink from McMaster University (Rink 2016 pers. comm.). As such, the ESR ages should be considered as no more than a supporting data set, providing corroborative support for the remaining age estimates from this layer. The calibrated $^{14}C$ age ranges presented in Table 3.2 reveal considerable scatter in the age distributions for both sublayers. Although not calibrated, the two oldest age estimates accord well with our best estimate range. Even if future refinement and extension of the internationally accepted calibration curve allows these to be calibrated, they would still be consistent with our best estimate for Layer 3. The remaining age estimates, falling between ~44 and 46 cal ka B.P., either straddle the lower edge of our best estimate or are just marginally younger.

## Conclusions

The site of Pech IV is one of, if not the, most intensively geochronologically investigated Middle Palaeolithic sites in the Perigord region of southwest France. The four absolute dating techniques applied here have returned remarkably consistent age estimates despite measuring slightly different 'events,' i.e. deposition (OSL) versus occupation (TL, ESR, 14C). Four chronologically distinct packages of sediment can be identified based primarily upon the depositional ages determined using single-grain OSL. As such, the entire deposit is considered to have been deposited between ~99 and 45 ka ago. This time period can be subdivided into four separate blocks according to the stratigraphy, where Layer 8 was formed between ~99 and 90 ka ago, Layers 6 and 5 from ~81 to 72 ka, Layer 4 between 72 and 54 ka, and Layer 3 between ~54 and 45 ka. With this chronological framework in place, it is now possible to synergize the geological, climatic, faunal, and archaeological records in their correct temporal context.

## References

Aitken, M. J. (1985). *Thermoluminescence dating*. London: Academic Press.

Aitken, M. J. (1998). *An introduction to optical dating: The dating of quaternary sediments by the use of photon-stimulated luminescence*. Oxford: Oxford University Press.

Aldeias, V., Dibble, H., Sandgathe, D., Goldberg, P., & McPherron, S. (2016). How heat alters underlying deposits and implications for archaeological fire features: A controlled experiment. *Journal of Archaeological Science, 67,* 64–79.

Bowman, S. G. E., Loosemore, R. P. W., Sieveking, Gd G, & Bordes, F. (1982). Preliminary dates for Pech de l'Aze IV. *PACT, 6,* 362–369.

Chappell, J. (1991). Late quaternary environmental changes in eastern and central Australia, and their climatic interpretation. *Quaternary Science Reviews, 10,* 377–390.

Dibble, H., Berna, F., Goldberg, P., McPherron, S., Mentzer, S., Niven, L., et al. (2009). A preliminary report on Pech de l'Azé IV, Layer 8 (Middle Paleolithic, France). *PaleoAnthropology, 2009,* 182–219.

Flint, R., & Deevey, E. S. (1962). Editorial statement. *Radiocarbon, 4,* i–ii.

Godwin, H. (1962). Half-life of radiocarbon. *Nature, 195,* 984.

Grün, R. (1989). Electron spin resonance (ESR) dating. *Quaternary International, 1,* 65–109.

Grün, R. (2007). Electron spin resonance dating. In S. A. Elias (Ed.), *Encyclopedia of quaternary science* (pp. 1505–1516).

Grün, R., Aubert, M., Joannes-Boyau, R., & Moncel, M.-H. (2008). High resolution analysis of uranium and thorium concentration as well as U-series isotope distributions in a Neanderthal tooth from Payre (Ardèche, France) using laser ablation ICP-MS. *Geochimica et Cosmochimica Acta, 72,* 5278–5290.

Grün, R., Eggins, S., Aubert, M., Spooner, N., Pike, A. W. G., & Müller, W. (2010). ESR and U-series analyses of faunal material from Cuddie Springs, NSW, Australia: Implications for the timing of the extinction of the Australian megafauna. *Quaternary Science Reviews, 29,* 596–610.

Guérin, G., Discamps, E., Lahaye, C., Mercier, N., Guibert, P., Turq, A., et al. (2012). Multi-method (TL and OSL), multi-material (quartz and flint) dating of the Mousterian site of Roc de Marsal (Dordogne, France): Correlating Neanderthal occupations with the climatic variability of MIS 5–3. *Journal of Archaeological Science, 39,* 3071–3084.

Higham, T. F. G., Jacobi, R. M., & Ramsey, C. B. (2006). AMS radiocarbon dating of ancient bone using ultrafiltration. *Radiocarbon, 48,* 179–195.

Huntley, D. J., Godfrey-Smith, D. I., & Thewalt, M. L. W. (1985). Optical dating of sediments. *Nature, 313,* 105–107.

Jacobs, Z. (2010). An OSL chronology for the sedimentary deposits from Pinnacle Point Cave 13B: A punctuated presence. *Journal of Human Evolution, 59,* 289–305.

Jacobs, Z., Jankowski, N. R., Dibble, H., Goldberg, P., McPherron, S., Sandgathe, D., et al. (2016). The age of three Middle Palaeolithic sites: Single-grain optically stimulated luminescence chronologies for Pech de l'Azé I, II and IV in France. *Journal of Human Evolution, 95,* 80–103.

Jacobs, Z., Li, B., Jankowski, N., & Soressi, M. (2015). Testing of a single grain OSL chronology across the Middle to Upper Palaeolithic transition at Les Cottés (France). *Journal of Archaeological Science, 54,* 110–122.

Jacobs, Z., & Roberts, R. G. (2007). Advances in optically stimulated luminescence dating of individual grains of quartz from archeological deposits. *Evolutionary Anthropology, 16,* 210–223.

Libby, W. F., Anderson, E. C., & Arnold, J. R. (1949). Age determination by radiocarbon content: World-wide assay of natural radiocarbon. *Science, 109,* 227.

Lisiecki, L. E., & Raymo, M. E. (2005). A Pliocene-Pleistocene stack of 57 globally distributed benthic δ18O records. *Paleoceanography, 20,* 1–17.

McPherron, S., Talamo, S., Goldberg, P., Niven, L., Sandgathe, D., Richards, M. P., et al. (2012). Radiocarbon dates for the late middle palaeolithic at Pech de l'Azé IV, France. *Journal of Archaeological Science, 39,* 3436–3442.

Mercier, N., & Valladas, H. (2003). Reassessment of TL age estimates of burnt flints from the Paleolithic site of Tabun Cave, Israel. *Journal of Human Evolution, 45,* 401–409.

Murray, A. S., & Wintle, A. G. (2000). Luminescence dating of quartz using an improved single-aliquot regenerative-dose protocol. *Radiation Measurements, 32,* 57–73.

Prescott, J. R., & Hutton, J. T. (1994). Cosmic ray contributions to dose rates for luminescence and ESR dating: Large depths and long-term time variations. *Radiation Measurements, 23,* 497–500.

Ramsey, C. B., Buck, C. E., Manning, S. W., Reimer, P., & Hvd, Plicht. (2006). Developments in radiocarbon calibration for archaeology. *Antiquity, 80,* 783–798.

Reimer, P. J., Bard, E., Bayliss, A., Beck, J. W., Blackwell, P. G., Ramsey, C. B., et al. (2013). Intcal13 and Marine13 radiocarbon age calibration curves 0–50,000 years cal BP. *Radiocarbon, 55,* 1869–1887.

Rhodes, E. J. (2011). Optically stimulated luminescence dating of sediments over the past 200,000 years. *Annual Review of Earth and Planetary Sciences, 39,* 461–488.

Richter, D., Dibble, H., Goldberg, P., McPherron, S., Niven, L., Sandgathe, D., et al. (2013). The late middle palaeolithic in southwest France: New TL dates for the sequence of Pech de l'Azé IV. *Quaternary International, 294,* 160–167.

Richter, D., McPherron, S., Dibble, H., Goldberg, P., Sandgathe, D., Turq, A. (In Press). Additional chronometric data for the small flake assemblages ('Asinipodian') from Pech de l'Azé IV (France) and a comparison with similar assemblages at the nearby site of Roc de Marsal.

Soressi, M., Jones, H. L., Rink, W. J., Maureille, B., & Tillier, A. (2007). The Pech-de-l'Azé I Neandertal child: ESR, uranium-series, and AMS 14C dating of its MTA type B context. *Journal of Human Evolution, 52,* 455–466.

Taylor, R. E. (1987). *Radiocarbon dating: An archaeological prespective*. London: Academic Press, INC.

Turq, A., Dibble, H., Goldberg, P., McPherron, S., Sandgathe, D., Jones, H., et al. (2011). Les fouilles récentes du Pech de l'Azé IV (Dorgodne). *Gallia Préhistoire, 53,* 1–58.

Valladas, H. (1983). Estimation of heating temperature of prehistoric burnt flints by thermoluminscence. *PACT, 9,* 251–253.

Valladas, H., Mercier, N., Joron, J. L., McPherron, S., Dibble, H., & Lenoir, M. (2003). TL dates for the middle paleolithic site of Combe-Capelle Bas, France. *Journal of Archaeological Science, 30,* 1443–1450.

# Taphonomical and Zooarchaeological Analysis of Bordes' Excavated Material from Levels I2 and Y-Z

4

Jamie Hodgkins

## Introduction

This chapter explores the taphonomy and zooarchaeology of two archaeological levels from Pech IV (Levels I2 and Y-Z, corresponding to Layers 4B–4C and Layer 8, respectively, in the current excavations) that were excavated by Bordes from 1970 to 1977. The analyses described here were specifically geared toward (1) documenting the extent of post-depositional destruction that occurred in each level analyzed and; (2) determining the relative contribution by Neanderthals and carnivores to the accumulation and nutrient extraction of faunal remains found in Levels I2 and Y-Z. In the case that Neanderthals were responsible for transporting and butchering faunal remains in both levels, an additional set of analyses will be run with the goal of interpreting transport behaviors. For this particular analysis, herbivore remains ranging from small roe-deer-sized animals to those as large as European red deer were the focus. Ethnographic data indicate that an animal larger than a red deer is typically field butchered by hunter-gatherers, and a majority of the skeleton is not transported to cave sites (Binford 1978; Bunn et al. 1988; Monahan 1998; O'Connell et al. 1998, 1990). That being said, animals that are the size of red deer may also be butchered partially in the field, with parts of the skeleton being left behind. Therefore, when interpreting Neandertal subsistence behaviors it is important to try to document the transport decisions made by Neanderthals in terms of whether or not they were transporting whole carcasses for butchering at the cave site or transporting portions of carcasses.

## Excavation Background

### Sample Preservation and Curation

As described in Chap. 1, Bordes excavated in meters with stratigraphic subdivisions within each unit corresponding to the natural geological contours and features as he observed them. Many of the lithic and faunal remains recovered were point-provenienced. In their study of the excavator bias apparent in Bordes' excavation, Dibble et al. (2005) found few artifacts greater than 5 mm in size in his back dirt, providing evidence that most of the material excavated from the site by Bordes was retained. During his excavation, fragments deemed large enough (no actual size was specified) by excavators, and especially tooth fragments and epiphyseal ends, were piece-plotted and assigned a number, which was subsequently written on each specimen. The piece-plotted and numbered material was then stored in bags organized by square. Smaller specimens were placed in small finds bags with the elevations at which they were found written on the bag. Beginning in the 1990s, a database was created using Bordes' field notes, which lists all stratigraphic levels and units from the site, along with the minimum and maximum elevation coordinates of each. Thus, this database allowed all bone fragments, even small finds, to be assigned to stratigraphic level, mitigating the biasing effect of preferentially numbering teeth and epiphyseal ends. As described in Chap. 2, it was also determined that stratigraphic layers identified by Bordes aligned well with layers identified in the recent excavation of the site. Therefore, because the faunal material analyzed for this study comes from Bordes' original excavations, and because stratigraphic layers from both Bordes' excavation and Dibble and McPherron's excavation align, Bordes' original level designations will be used throughout the Chapter (specifically Level Y-Z, corresponding to Layer 8, and Level I2, corresponding to lower Layer 4B and Layer 4C).

J. Hodgkins (✉)
Department of Anthropology, University of Colorado, Denver, CO, USA
e-mail: Jamie.Hodgkins@UCDenver.edu


H.L. Dibble et al. (eds.), *The Middle Paleolithic Site of Pech de l'Azé IV*,
Cave and Karst Systems of the World, https://doi.org/10.1007/978-3-319-57524-7_4

## Methods

### Sample Selection

Faunal material was abundant in Levels I2 ($N$ = 9374) and Y-Z ($N$ = 4847). It was therefore necessary to randomly sample specimens from each archaeological layer. In Level I2, more than 10% of the total number of specimens was sampled ($N$ = 910 fragments), and in Level Y-Z greater than 15% of the total number of specimens was sampled ($N$ = 727).

For this analysis, artiodactyls in the size 2 (21–113 kg) and 3 (114–340 kg) body size range were chosen for analysis (Bunn et al. 1988). Among the animals in this size range present during the Late Pleistocene of southwestern France, three genera of deer dominated: *Cervus*, *Rangifer*, and *Capreolus* (Delpech 1983; Dibble et al. 2009; Kurtén 1968; Laquay 1981). Bone fragments were identified in as much detail as possible, with categories that included size 2 or 3 mammal, size 2 or 3 bovid/cervid, genus, and species (when possible). Because Pech IV is a cave site to which animals were transported for processing, it was prudent to focus on small to moderately large-sized animals as they are often transported completely or nearly completely to habitation sites by both hunter-gatherers and large carnivores (Bunn et al. 1988; O'Connell et al. 1998, 1990; Lyman 1995; Monahan 1998; Pokines and Kerbis Peterhans 2007). Documenting the presence or absence of skeletal elements belonging to size 2/3 mammals will provide detailed insight into patterns of Neandertal and/or carnivore prey carcass transport behaviors.

### Skeletal Element Analysis

In addition to the taxonomic and size identification procedures described above, several other methods were used to analyze faunal fragments from both levels. Data generated by these methods were entered into Microsoft Access, Bone Entry GIS, and Bone Sorter (Abe et al. 2002; Marean et al. 2001). All fragments were identified as specifically as possible to skeletal element and element side. Skeletal elements from the site were randomly selected from each level. Long bone shaft fragments were of particular interest in this analysis because dense cortical bone is more resistant to taphonomic destruction and carnivore modification than epiphyseal ends and elements of the postcranial axial skeleton, which have lower density and contain more bone grease (Cleghorn and Marean 2004; Lam et al. 1999; Marean 1998; Marean and Cleghorn 2003; Marean and Frey 2013; Marean and Kim 1998; Marean and Spencer 1991; Marean et al. 1992; Pickering et al. 2003). Due to their higher probability of preservation, limb bone shaft fragments may be more indicative of the original number of skeletal elements transported to a cave site (Cleghorn and Marean 2004; Faith and Gordon 2007; Lam et al. 1999; Marean 1998; Marean and Cleghorn 2003; Marean and Frey 2013; Marean and Kim 1998; Marean and Spencer 1991; Marean et al. 1992; Pickering et al. 2003; Yravedra and Domínguez-Rodrigo 2009; Cleghorn and Marean 2007). These elements have been dubbed the "high survival set" (Marean and Cleghorn 2003), which includes crania, mandibles, humeri, radioulnae, metacarpals, femora, tibiae, and metatarsals. The high survival set can be used in calculations of minimum number of skeletal elements (MNE) and individuals (MNI). When bone fragments of skeletal elements from the high survival set are used in calculations of MNE, taphonomic factors that affect bone representations, such as density-mediated destruction, and carnivore consumption of epiphyseal ends may be controlled.

During analysis, the presence and completeness of diagnostic landmarks on each skeletal fragment were recorded (e.g., "greater trochanter 10% present"). Teeth were identified as specifically as possible to their anatomical position and taxon. The age of specimens was noted when possible by coding skeletal elements as "fused" or "unfused" and noting the wear pattern on teeth. Bone and tooth fragments that were identifiable to skeletal element and side were then drawn onto a bone template in ArcView GIS 3.3 (see Marean et al. 2001; Abe et al. 2002). Drawing fragments into ArcView allowed the MNE and MNI to be calculated by overlapping all fragments drawn onto a template and counting the number of times fragments overlapped. It also provided a record of the shape and placement of each fragment (Abe et al. 2002).

Once the MNE was known, it was used to calculate the minimum animal units (MAU) and the (%MAU) found in each archaeological layer. The MAU is the number of times the element appears in the complete skeleton of an animal (Binford 1984; Grayson 1984). The MAU was calculated by taking the MNE for each element and dividing it by the number of times that element appears in the body (e.g., MNE of 4 femora/2 femora in the body = MAU of 2). The element with the highest MAU was used as the standard for a normalized scale called the %MAU: each skeletal element's MAU was divided by the highest MAU found in the sample and multiplied by 100 (Binford 1978). It should be noted that for cranial and mandibular remains, the MNE and %MAU were calculated both with and without identifiable teeth included. Providing both values make the data broadly comparable for researchers who may or may not choose to include dentition in their calculations. However, in this study the %MAU was calculated with identifiable teeth included, and these %MAU values were used in analyses of skeletal element abundances to estimate the number of cranial remains and mandibles that were transported in each archaeological level. Once calculated, the %MAU was plotted for each high survival skeletal element to visually determine the transport patterns utilized at each site. The % MAU was also compared to the whole bone standard food utility index (SFUI), which is a calculation of the meat, marrow, and bone grease associated with each skeletal element (Metcalfe and Jones 1988). Plotting the %MAU versus the SFUI for each archaeological level allowed a depiction of the relationship between bone utility and the transport patterns utilized. This relationship was then compared to Binford's theoretical utility curves to determine the transport decisions made by Neanderthals at Pech IV (Metcalfe and Jones 1988; Binford 1978). These utility curves include (1) the bulk strategy, where all bones are transported except those of the lowest utility; (2) the gourmet strategy, where only the highest quality bones are transported; and (3) the unbiased strategy, where bones are transported in direct correlation to their utility. Faith and Gordon (2007) have added the unconstrained strategy to this suite of utility curves in which full skeletons are transported, and thus all bones are transported equally.

## Taphonomic Analysis of Nutrient Extraction Behaviors

One of the most important taphonomic analyses run at the site was to establish the frequency of skeletal elements that were intentionally broken for the extraction of nutrients (by Neanderthals or carnivores) versus the number of elements unintentionally broken by processes such as trampling, roof collapse, etc. The purpose of this analysis was to evaluate the extent to which hominins and carnivores extracted nutrients from skeletal elements at the site and to evaluate the role that post-depositional processes played in the formation of the assemblage. This analysis was conducted following methods outlined by Villa and Mahieu (1991) who established criteria for determining if long bones had been broken in a fresh state, in which internal bone nutrients such as bone marrow and grease were still present in the element, versus if a long bone was broken in a dry state when nutrients were no longer preserved. Experimental studies of breakage patterns caused by hominins and carnivores (Marean et al. 2000) can now be used to calibrate data generated by bone breakage analyses allowing them to be more effectively applied to zooarchaeological remains. For example, based on actualistic experiments in which skeletal elements were broken for the express purpose of marrow extraction, the expected frequency of elements with fresh-bone breaks is 85 ± 3% (Blumenschine 1988; Capaldo 1995, 1998; Marean et al. 2000). It should be noted that data from the experimental assemblages listed here as "carnivore only", "hominid to carnivore", and "hominid only" represent the limits of a range of variation that exists for the accumulating processes for large mammals in caves. Within the experimental data comparisons "carnivore only" includes experiments in which carnivores were fully responsible for breaking bones, "hominid only" refers to experiments for which Neanderthals were fully responsible for breaking bones, and "hominid to carnivore" is a situation where Neanderthals broke the bones first and then carnivores later fed on them.

## Surface Modification Analysis

In addition to the analysis of nutrient extraction behaviors, an analysis of percussion marks, tooth marks, and marks resulting from other taphonomic processes (e.g., chemical etching, trampling) was undertaken for this study. Observing percussion and tooth marks on skeletal fragments enables the intensity of hominin butchering behaviors to be quantified, and to highlight the role that carnivores, other animals and/or taphonomic processes played in accumulating and modifying skeletal elements in each assemblage. Experimental work has established criteria for identifying distinctive surface modification marks (Binford 1981, 1988; Blumenschine 1995; Blumenschine et al. 1996; Blumenschine and Selvaggio 1988; Brain 1984; Domínguez-Rodrigo and Barba 2006; Marean et al. 2004; Pickering et al. 2004; Pickering and Egeland 2006; Potts and Shipman 1981; Shipman 1981). Such work has included criteria for differentiating among tooth marks created by different carnivores (Domínguez-Rodrigo and Piqueras 2003; Pickering et al. 2004), chemical etching marks versus tooth marks (Domínguez-Rodrigo and Barba 2006), and trampling marks versus butchering marks (Domínguez-Rodrigo et al. 2009; Fiorillo 1989; Olsen and Shipman 1988). The use of incident lighting and low-powered magnification has been shown to greatly enhance the correct identification of these features (Blumenschine et al. 1996). These criteria were used here to examine each fragment from Pech IV. Every surface modification mark observed on a bone fragment was then drawn onto the image of that fragment within ArcView GIS. The surface modification marks were labeled with the specimen number of the fragment allowing for the number of marks per fragment to be quantified. As with the analysis of bone breakage, these data were then compared to baseline values established from the actualistic experiments (Blumenschine 1988; Capaldo 1998; Marean et al. 2000).

## Results and Discussion

### Taphonomy and Post-depositional Destruction

Dry bone breaks generally occur after skeletal elements have been deposited and all nutrients have been extracted from them, either actively through hominin and/or carnivore consumption or passively through drying, degradation, leaching, etc. However, a small frequency of right angle breaks and transverse outlines occur when hominins and carnivores break elements open (Marean et al. 2000). Results from this analysis show that long bone fragments from both levels at Pech IV were in line with expected frequencies of right angle breaks caused by hominin and carnivore nutrient extraction behaviors (Fig. 4.1, for raw data from Pech IV see Table 4.1, for data from actualistic experiments see Marean et al. 2000). However, the number of long bone fragments with transverse outlines was lower than predicted by experimental data: results that suggest post-depositional destructive processes that result in bone breakage, such as trampling, sediment compaction, and roof fall, did not play a significant role in modifying skeletal material at the site.

The frequency of long bone fragments with oblique angles from Levels I2 and Y-Z at Pech IV fall just outside the lower 95% confidence interval for assemblages modified by hominins only, but the frequencies are well below those produced when carnivores alone and both carnivores and hominins extract nutrients from long bones (Fig. 4.2). The frequency of long bone fragments with curved and v-shaped outlines fits well within expected values for assemblages in which hominins alone, and carnivores alone extracted all internal nutrients from skeletal elements. These results suggest that most skeletal material at the site was broken while elements were fresh. The pattern of breakage aligns most closely with patterns produced when hominins extract marrow and bone grease from long bones. However, carnivore activity cannot be ruled out based on the results from this analysis alone.

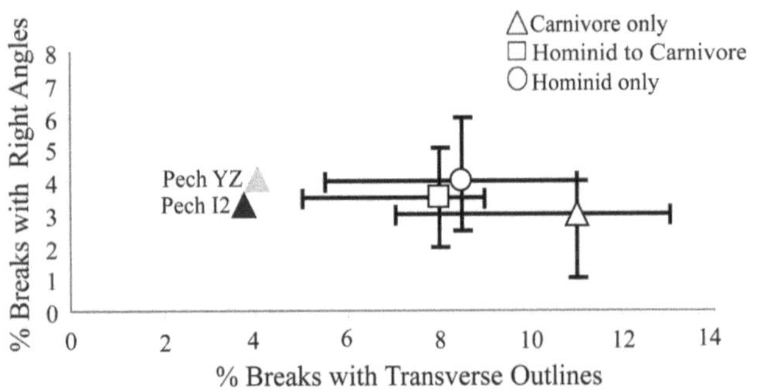


**Fig. 4.1** The frequency of non-nutritive bone breaks recorded on all size 2/3 long bone fragments from Pech IV. Data are compared to experimental and naturalistic assemblages where nutrients were extracted by hominins only, hominins then modified by carnivores, and carnivores only (Marean et al. 2000)

It has been documented that carnivores will preferentially chew and consume the spongy epiphyseal ends of long bones, often removing epiphyses from the assemblage, moving near-epiphysis shaft fragments away from their original place of deposition by hominins, but leaving mid-shaft fragments in place (Marean and Bertino 1994; Marean and Spencer 1991; Marean et al. 1992; Bunn et al. 1988; Bunn and Kroll 1986; Blumenschine 1988). Therefore, another method that can be used to investigate the extent to which carnivores modified skeletal elements is to look at frequency of long bone epiphyseal ends compared to long bone mid-shaft fragments. Another benefit of this approach is that it can highlight if an archaeological assemblage has been subjected to excavation and curation bias. As mentioned in the excavation background section, Bordes preferentially gave specimen numbers to identifiable epiphyseal ends, while placing unidentifiable small finds into bags with minimum and maximum elevation coordinate data. For the current analysis, identifiable specimens with specimen numbers and small finds where combined and reorganized by level. Specimens were then randomly sampled from each level in an attempt to avoid bias in the assemblage. If bias is still present in the sampled assemblage, then the frequency of epiphyseal ends compared to mid-shaft fragments will not match experimentally derived frequencies created when hominins alone, carnivores alone, and both hominins and carnivores extract nutrients from skeletal elements (Blumenschine 1988; Capaldo 1995; Marean et al. 2000).

Table 4.2 provides the raw data for the frequency of epiphyseal fragments and shaft fragments from each level. Figure 4.3 then shows the frequency of shaft fragments and epiphyseal fragments plotted against expected frequencies based on experimental assemblages. The results show that epiphyseal ends are rare in Levels I2 and Y-Z, matching experimentally derived frequencies from carnivore-only and hominid-to-carnivore assemblages. Results from this analysis, therefore, cannot help to determine whether Neanderthals or carnivores played a more active role in modifying skeletal elements at the site, but they do provide strong evidence that the sampled assemblage does not reflect excavator bias that favored epiphyseal ends.

The best way to determine whether carnivores were responsible for accumulating skeletal elements, and/or the extent to which they consumed portions of skeletal elements left by hominins in Levels I2 and Y-Z at Pech IV, is to compare the frequency of carnivore tooth marks and hominin produced percussion marks on long bone fragments in each level (Marean et al. 2000). When the frequency of these surface modification marks is compared to established 95% confidence intervals calculated from actualistic studies

**Table 4.1** Number of identifiable long bone ends with nutritive and non-nutritive bone breaks

| Level | Oblique | Curved | Total long bone ends |
|---|---|---|---|
| NISP nutritive bone breaks | | | |
| I2 | 603 | 600 | 728 |
| Y-Z | 477 | 477 | 568 |

| Level | Right | Transverse | Unbroken | Total long bone ends |
|---|---|---|---|---|
| NISP non-nutritive bone breaks | | | | |
| I2 | 24 | 27 | 17 | 728 |
| Y-Z | 23 | 23 | 8 | 568 |

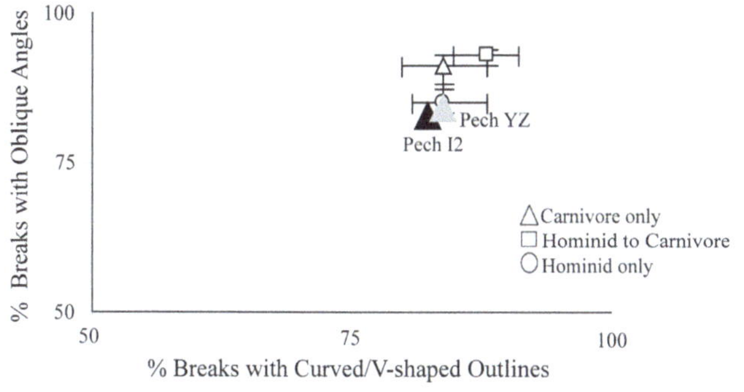


**Fig. 4.2** The frequency of nutritive bone breaks recorded on all size 2/3 long bone fragments from Pech IV Levels Y-Z and I2. Data are compared to experimental and naturalistic assemblages where nutrients were extracted by hominins only, hominins then modified by carnivores, and carnivores only (Marean et al. 2000)

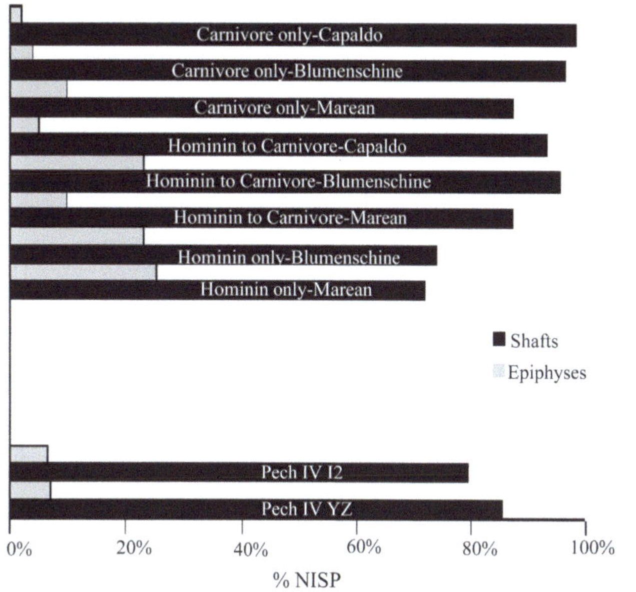


**Fig. 4.3** The frequency of size 2/3 long bone shaft fragments compared to epiphyseal fragments at Pech IV Levels Y-Z and I2 compared to actualistic data (Blumenschine 1988; Marean and Spencer 1991; Marean et al. 2000, 2004; Capaldo 1995)

**Table 4.2** Raw number of long bone shaft and epiphyseal fragments

| Level | Long bone shaft fragments | Long bone epiphyses |
|---|---|---|
| I2 | 364 | 30 |
| Y-Z | 284 | 24 |

(Marean et al. 2000; Marean and Bertino 1994; Marean and Spencer 1991; Marean et al. 1992; Blumenschine 1988, 1995; Capaldo 1995, 1997) the results demonstrate that carnivore activity was rare to nonexistent in levels I2 and Y-Z (Table 4.3, Fig. 4.4). Thus, Neanderthals appear to have been responsible for collecting skeletal elements at the site and extracting nutrients from them.

## Taxonomic Analysis

As discussed previously, taxonomic analysis for this study was focused on the identification of animals in the size 2 and 3 body size categories, which is largely dominated by cervids in the Pleistocene of southwestern France. Larger species were randomly sampled, but for present purposes, these mammals have been labeled "other taxa". Table 4.4 shows the number of skeletal elements (NISP) belonging to size 2/3 bovid/cervids that were identifiable to a specific skeletal element, while Table 4.5 provides raw data for the number of skeletal remains that were identifiable to a specific taxonomic group.

When an MNE analysis using ArcView GIS is run on shaft fragments identifiable to specific taxa (Fig. 4.5) the results show that while each species of deer (*R. tarandus*, *C. capreolus*, and *C. elaphus*) was represented in each level, *R. tarandus* was the dominate species transported to the cave in Level I2 (Layer 4B, 4C), while *Cervus elaphus* dominated the assemblage in level Y-Z (Layer 8) (Table 4.6).

When the combined MNE for all size 2/3 bovid/cervids is used to calculate the %MAU of each high survival element (Tables 4.7 and 4.8), the results show that the abundance of each element remained fairly consistent over time, which

**Table 4.3** Raw number of skeletal elements with at least one percussion mark (PM) or one tooth mark (TM)

| Element | Level I2 PM | Level Y-Z PM | Level I2 TM | Level Y-Z TM |
|---|---|---|---|---|
| Humerus | 18 | 9 | 0 | 0 |
| Radioulna | 12 | 6 | 1 | 0 |
| Metacarpal | 11 | 7 | 0 | 0 |
| Femur | 11 | 7 | 0 | 0 |
| Tibia | 40 | 13 | 0 | 0 |
| Metatarsal | 44 | 22 | 0 | 0 |
| Metapodial | 7 | 2 | 0 | 0 |
| Prox phalanges | 1 | 1 | 0 | 0 |
| Mid. phalanges | 0 | 0 | 0 | 0 |
| Long bones | 12 | 23 | 0 | 0 |

No specimens were found to have both percussion marks and tooth marks

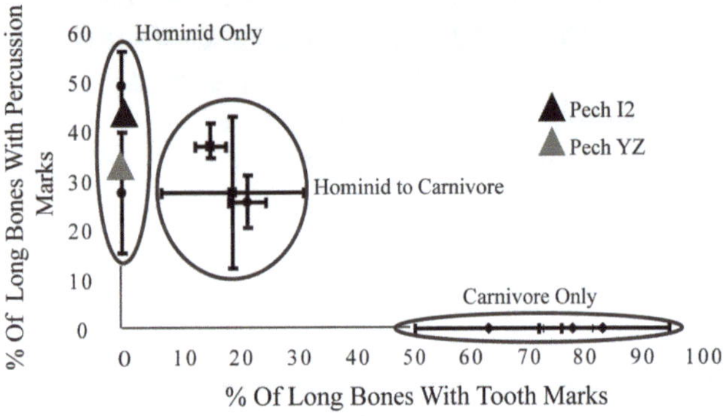


**Fig. 4.4** The percentage of long bone shafts with percussion marks versus the percentage with tooth marks. Data are compared to experimental and naturalistic assemblages broken by hominins only, hominins then modified by carnivores, and carnivores only (Blumenschine 1995, 1988; Capaldo 1997, 1995; Marean and Spencer 1991; Marean et al. 2000, 1992; Marean and Bertino 1994). The *ellipses* are used for emphasis, but have no statistical meaning

suggests Neanderthals used similar transport strategies through time (Fig. 4.6).

Finally, when the %MAU of each high survival element is plotted against the standard food utility index (SFUI), which is a measure of the quantity of nutrients (e.g., meat, marrow, and bone grease) associated with different body parts (Metcalfe and Jones 1988), the resulting plot aligns most closely (although not perfectly) with Binford's bulk transport strategy (Fig. 4.7). This strategy is characterized by the transport of all high utility (i.e., high nutritional value) elements, with elements of lower nutritional value left behind (Binford 1978). The main difference between Binford's proposed utility strategies and skeletal representation at Pech IV is that in all levels femora are not as well represented as would be predicted by the bulk transport strategy. Binford calculates the femur to be the bone with the highest utility (SFUI = 100). It is also interesting to note that transport strategies employed in Level Y-Z at Pech IV

**Table 4.4** Number of specimens identified to specific skeletal elements for size 2/3 bovid/cervids

| NISP skeletal element | Level I2 | Level Y-Z |
|---|---|---|
| Cranial bones | 8 | 6 |
| Maxillary single teeth | 18 | 20 |
| Mandible | 31 | 16 |
| Mandibular single teeth | 43 | 50 |
| Humerus | 34 | 29 |
| Radioulna | 35 | 24 |
| Metacarpal | 24 | 30 |
| Femur | 31 | 18 |
| Tibia | 74 | 37 |
| Metatarsal | 104 | 69 |
| Metapodial | 22 | 30 |
| Proximal phalanges | 8 | 5 |
| Middle phalanges | 3 | 5 |
| Distal phalanges | 2 | 1 |
| Long bone fragments | 29 | 37 |
| Atlas | 0 | 0 |
| Axis | 0 | 0 |
| Cervical | 0 | 0 |
| Thoracic | 1 | 1 |
| Lumbar | 2 | 1 |
| Sacrum | 0 | 0 |
| Rib | 15 | 3 |
| Scapula | 9 | 3 |
| Sternabra | 0 | 0 |
| Carpals | 9 | 8 |
| Tarsals | 5 | 4 |
| Pelvis | 5 | 1 |
| Patella | 0 | 0 |

Note that this table does not include fragments that could not be identified to a specific element

**Table 4.5** Number of dental and skeletal remains identified to each taxonomic group

| Taxon | Level I2 | Level Y-Z |
|---|---|---|
| Bovid/cervid | 143 | 148 |
| Cervid | 21 | 13 |
| *Capreolus capreolus* | 21 | 29 |
| *Cervus elaphus* | 69 | 208 |
| *Rangifer tarandus* | 253 | 1 |
| Mammal | 291 | 288 |
| Other taxa | 112 | 40 |
| Total bones analyzed | 910 | 727 |

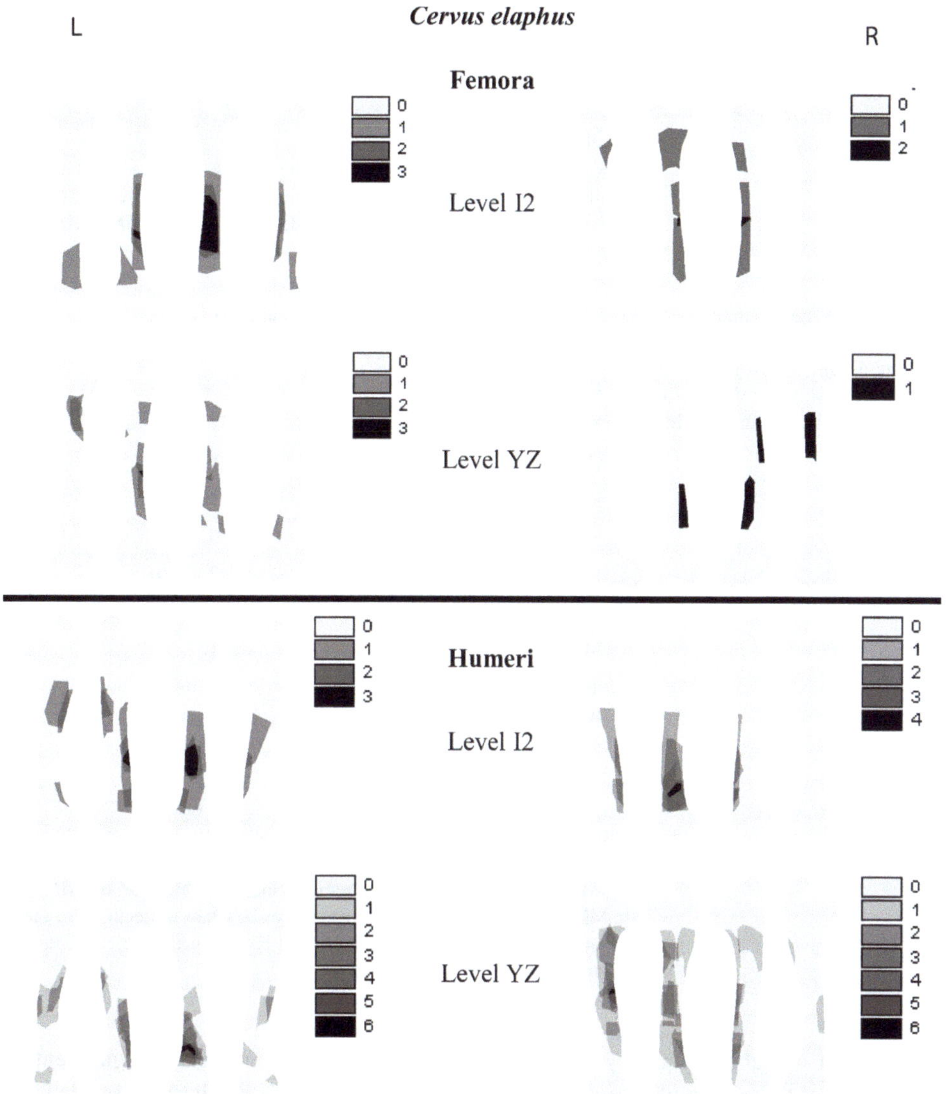


**Fig. 4.5** Shaft fragments of *Cervus elaphus* femora (*top*) and humeri (*bottom*) from Bordes' excavation of Pech IV. Each fragment was drawn into ArcView GIS where it was then mapped onto the same template and fragment overlaps were used to identify elements from different individuals. The *darkest areas* on each element indicate the highest number of overlaps. Similarly, the *color bars* to the right of each element indicate the number of overlaps along each element with the *bottom black bar* indicating the highest number of overlaps [minimum number of elements (MNE)]. The MNI of a species is equal to the highest MNE value (between *lefts* and *rights*) found out of all bones analyzed of a particular species

**Table 4.6** Minimum number of individuals calculated from arc view GIS for each species of cervid

| Taxon | Level I2 | Level Y-Z |
|---|---|---|
| *Rangifer tarandus* | 9 | 0 |
| *Cervus elaphus* | 4 | 10 |
| *Capreolus capreolus* | 2 | 1 |

**Table 4.7** The minimum number of skeletal elements and the minimum animal units calculated using ArcView GIS of size 2/3 bovid/cervids from Level I2

| Element | MNE L | With teeth | MNE R | With teeth | MAU | MAU teeth | SFUI | %MAU with teeth |
|---|---|---|---|---|---|---|---|---|
| Cranial | 1 | 2 | N/A | N/A | 0.5 | 1 | 9.1 | 11 |
| Mandible | 5 | 7 | 6 | Same | 5.5 | 6.5 | 11.5 | 72 |
| Humerus | 6 | | 6 | | 6 | | 36.8 | 67 |
| Radioulna | 5 | | 3 | | 4 | | 25.8 | 44 |
| Metacarpal | 2 | | 0 | | 1 | | 5.2 | 11 |
| Femur | 5 | | 4 | | 4.5 | | 100 | 50 |
| Tibia | 9 | | 9 | | 9 | | 62.8 | 100 |
| Metatarsal | 1 | | 6 | | 3.5 | | 37 | 39 |
| Total sample size without teeth | | | 68 | | | | | |
| Total sample size with teeth | | | 71 | | | | | |
| Evenness value | | | 0.91 | | | | | |
| Spearman's rho | | | 0.56 | | | | | |

Note that the MNE of cranial remains increases by one and the MNE of left mandibles increase by two when all teeth were included in the analysis, increasing the total sample size by 3. The Evenness and Rho values were calculated using the %MAU derived from the MNE with teeth

**Table 4.8** The minimum number of skeletal elements and the minimum animal units calculated using ArcView GIS of size 2/3 bovid/cervids from level Y-Z

| Element | MNE L | With teeth | MNE R | With teeth | MAU | MAU teeth | SFUI | %MAU with teeth |
|---|---|---|---|---|---|---|---|---|
| Cranial | 0 | 3 | N/A | N/A | 0 | 1.5 | 9.1 | 21 |
| Mandible | 2 | 4 | 4 | 10 | 3 | 7 | 11.5 | 100 |
| Humerus | 6 | | 6 | | 6 | | 36.8 | 86 |
| Radioulna | 2 | | 3 | | 2.5 | | 25.8 | 35 |
| Metacarpal | 1 | | 0 | | 0.5 | | 5.2 | 7 |
| Femur | 3 | | 2 | | 2.5 | | 100 | 36 |
| Tibia | 4 | | 4 | | 4 | | 62.8 | 57 |
| Metatarsal | 3 | | 1 | | 2 | | 37 | 29 |
| Total sample size without teeth | | | 41 | | | | | |
| Total sample size with teeth | | | 52 | | | | | |
| Evenness value | | | 0.90 | | | | | |
| Spearman's rho | | | 0.14 | | | | | |

Note that the MNE of cranial remains increases by three, the MNE of left mandibles increases by two, and the MNE of right mandibles increases by six when all teeth were included in the analysis, increasing the total sample size by 11. The Evenness and Rho values were calculated using the %MAU derived from the MNE with teeth

resulted in a higher frequency of mandibles and a lower frequency of tibiae than observed in Level I2 (Figs. 4.6 and 4.7).

## Conclusions

The results from this study show that post-depositional processes that result in fracturing of skeletal elements after all nutrients have been removed from the bone did not play a significant role in altering the assemblage. This result suggests that sediments from Levels I2 and Y-Z were well protected even from the effects of trampling, a process that often causes fragmentation of skeletal material (cf. Chap. 2). The high frequency of percussion marks and the low frequency of tooth marks on long bone shaft fragments demonstrate that Neanderthals were the primary accumulators of faunal material in both Levels I2 and Y-Z. The exceptionally low number of tooth marks

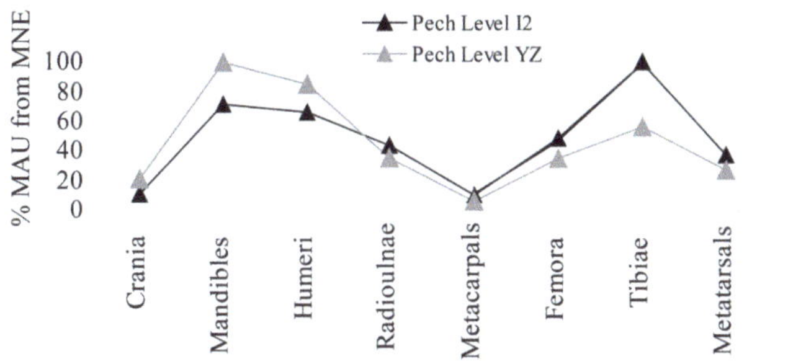


**Fig. 4.6** The abundance (measured in %MAU) of high survival skeletal elements including teeth analyzed from Levels I2 and Y-Z at Pech IV

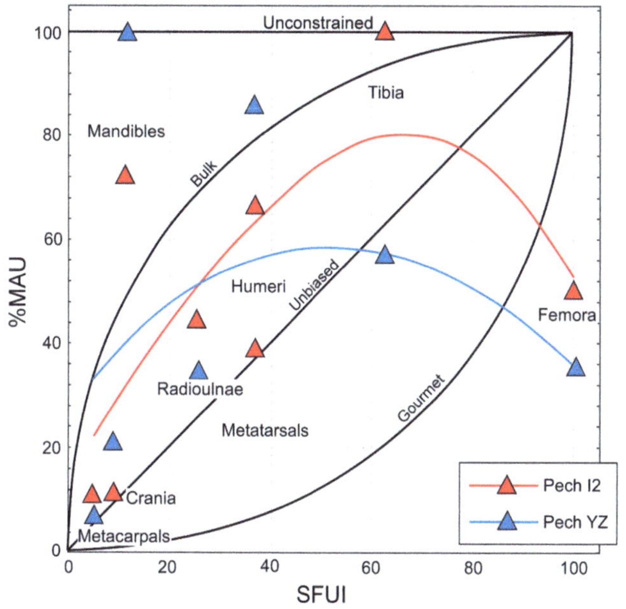


**Fig. 4.7** The relationship between the relative abundance of high survival skeletal elements and their Standardized Food Utility Index (SFUI) for Levels I2 and Y-Z at Pech IV. A distance-weighted least squares curve was fitted to the data points from each level. Utility curves proposed by Binford (1978) are plotted with the data for comparison

on skeletal elements from each level suggests that carnivores did not gain access to faunal material after it was discarded by Neanderthals. Finally, results from the current study indicate that the abundance of skeletal elements from each level most closely approximate Binford's bulk transport strategy in which all bones are transported except those of the lowest utility, although skeletal elements from both levels deviated from the bulk strategy in the lack femoral fragments. In addition, mandibles were more abundant in both levels at Pech IV than would be predicted based on their calculated utility. Level Y-Z (Layer 8) also stood out as having a lower than expected abundance of tibia. The lack of femora and abundance of mandibles has also been reported from the Middle Paleolithic site of Les Pradelles (Costamagno et al. 2006), and those researchers suggested that the deficit of femora was caused by the transport of these elements to another site. While this interpretation is possible, it seems likely that femora were never transported to the site to begin with. For example, Binford (1978) observed femora being processed at a kill site for the purposes of snacking, and these femora were then left behind.

Perhaps the most important result from this analysis is that minimal evidence of excavator or curation bias exists. Biases that existed at the time of excavation appear largely correctable, which means that all data collected on the older assemblage are directly comparable to analyses conducted on faunal remains from the recent excavation.

## References

Abe, Y., Marean, C. W., Nilssen, P. J., Assefa, Z., & Stone, E. C. (2002). The analysis of cutmarks on archaeofauna: A review and critique of quantification procedures, and a new image-analysis GIS approach. *Society for American Archaeology, 67*(4), 643–663.

Binford, L. R. (1978). *Nunamiut ethnoarchaeology*. New York: Academic Press Inc.

Binford, L. R. (1981). *Bones: Ancient men and modern myths*. New York: Academic Press Inc.

Binford, L. R. (1984). *Faunal remains from Klasies River Mouth (Studies in Archaeology)*. New York: Academic Press Inc.

Binford, L. R. (1988). Etude taphonomique des restes fauniques de la grotte Vaufrey. In J.-P. Rigaud (Ed.), *La Grotte Vaufrey à Cénac et Saint-Julien (Dordogne): Paléoenvironments, Chronologie et Activités Humaines* (Vol. 19, pp. 535–564).

Blumenschine, R. J. (1988). An experimental model of the timing of hominid and carnivore influence on archaeological bone assemblages. *Journal of Archaeological Science, 15*(5), 483–502.

Blumenschine, R. J. (1995). Percussion marks, tooth marks, and experimental determinations of the timing of hominid and carnivore access to long bones at FLK Zinjanthropus, Olduvai Gorge, Tanzania. *Journal of Human Evolution, 29*(1), 21–51.

Blumenschine, R. J., Marean, C. W., & Capaldo, S. D. (1996). Blind tests of inter-analyst correspondence and accuracy in the identification of cut marks, percussion marks, and carnivore tooth marks on bone surfaces. *Journal of Archaeological Science, 23*(4), 493–507.

Blumenschine, R. J., & Selvaggio, M. M. (1988). Percussion marks on bone surfaces as a new diagnostic of hominid behaviour. *Nature, 333*(6175), 763–765. doi:10.1038/332141a0.

Brain, C. K. (1984). *The hunters and the hunted*. Chicago: University of Chicago Press.

Bunn, H. T., Bartram, L. E., & Kroll, E. M. (1988). Variability in bone assemblage formation from Hadza hunting, scavenging, and carcass processing. *Journal of Anthropological Archaeology, 457*, 412–457.

Bunn, H. T., & Kroll, E. M. (1986). Systematic Butchery by Plio/Pleistocene Hominids at Olduvai Gorge, Tanzania. *Current Anthropology, 27*(5), 431–452.

Capaldo, S. D. (1995). *Inferring Hominid and carnivore behavior from dual-patterned archaeofaunal assemblages*. New Jersey: Rutgers University.

Capaldo, S. D. (1997). Experimental determinations of carcass processing by Plio-Pleistocene hominids and carnivores at FLK 22 (Zinjanthropus), Olduvai Gorge, Tanzania. *Journal of Human Evolution, 33*(5), 555–597.

Capaldo, S. D. (1998). Simulating the formation of dual-patterned archaeofaunal assemblages with experimental control samples. *Journal of Archaeological Science, 25*(4), 311–330.

Cleghorn, N. E., & Marean, C. W. (2004). Distinguishing selective transport and in situ attrition: A critical review of analytical approaches. *Journal of Taphonomy, 2*(2), 43–68.

Cleghorn, N. E., & Marean, C. W. (2007). The destruction of skeletal elements by carnivores: the growth of a general model for skeletal element destruction and survival in zooarchaeological assemblages. In T. R. Pickering, N. Toth, & K. Schick (Eds.), *African taphonomy: A tribute to the career of C.K. "Bob" Brain* (pp. 13–42). Bloomington: Stone Age Press.

Costamagno, S., Liliane, M., Cédric, B., Bernard, V., & Bruno, M. (2006). Les Pradelles (Marillac-le-Franc, France): a mousterian reindeer hunting camp? *Journal of Anthropological Archaeology, 25*(4), 466–484.

Delpech, F. (1983). *Les faunes du Paleolithique Superieur dans le sud-ouest de la France*. Bordeaux: University of Bordeaux I.

Dibble, H., Berna, F., Goldberg, P., McPherron, S., Mentzer, S., Niven, L., et al. (2009). A preliminary report on Pech de l'Azé IV, Layer 8 (Middle Paleolithic, France). *Paleoanthropology, 2009*, 182–219.

Dibble, H., Raczek, T. P., & Mcpherron, S. (2005). Excavator bias at the site of Pech de l'Azé IV, France. *Journal of Field Archaeology, 30*(3), 317–328.

Domínguez-Rodrigo, M., & Barba, R. (2006). New estimates of tooth mark and percussion mark frequencies at the FLK Zinj site: The carnivore-hominid-carnivore hypothesis falsified. *Journal of Human Evolution, 50*(2), 170–194.

Domínguez-Rodrigo, M., de Juana, S., Galán, A. B., & Rodríguez, M. (2009). A new protocol to differentiate trampling marks from butchery cut marks. *Journal of Archaeological Science, 36*(12), 2643–2654.

Domínguez-Rodrigo, M., & Piqueras, A. (2003). The use of tooth pits to identify carnivore taxa in tooth-marked archaeofaunas and their relevance to reconstruct hominid carcass processing behaviours. *Journal of Archaeological Science, 30*(11), 1385–1391.

Faith, J. T., & Gordon, A. D. (2007). Skeletal element abundances in archaeofaunal assemblages: Economic utility, sample size, and assessment of carcass transport strategies. *Journal of Archaeological Science, 34*(6), 872–882.

Fiorillo, A. R. (1989). An experimental study of trampling: implications for the fossil record. In R. Bonnichsen & M. H. Sorg (Eds.), *Bone modification*. Orono: The Center for the Study of First Americans.

Grayson, D. K. (1984). *Quantitative zooarchaeology*. New York: Academic Press.

Kurtén, B. (1968). *Pleistocene mammals of Europe*. London: Weidenfeld & Nicolson.

Lam, Y. M., Xingbin, C., & Pearson, O. M. (1999). Intertaxonomic variability in patterns of bone density and the differential representation of bovid, cervid, and equid elements in the archaeological record. *American Antiquity, 64*(2), 343–362.

Laquay, G. (1981). *Recherches sur les faunes du Wurm I en Perigord*. Bordeaux: University of Bordeaux.

Lyman, R. L. (1995). A study of variation in the prehistoric butchery of large artiodactyls. In E. Johnson (Ed.), *Ancient peoples and landscapes* (pp. 233–253). Lubbock: Museum of Texas Tech University.

Marean, C. W. (1998). A critique of the evidence for scavenging by Neanderthals and early modern humans: new data from Kobeh Cave ( Zagros Mountains, Iran) and Die Kelders Cave 1 Layer 10 ( South Africa). *Journal of Human, 35*(2), 111–136.

Marean, C. W., Abe, Y., Frey, C. J., & Randall, R. C. (2000). Zooarchaeological and taphonomic analysis of the Die Kelders Cave 1 layers 10 and 11 middle stone age larger mammal fauna. *Journal of Human Evolution, 38*(1), 197–233.

Marean, C. W., Abe, Y., Nilssen, P. J., & Stone, E. C. (2001). Estimating the minimum number of skeletal elements (MNE) in zooarchaeology: A review and a new image-analysis GIS approach. *American Antiquity, 66*(2), 333–348.

Marean, C. W., & Bertino, L. (1994). Intrasite spatial analysis of bone: Subtracting the effect of secondary carnivore consumers. *American Antiquity, 59*(4), 748–768.

Marean, C. W., & Cleghorn, N. (2003). Large mammal skeletal element transport: Applying foraging theory in a complex taphonomic system. *Journal of Taphonomy, 1*(1), 15–42.

Marean, C. W., Domínguez-rodrigo, M., & Pickering, T. R. (2004). Skeletal element equifinality in zooarchaeology begins with method: The evolution and status of the "Shaft Critique". *Journal of Taphonomy, 2*(2), 69–98.

Marean, C. W., & Frey, C. J. (2013). Animal bones from Caves to Cities: Reverse utility curves as methodological artifacts. *American Antiquity, 62*(4), 698–711.

Marean, C. W., & Kim, S. Y. (1998). Mousterian large-mammal remains from Kobeh Cave: Behavioral implications for Neanderthals and early modern humans. *Current Anthropology, 39*, S79–S109.

Marean, C. W., & Spencer, L. M. (1991). Impact of carnivore ravaging on zooarchaeological measures of element abundance. *American Antiquity, 56*(4), 645–658.

Marean, C. W., Spencer, L. M., Blumenschine, R. J., & Capaldo, S. D. (1992). Captive hyaena bone choice and destruction, the Schlepp effect and olduvai archaeofaunas. *Journal of Archaeological Science, 19*(1), 101–121.

Metcalfe, D., & Jones, K. T. (1988). A reconsideration of animal body-part utility indices. *American Antiquity, 53*(3), 486–504.

Monahan, C. M. (1998). The Hadza carcass transport debate revisited and its archaeological implications. *Journal of Archaeological Science, 25*(5), 405–424.

O'Connell, J. F., Hawkes, K., & Blurton Jones, N. (1990). Reanalysis of large mammal body part transport among the Hadza. *Journal of Archaeological Science, 17*(3), 301–316.

O'Connell, J. F., Hawkes, K., & Blurton Jones, N. (1998). Hadza hunting, butchering, and bone transport and their archaeological implications. *Journal of Anthropological Research, 44*(2), 113–161.

Olsen, S. L., & Shipman, P. (1988). Surface modification on bone: Trampling versus butchery. *Journal of Archaeological Science, 15* (5), 535–553.

Pickering, T. R., Domínguez-Rodrigo, M., Egeland, C. P., & Brain, C. K. (2004). Beyond leopards: Tooth marks and the contribution of multiple carnivore taxa to the accumulation of the Swartkrans Member 3 fossil assemblage. *Journal of Human Evolution, 46*(5), 595–604.

Pickering, T. R., & Egeland, C. P. (2006). Experimental patterns of hammerstone percussion damage on bones: Implications for inferences of carcass processing by humans. *Journal of Archaeological Science, 33*(4), 459–469.

Pickering, T. R., Marean, C. W., & Domínguez-Rodrigo, M. (2003). Importance of limb bone shaft fragments in zooarchaeology: A response to "On in situ attrition and vertebrate body part profiles" (2002), by M. C. Stiner. *Journal of Archaeological Science, 30*, 1469–1482.

Pokines, J. T., & Kerbis Peterhans, J. C. (2007). Spotted hyena (*Crocuta crocuta*) den use and taphonomy in the Masai Mara National Reserve, Kenya. *Journal of Archaeological Science, 34*(11), 1914–1931.

Potts, R., & Shipman, P. (1981). Cutmarks made by stone tools on bones from Olduvai Gorge, Tanzania. *Nature, 291*(5816), 577–580.

Shipman, P. (1981). Applications of scanning electron microscopy to taphonomic problems. *Annals of the New York Academy of Sciences, 376,* 357–385.

Villa, P., & Mahieu, E. (1991). Breakage pattern of human long bones. *Journal of Human Evolution, 21*(1), 27–48.

Yravedra, J., & Domínguez-Rodrigo, M. (2009). The shaft-based methodological approach to the quantification of long limb bones and its relevance to understanding hominid subsistence in the Pleistocene: Application to four Palaeolithic sites. *Journal of Quaternary Science, 24*(81), 85–96.

# Zooarcheological Analysis of the Assemblage from the 2000–2003 Excavations

# 5

Laura Niven and Hélène Martin

## Introduction

The dynamics between Neandertal hunters, their prey, and their natural and social landscapes have long been the focus of Paleolithic zooarcheological research. Investigations over the past several decades have yielded exceptional insight on their hunting and exploitation of large mammals, particularly ungulates, which allows us to better understand their broader subsistence strategies, site use, mobility, and life history. Sites with multi-layer stratigraphic sequences of Middle Paleolithic deposits containing associated Mousterian artifacts and archeofaunas spanning long ranges of time are particularly valuable sources for evaluating change and stability in subsistence during the Neandertals' long presence in Eurasia (e.g., Chase 1986, 1999; Stiner 1994, 2005; Gaudzinski 1995; Blasco 1997; Pike-Tay et al. 1999; Burke 2000; Conard and Prindiville 2000; Beauval 2004; Fiore et al. 2004; Morin 2004; Rabinovich and Hovers 2004; Speth and Clark 2006; Speth and Tchernov 2007; Rivals et al. 2009; Niven et al. 2012).

Pech IV contains roughly three meters of deposits spanning much of the Late Pleistocene, representing approximately 60 ka of Neandertal life to investigate. This chapter presents the results of the zooarcheological analysis of all faunal remains recovered from the recent excavations. Not all stratigraphic layers provided adequate faunal samples for robust analyses; therefore, the bulk of this chapter focuses on the larger assemblages. The results indicate that both large-scale and more local environmental conditions influenced the abundance of the two dominant prey species—red deer and reindeer—across the sequence. Site occupations may have been partly based on availability and abundance of these prey as well. When the zooarcheological results are evaluated in conjunction with data from the lithics and other features, a clearer picture emerges of site use over time and through changing climate regimes.

This archeozoological study involves the material from the recent excavations between 2000 and 2003. With the exception of a sample from the coarse water-screen material recovered from Layer 8, this analysis is limited to the piece-plotted finds. Laquay (1981) previously studied the fauna from Bordes' excavation at Pech IV, and his analysis covered much of the same sequence as ours, excluding the uppermost layers. Overall, however, the study presented here reflects the same pattern of species representation as presented by Laquay. The faunal spectrum published by Laquay provided a valuable framework for understanding the climatic conditions under which the Pech de l'Azé caves and neighboring sites were occupied by Neandertals (e.g., Mellars 1996). An additional archeozoological study by Hodgkins, using methods different from Laquay, of faunal material from the Bordes' excavations is presented in Chap. 4.

## Materials and Methods

Approximately 23,000 piece-plotted faunal remains were recovered from the 2000–2003 excavations. Although the size cutoff for piece plotting was >25 mm, identifiable bones or teeth below this size were also mapped in place whenever possible. Except for a small sample from Layer 8, coarse water-screen remains were not included in this study.

The entire assemblage was heavily fragmented, although the degree and causes of this varied through the sequence. Density-mediated attrition also influenced the preservation of certain elements or parts thereof. Both factors strongly affected the ability to identify remains to species, and consequently many specimens could be placed only in general categories by family or body size (following Brain 1981:

L. Niven (✉)
Annual Reviews, Palo Alto, CA, USA
e-mail: lniven@annualreviews.org

H. Martin
INRAP-Centre Archéologique de Saint-Orens, St.-Orens de Gameville, France
e-mail: helene.martin@inrap.fr


H.L. Dibble et al. (eds.), *The Middle Paleolithic Site of Pech de l'Azé IV*,
Cave and Karst Systems of the World, https://doi.org/10.1007/978-3-319-57524-7_5

Appendix Table 1). Examples include cancellous bone (spongy: articular ends, vertebrae), flat bone (scapula, rib, pelvis), or cranial vault fragments, which were grouped by body size when possible. Due to the attrition of fragile and less dense cancellous bone, the assemblages were dominated by long bone shaft fragments that are often small. Thus, a conservative approach was taken in identifying fragments of some ungulate long bones, especially metapodials; unless they could be identified to taxon with 100% certainty, they were classified as large or small cervid or bovid.

A combination of primary and derived data is used to quantify the Pech IV fauna. The number of identified specimens (NISP) includes specimens identified to taxon, genus, family, and order (Lyman 1994). Although it is possible that lumping "large cervid" (i.e., red deer or reindeer) remains is not ideal and may introduce errors in the data, the small sample sizes of many assemblages necessitated such an approach. Moreover, this will not obscure the general pattern of faunal exploitation since this lumped group cannot include any other taxa. Also used in this study is %NISP, which allows us to evaluate species abundance overall or to quantify proportions of bone modifications by element.

The minimum number of elements (MNE) includes all complete or fragmentary specimens observed for that skeletal element, by taxon (Binford 1984), considering side, sex, or age when possible. MNEs can be converted to minimum animal units (MAU) and %MAU, which allow for comparing element frequencies to a standard (i.e., complete skeleton; Binford 1984). Because metapodial MNEs are so difficult to quantify from midshaft fragments, this study follows Castel's (personal communication) approach that measures (in mm) the anterior groove length from each archeological fragment and sums them by element per layer. That number is divided by a standard length averaged from multiple modern specimens of that species (in this case, red deer, roe deer, and reindeer). This number provides an estimate of how many metacarpals and metatarsals are represented by what may be hundreds of shaft fragments.

The minimum number of individuals (MNI) is based on the minimum number of individuals represented by each skeletal element by taxon. MNI counts are only based on specimens identified to species; they also consider side, sex, and age.

Fragmentation affected the ability to identify long bone shaft splinters, the most numerous specimens in all assemblages. However, external and internal morphological features were used to identify them whenever possible. The length of these shaft fragments was also measured and quantified to evaluate breakage patterns among these fat-rich skeletal parts across taxa and layers.

The type of bone breakage was recorded to evaluate aspects of site formation and exploitation of within-bone nutrients by the Neandertals. Green or fresh breakage occurs when the bone still retains its organic content. Fracture outlines are helical and surfaces are oblique and smooth. Such damage may result from intentional breakage of bones by humans to access the marrow cavity, which can often be confirmed through associated percussion marks. Dry breakage occurs after the bone's organic content is lost, resulting in longitudinal or right angle fractures. Curational breakage is related but distinctive, primarily by the color of the breakage surface.

Despite the extensive fragmentation, bone surface preservation is generally good, except for in Layers 3A and 3B. This allowed for the recognition and analysis of various traces such as stone tool cutmarks, striations, hammerstone impacts, and carnivore tooth marks. Every bone specimen was thoroughly examined for surface modifications using a strong primary light source and 10× magnification; specific examples were examined under the microscope if necessary. Only clear and unambiguous evidence was recorded. Bone modifications were quantified overall (i.e., including the "lumped" specimens) and by taxon or element.

Burning was identified through several methods, including field observations, context, geomorphological thin sections, and macroscopic evaluation of bone surface or structure. It can be difficult to distinguish burning from mineral (e.g., manganese oxide) staining (Shahack-Gross et al. 1997; Stiner 2005), so additional microscopic methods are needed to confirm that archeological bone was, in fact, heated (e.g., Bellamo 1993; Hanson and Cain 2007; Nicholson 1993; Schiegl et al. 2003; Shipman et al. 1984; Stiner et al. 1995).

Documentation of burned bone follows the six burning stages outlined by Stiner et al. (1995: Table 3), though its quantification is based on Costamagno et al. (2009), whose method was designed specifically for interpreting the origin of burned bone in archeological contexts.

## Coarse Water-Screen Material

A sample of coarse (6 mm mesh) water-screen material from Layer 8 was evaluated to address several questions: (1) the degree and type of burning; (2) to what degree were fragile skeletal parts preserved (e.g., cancellous, fetal, or axial bone); (3) to what degree were the smallest skeletal specimens present. Because the size cutoff for piece-plotting was 25 mm, many of the smallest specimens went to the water-screen buckets, even if they were identifiable. Thus, a survey of water-screen material was essential for assessing the true abundance of elements such as phalanges, tarsals, carpals, sesamoids, and even teeth of the main ungulate prey, as well as remains of small mammals and birds.

Detailed results of this survey are presented in the following sections. To briefly summarize the findings, analysis and quantification of burned bone ($N = 23{,}070$) revealed valuable insights on the actual preservation of spongy bone, the degree of burning across the bone assemblage in general, and the role of bone in the use of fire in Layer 8 (see full review in Dibble et al. 2009). The survey indicated that the small skeletal parts of ungulates and bones of small mammals and birds were indeed present and quite abundant. Fragmented portions of red deer petrous bone were also numerous, which provided important evidence for the bony parts of the cranium that were not represented by piece-plotted finds. In other words, several anatomical portions that were seemingly "missing" in the main assemblage were in fact represented. Lastly, evidence for small mammal and bird remains was found, yielding insight on the exploitation of these resources by Neandertals.

## Anthropogenic or Carnivore Accumulation?

The primary taphonomic question involves which agent accumulated the bone in Pech IV: Neandertals, carnivores (e.g., hyena, bear, lion, wolf), or a combination of both. Because caves served the needs of both humans and animals, it is quite common that sites in karstic areas contain the food remains of both groups. Evaluating the bone accumulator(s) of assemblages is based on the presence/absence of carnivore skeletal remains or coprolites; the presence/absence of carnivore traces on bones (e.g., tooth punctures or scores, gnawing, characteristic breakage); and attrition of spongy and fragile skeletal parts.

Carnivores are extremely rare in all layers of the Pech IV sequence, both in terms of physical remains and traces such as toothmarks or coprolites. Wolf (*Canis lupus*) was found in Layers 4C and 8, and unidentified bear (*Ursus* sp.) was present in Layers 6A and 6B. Fox (*Vulpes* or *Alopex*) was found only in Layer 6A. The number of toothmarks on any bone surface is correspondingly small (0.002%). Both lines of evidence are in complete agreement with the data from Bordes' collection (Chap. 4). Although the lack of carnivores and their signatures is surprising in a Pleistocene cave, in this case, the absence of evidence is not due to issues of identification or preservation (except perhaps in terms of coprolites). Thus, bone collecting carnivores were simply not involved in the faunal accumulation here, and the assemblages are presumed to be completely anthropogenic.

Fragile cancellous bone portions and axial skeletal parts are strongly underrepresented in every layer, but this bias cannot be attributed to carnivore feeding. The denser bone elements and teeth were consistently better preserved and long bones are overwhelmingly represented by midshaft portions and unidentifiable shaft splinters. Statistically significant relationships between bone density and skeletal abundance are evident for several assemblages. However, an evaluation of the coarse water-screen sample from Layer 8 showed that cancellous and axial bone is, in fact, present to some degree, albeit in tiny fragments. A variety of taphonomic factors could have contributed to the deterioration of these skeletal portions, including geologic and mechanical factors specific to limestone cave deposits. Overall, these bones are poorly represented among the piece-plotted specimens across all layers in the new collection and in the sample from Bordes' collection (Chap. 4).

## The Faunal Assemblages

The NISP and MNI per layer reflect the changing abundance and diversity of fauna across the sequence of Pech IV (Table 5.1; Fig. 5.1).

To briefly set the stage for the results presented below, the basal Layer 8's faunal spectrum, while rather small, is dominated by red deer and roe deer. Smaller numbers of wild boar (*Sus scrofa*) and beaver (*Castor fiber*) were also represented, collectively indicating temperate, wooded environs. Horse and *Bos/Bison* are also present in small numbers. Most notable here is the use of fire that affected the bone and stone assemblages significantly.

Although some fauna was recovered from layer 7, it was severely damaged by post-depositional processes (specifically cryoturbation) and cannot be properly analyzed.

Layer 6 yielded the largest assemblage in the Pech IV sequence. Smaller fragmentation size (likely from butchery) of long bone shaft fragments made these assemblages more difficult to identify, but the faunal spectrum is essentially the same as Layer 8.

Layer 5B is dominated by large ungulates, such as horse and bison. The large cervids are mixed, with red deer (*Cervus elaphus*) and reindeer (*Rangifer tarandus*) present. A similar scenario is evident in Layer 5A. In both layers, it was not always possible to identify long bone shaft fragments to species; these were grouped as "large cervid." These "mixed" faunas indicate the shifting climate conditions occurring over the course of the deposition of Layer 5.

Reindeer clearly and overwhelmingly dominate all three sublayers of Layer 4. A few horse specimens were present but little else. Preservation was good, resulting in better identifiability to species and element.

In the uppermost part of the sequence (Layer 3), the Mousterian of Acheulian Tradition (MTA) industries were associated with faunal assemblages containing primarily reindeer and *Bos/Bison*. Preservation was poor and indeterminate long bone shaft fragments were most common.

In summary, the faunal sequence in this site was clearly dominated by large cervids (reindeer or red deer), with other

**Table 5.1** Summary of faunal remains from Pech IV expressed as NISP, MNI (in brackets), or N

| | 3A | 3B | 4A | 4B | 4C | 5A | 5B | 6A | 6B | 8[a] |
|---|---|---|---|---|---|---|---|---|---|---|
| *Lepus* sp. (hare) | – | – | – | – | – | – | – | – | – | 1 [1] |
| *Castor fiber* (beaver) | – | – | – | – | – | – | – | 3 [1] | 1 [1] | 3 [1] |
| *Canis lupus* (wolf) | – | – | – | – | 2 [1] | – | – | – | – | 2 [1] |
| *Vulpes/Alopex* (fox) | – | – | – | – | – | – | – | 1 [1] | – | – |
| Mustelidae | – | – | – | – | – | – | 1 [1] | – | – | – |
| *Ursus* sp. (indeterminate bear) | – | – | – | – | – | – | – | 1 [1] | 4 [1] | – |
| Rhinocerotidae | – | – | – | – | – | – | – | – | 5 [1] | 1 [1] |
| *Equus ferus* (horse) | 1 [1] | 4 [1] | 6 [1] | 2 [1] | 5 [1] | 2 [1] | 4 [1] | 79 [3] | 104 [3] | 8 [1] |
| *Cervus elaphus* (red deer) | 1 [1] | 2 [1] | 6[1] | 4 [1] | 7 [1] | 28 [2] | 16 [1] | 2197 [13] | 1834 [19] | 495 [10] |
| *Rangifer tarandus* (reindeer) | 2 [1] | 8 [1] | 507 [4] | 195 [2] | 1495 [19] | 45 [4] | 23 [2] | 22 [1] | 3 [1] | 4 [1] |
| *Capreolus capreolus* (roe deer) | 3 [1] | 13 [1] | 9 [1] | 13 [1] | 35 [2] | 42 [5] | 13 [1] | 353 [7] | 948 [7] | 135 [2] |
| *Bos/Bison* (aurochs/bison) | 16 [1] | 4 [1] | 1 [1] | – | 1 [1] | 25 [1] | 14 [1] | 26 [1] | 17 [1] | – |
| *Capra ibex* (ibex) | – | – | – | – | – | – | 1 [1] | 1 [1] | – | 1 [1] |
| *Sus scrofa* (boar) | – | – | – | – | 1 [1] | – | 1 [1] | 17 [3] | 26 [2] | 20 [3] |
| Aves | – | – | – | – | – | – | – | – | – | 1 [1] |
| Microfauna | – | – | – | – | – | – | – | 5 [1] | 5 [1] | – |
| Large carnivore | – | – | 1 | – | 1 | – | – | 2 | 4 | – |
| Small carnivore | – | – | – | – | – | – | – | 1 | 1 | – |
| Large cervid | 405 | 724 | 232 | 111 | 2141 | 2091 | 799 | 443 | 270 | 757 |
| Small cervid | 3 | 3 | 5 | – | – | 9 | 2 | 83 | 113 | 375 |
| Large bovid/horse | 70 | 60 | 6 | – | 22 | 80 | 31 | 286 | 321 | 62 |
| Body size 2 (30–100 kg) | – | – | – | – | – | – | – | – | 3 | 5 |
| Body size 3 (100–300 kg) | 39 | 16 | 12 | 5 | 4 | 59 | 18 | 210 | 411 | 89 |
| Body size 4 (300–1000 kg) | 35 | 3 | – | 1 | 8 | 20 | 11 | 188 | 380 | 36 |
| Indeterminate | 170 | 112 | 24 | 78 | 88 | 218 | 90 | 931 | 837 | 43 |
| Total | 745 | 949 | 809 | 409 | 3810 | 2619 | 1024 | 4849 | 5287 | 2038 |

[a]Numbers include piece-plotted bone plus sample of coarse water-screen fraction

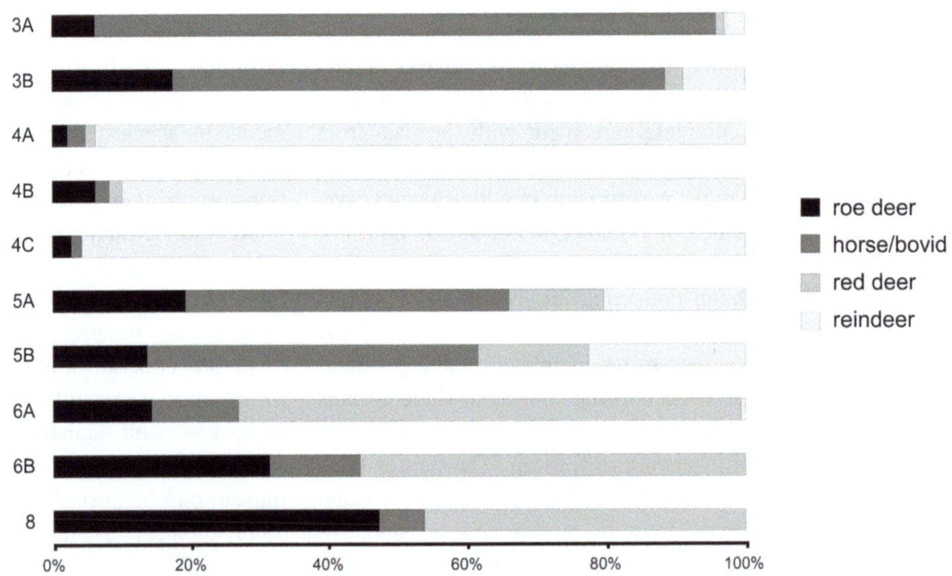


**Fig. 5.1** Relative abundances (%NISP) of primary prey species through the Pech IV sequence

ungulates represented by much smaller numbers. Considering the predominance of cervids, interpretations of the Pech IV archeofauna in terms of Neandertal subsistence focuses primarily on those assemblages (see Niven 2013). Samples of horse and large bovids were generally too small to provide insight on their exploitation as prey, but aspects of these assemblages are presented when the data were available.

Small mammals and birds are infrequent in this assemblage, but this is partly a factor of excavation methods (i.e., the 25-mm size cutoff for piece-plotting). Carnivores are also extremely rare.

## Small Mammals and Birds

Remains of small mammals and birds are rare in the Pech IV assemblage overall. This scarcity may be largely due to the size cutoff for collecting bone during excavation. Supporting this argument is the evidence of these species in the coarse water-screen material that was sampled from Layer 8. As mentioned earlier, one goal of examining that material was to assess the true presence or absence of small fauna, and results showed that they were indeed present, though not abundant.

Identified small mammals included hare (*Lepus* sp.), beaver (*Castor fiber*), and an indeterminate medium-sized raptor. All were found in Layer 8, although Layers 6A ($N$ = 3) and 6B ($N$ = 1) also yielded beaver remains. These animals are rarely recovered in Mousterian contexts, suggesting that they played little role in subsistence. However, an increasing number of Neandertal sites in Europe are producing remains of small game and birds. Whether this increase relates to better excavation techniques, preservation, or other factors remains to be seen.

The beaver remains are of particular interest. First, this animal's presence is an excellent indicator of wooded environs, as trees are essential for providing beavers with food and shelter. Not surprisingly, the fossil record of *Castor* is tightly associated with temperate climate regimes (Stuart 1982). Along with the abundance of other temperate taxa such as roe deer and boar, the presence of beaver in Layers 6 and 8 are consistent with the paleobotanical results and date ranges indicating warmer wooded environments.

In Layer 8, the second and third phalanges from beaver do not articulate but are presumably from the same individual. Both specimens are burned and cutmarks were recognized on the lateral aspect of the second phalanx. This may relate to skinning the animal, which could have represented a good source of warm fur but also a fat-rich food. The beaver remains from Layer 6 were not modified in any way but are presumed to have been acquired for the same reasons.

The raptor from Layer 8 is represented by a single third phalanx from the talon. Cutmarks are evident on the articular surface and visible with the naked eye, although a microscope was used to verify their presence on this unusual specimen. The cutmarks resemble those shown by Fiore et al. (2004: Fig. 2) on an eagle specimen from a Mousterian horizon at Fumane Cave, Italy. Similar specimens were found at the neighboring Pech I but in later (MTA Type B) deposits (Soressi et al. 2008; Rendu 2010). Morin and Laroulandie (2012) also presented examples of cutmarked eagle phalanges (talons) from the nearby Dordogne site of Combe Grenal and other regional Mousterian sites (Morin and Laroulandie 2012: Table 1). These authors suggested that the cutmarks resulted from removing the talon's sheath, though the purpose of doing so is unclear. Because no meat or other edible tissues are found in this part of the foot, removing the sheath might relate to other activities (see Peresani et al. 2011; Morin and Laroulandie 2012).

Although the evidence for small fauna and birds at Pech IV is not extensive, it nonetheless contributes to the growing number of Mousterian sites showing that Neandertals did occasionally exploit these animals for food and, perhaps, other purposes.

## Sex Ratios of Prey

Diagnostic skeletal parts or articular ends suitable for osteometrics are overall lacking across the sequence. Nonetheless, specimens from Layers 6 and 8 yielded limited information on the sex ratios of red deer. For example, one neonate and one fetal red deer represent at least two females among the 10 adult individuals in Layer 6A, and several antler fragments indicate at least one male. A small number of antler fragments but no fetal remains were recovered from Layer 6B. Skeletal elements in this layer were generally more robust than in the surrounding layers, likely reflecting more male red deer in 6B, but measurable bones and hence, quantifiable data, are lacking to say anything more precise. The best results come from Layer 8, where red deer canine teeth indicate six females and one male (following methods of d'Errico and Vanhaeren 2002). Remains of at least three fetal red deer also support an argument for a predominance of females.

## Seasonality and Age

Information on the season of occupation at Pech IV comes from fetal bone and teeth that can be aged from all layers except for 3A and 5B (Table 5.2). A combination of methods was used to address questions about the season of death from red deer, reindeer, and wild boar. An additional study using cementochronological methods was conducted by Martin (see below) on a select sample of teeth from red deer, roe deer, and reindeer from Layer 6 and Layer 5A. It should be noted that all attempts were made to associate specimens from the cementochronological study with those in the main assemblage to produce accurate MNIs. In other words, in many cases, a tooth used in one study (cementochronology) could be confidently associated with teeth used the other study.

## Analyzing Fetal Remains

Detailed data on fetal bone growth in domestic horses are found in Habermehl (1975). Precise measurements of long bone diaphyses for each week of growth allow us to determine the season of death from well-preserved archeological specimens. Similar osteometric data from domestic pig long bones at various stages of gestation are also available (Habermehl 1975: Table 14). Both wild boar and domestic pig have gestation periods of 110–120 days and give birth to 5–6 young in spring (Habermehl 1975, 1985). Red deer fetal specimens were aged using two sets of comparative fetal skeletons of modern *Cervus*. One skeleton was approximately half-term (4.5–5.0 months) and the other nearly full-term (7–8 months). Based on the gestation schedule of ~250 days for modern *Cervus*, with the mating season taking place in September–October and birth in May–June (Habermehl 1985), the comparative skeletons allowed age estimations of the Pech IV specimens.

## Dental Remains

Dental eruption and wear stages of partial tooth rows were evaluated for a season of death and age. These deciduous and permanent teeth from the maxilla and mandible were compared to the extensive sample of precisely aged dentition from modern caribou found in Miller (1974). Because *Rangifer* have seasonally restricted mating (autumn) and birthing (late May to early June) periods, their season of death can be determined using comparative specimens of known age.

## Season of Death Estimates

Season of death information from analysis of fetal bone and/or dentition are presented below in chronological order from earliest to latest (Table 5.2).

Red deer fetal remains ($N = 19$) from Layer 8 indicate hunting and site occupation during summer, fall, and winter. A single fetal boar radius is from a full-term or neonate who died in spring/early summer. These results show the site being visited during all seasons of the year during the Layer 8 occupations, with locally available red deer being hunted throughout.

Seasonal information from Layer 6 was estimated from four fetal red deer (6A: $N = 4$, 6B: $N = 1$) and two boar specimens (one each from 6A and 6B). Based on the methods outlined above, the red deer were hunted during spring and winter. The boar from 6B could not be aged with any confidence due to poor preservation, but the specimen from 6A was full-term or neonate, indicating death in spring/early summer. Overall, occupations represented in both sublayers of Layer 6 took place in all seasons except fall.

One reindeer mandibular third molar from Layer 5A could be aged to 18–22 months, indicating death in the winter/spring. This corresponds to three fetal reindeer bones from this layer that were assigned to late winter and spring. These individuals may have migrated early or simply been resident in the local area.

Layer 4 yielded robust samples for a season of death and age of reindeer. One individual aged 2–3 months old indicates death in summer, and one 18–24 months old points to winter/spring for Layer 4A. Two ca. 24-month olds in Layer 4B were hunted in spring/early summer. Layer 4C contained

**Table 5.2** Seasonality evidence across the Pech IV sequence, based on eruption and wear of teeth and fetal bone

| | Spring | Summer | Fall | Winter |
|---|---|---|---|---|
| 3A | No data | | | |
| 3B | | | | *Equus* sp. |
| 4A | *Rangifer* | *Rangifer* | | |
| 4B | *Rangifer* | | | |
| 4C | *Rangifer* | *Rangifer* | | |
| 5A | *Rangifer* | | | *Rangifer* |
| 5B | No data | | | |
| 6A | *Sus/Cervus* | | | *Cervus* |
| 6B | *Sus/Cervus* | *Sus scrofa* | | *Cervus* |
| 8 | *Sus scrofa* | *Cervus* | *Cervus* | *Sus/Cervus* |

a yearling (spring/early summer) and two reindeer slightly younger (winter/spring).

These results show that cave occupations and reindeer hunting were limited to a seasonally restricted window, which likely reflects the local availability of this prey resource only during this time period in spring and perhaps early summer, i.e., during their migration. Pleistocene *Rangifer* migrated long distances like their modern counterparts (Britton et al. 2009) and results from reindeer assemblages from other Quina Mousterian sites (e.g., Costamagno et al. 2006; Britton et al. 2011) also indicate hunting episodes restricted to reindeer migration periods in spring or fall. Together, these sites suggest that Neandertals took advantage of the local abundance of reindeer during restricted seasonal windows. More specifically, this implies that Pech IV was occupied for shorter durations in the cold periods of Layer 4A.

A single left femur shaft from a fetal horse was identified in Layer 3B. Metric data from Habermehl (1975) indicate this animal was 23–25 weeks (ca. half of gestation) into term at death, which would indicate winter.

## Seasonality Based on Cementochronology of a Sample of Cervids (Hélène Martin)

In addition to the observations above, a sample project was conducted to investigate seasonal signatures using skeletochronological analysis of tooth cementum increments of selected ungulate specimens. This method is based on the hypothesis that cementum growth follows predictable seasonal cycles with an alternation of a fast-growth deposit during the warm season and a slow-growth deposit during the cold season (e.g., Klevezal' and Kleinenberg 1969; Lieberman and Meadow 1992; Burke and Castanet 1995; Klevezal' 1996). The outermost increment, forming at the time of death, is assumed to give a precise estimation of the season of death.

Fifty thin sections were made from 50 teeth of 22 individuals from Layer 6A, 6B, and 5A (Table 5.3). Of these 50 samples, 17 thin sections from eight red deer, nine roe deer, and one reindeer were readable. Isolated teeth, as well as tooth rows, were sampled, and it should be noted that associated teeth were likely remaining in the main assemblage analyzed by Niven. In other words, both studies examined teeth from the same individuals in some cases.

Specimens from both layers reflected a similar pattern: red deer and roe deer were hunted from early summer through fall/late fall (Fig. 5.2). These seasons comprise different biological cycles of these cervids that must have played a role in their capture by Neandertals. Females separate and isolate themselves to give birth in spring. After a month, the cows and calves would have moved around more freely. Fall is the mating season, and herds of males would have gathered. Both situations would have offered opportunities for hunters to prey on vulnerable animals.

The cementochronology data both correspond to and contradict the seasonality evidence obtained from red deer in the main assemblages in Layer 6. Those data pointed to hunting of *Cervus* during spring, summer, and winter; no fall hunts were indicated by fetal bone or dental analysis. This discrepancy may simply show that the cementochronology sample included the only animals hunted in fall, although the number of samples in this analysis consistently showing this signal makes this unlikely. The evidence for seasonality from the main assemblage was quite sparse, consisting mostly of fetal red deer and boar. Aging these specimens precisely is difficult due to a lack of comparative material and the results are estimates. It is possible that the fetal bones thought to represent summer or winter predation may actually reflect hunting that stretched into fall or were in late fall, respectively. The most important point is that cementochronology results indicate that red deer and roe deer were also hunted in fall, which means Layer 6 reflects occupations throughout most if not all of the year.

The single reindeer tooth from Layer 5A indicated death in spring. This corresponds to the evidence obtained from the full assemblage from this layer. As discussed above, the seasonality evidence from Layer 5A and Layer 4 suggests that reindeer were hunted during their spring migration.

In addition to assessing season of death on this sample of teeth, individual ages of red deer and roe deer were also estimated. These data indicate a predominance of prime aged animals, and this observation matches well with other estimates (see below). All individuals that yielded readable thin sections were between 3 and 6 years old (Table 5.3).

## Age Estimates

Material that could be aged is relatively sparse in the Pech IV assemblages, both in terms of teeth as well as a unfused skeletal bone of all prey taxa. The most robust datasets come from large samples of red deer and reindeer from Layers 4 and 6. Smaller numbers of these taxa, along with roe deer, horse, and large bovid, provided age information in all layers except for Layer 3, which contained no teeth that could be assigned a specific age.

Mandibular and maxillary teeth of red deer and reindeer were aged based on eruption state and/or wear stages (Miller 1974; Carter 2006) and placed into three age groups: juvenile (from birth to just before loss of deciduous teeth), prime adult (reproductive years of life), and old adult (tooth crown is >50% worn) (following Stiner 1990). This tripartite system allows us to construct a mortality profile for evaluating whether Neandertals targeted certain age groups of prey, and

**Table 5.3** Summary of results from cementochronological study of cervid teeth from Layers 6A, 6B, and 5A

| Specimen | Level | Taxon | ID | Age (year) | Season |
|---|---|---|---|---|---|
| F13-2397 | 6A | Roe | $LM_3$ | 3.5 | Spr–Fall |
| F14-3316 | 6A | Roe | $LM_{1-3}$ | 4 | Spr–Fall |
| F14-3318 | 6A | Roe | $LP_{2-4}$, $M_{1-2}$ | 3 | Late Fall–Early Winter |
| G14-1012 | 6A | Roe | $RM_3$ | 4.5 | Spr–Fall |
| H12-310 | 6A | Red | $LP^3$ | 3.5 | Spr–Fall |
| F14-3273 | 6A | Red | $LP^3$? | – | Late Fall–Early Winter |
| G14-951 | 6A | Red | $LP^2$ | – | Spr–Fall |
| E11-7418 | 6B | Roe | $LM_{2-3}$ | 4 | Spr–Fall |
| C13-1478 | 6B | Roe | $LM_{1-3}$ | 3.5 | Late Fall–Early Winter |
| G12-883 | 6B | Roe | $RM_{2-3}$ | – | Spr–Fall |
| G13-1831 | 6B | Roe | $RM_3$ | 4 | Spr–Fall |
| G14-1129 | 6B | Roe | $RM_2$ | – | Late Fall–Early Winter |
| D11-9341 | 6B | Red | $LM_{2-3}$ | 4.5 | Late Fall–Early Winter |
| G13-1236 | 6B | Red | $RM_3$ | 5.5 | Late Fall–Early Winter |
| E11-6907 | 6B | Red | $RM^{1-2}$ | – | Late Fall–Early Winter |
| G13-1166 | 6B | Red | $RI_1$ | 3.5 | Late Fall–Early Winter |
| F14-3388 | 6B | Red | $NM^?$ | – | Spr–Fall |
| E11-5134 | 5A | Rein | $LM_{1-2}$ | 4 | Spring |

See also Fig. 5.2

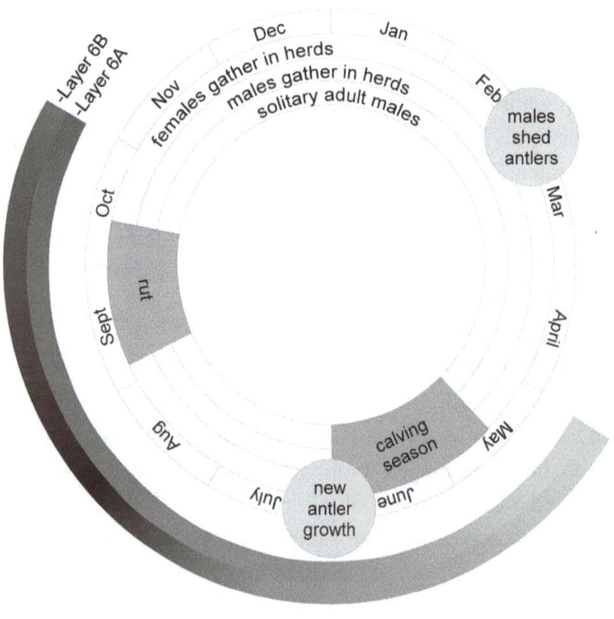


**Fig. 5.2** Illustration of the yearly behavioral cycle of cervids based on modern *Cervus elaphus* and *Capreolus capreolus*. *Shaded bands* along the exterior of the figure represent a season of death time spans for these cervids from Layers 6A and 6B based on cementochronological analysis. Illustration by H. Martin

in turn, whether their hunting strategies remained consistent or varied across the Pech IV sequence (Table 5.4).

Layer 8 yielded just three red deer teeth that could be assigned ages. Potentially from one individual, this animal was estimated to be a prime adult. However, if we consider the three fetal animals, there must have been at least three prime adults who carried them.

A predominance of prime adults is evident in the red deer assemblages from Layer 6, with five individuals each in 6A and 6B, and just a few juveniles and old adults (Table 5.4). This taxon was likely available locally in most or all seasons, considering our evidence for multi-season hunting in Layer 6. Strontium isotope data indicate that Pleistocene red deer were not highly migratory like reindeer, but made limited seasonal movements across the regional landscape (Pellegrini et al. 2008). A seasonally reliable prey animal near Pech IV may have influenced their frequent capture by Neandertals. A similar age pattern is seen among the roe deer from Layer 6, with prime adults dominating both assemblages. In 6A, one juvenile and six prime adults are represented, whereas 6B yielded seven prime adults. At least three horses were found in both Layers 6A and 6B, including three prime adults, and two prime adults and one juvenile, respectively. No seasonal data were obtained from these specimens. Finally, one prime adult large bovid was represented in 6A.

Layer 5 contained few ageable specimens, but a single reindeer and roe deer individuals from 5B were all prime adults. Two prime adult reindeer and one roe deer were represented in Layer 5A.

Interestingly, the data from Layer 4 are varied, with more juveniles in Layer 4A, but prime adults in 4B, and no clear pattern in 4C. Because the samples are limited

**Table 5.4** Summary of age groups for four prey taxa, based on teeth

| | 4A | 4B | 4C | 5A | 5B | 6A | 6B | 8 |
|---|---|---|---|---|---|---|---|---|
| *Cervus* | | | | | | | | |
| Juvenile | | | | | | 2 | 1 | 0 |
| Prime age | | | | | | 5 | 5 | 1 |
| Old adult | | | | | | 1 | 0 | 0 |
| *Rangifer* | | | | | | | | |
| Juvenile | 3 | | 2 | | | | | |
| Prime age | 1 | | 1 | 2 | 1 | | | |
| Old adult | | 3 | 1 | | | | | |
| *Capreolus* | | | | | | | | |
| Juvenile | | | | | | 1 | | |
| Prime age | | | | 1 | 1 | 6 | 7 | |
| Old adult | | | | | | | | |
| *Equus* | | | | | | | | |
| Juvenile | | | | | | | 1 | |
| Prime age | | | | | | 3 | 2 | |
| Old adult | | | | | | | | |

No ageable teeth were found in Layer 3

(3–4 individuals in each layer), we cannot assume any "focus" on one age group. Nonetheless, the different patterns in each layer might be unique to hunting this highly migratory taxon. Assuming that Pleistocene *Rangifer* migrations were of a similar character to those of today (see Britton et al. 2011), then they would have comprised several stages lasting one to two months each. These included a calving period, pulses of movement, pauses, the rut, and occasional splits in herd structure (i.e., males separating from females with calves) (Burch 1972; Enloe and David 1997). The spring and summer signals might reflect hunting of females and calves (4A) and mixed herds (4C). The prime adults hunted in summer from 4B may point to a migratory phase when Neandertals encountered barren females and males nearby, with the females and calves residing elsewhere. Overall, the variability seen in seasonal hunting in Layer 4 is unique to these cold, reindeer-dominated assemblages.

## Skeletal Element Frequencies

We assume that Pech IV was not the actual hunting location, but was instead the place to which Neandertals brought prey animals for processing and consumption. Being the end-point of prey transport (Stiner 1994), caves thus represent a unique opportunity for evaluating transport decisions by hominins. Ethnographic studies show that modern hunter–gatherers often selectively transport certain carcass portions or elements based on their nutritional yield (e.g., Binford 1978; Bunn et al. 1988; O'Connell et al. 1988; Bartram 1993), despite the potential desire to utilize as much of the carcass as possible. These decisions are influenced by various factors including body size, distance to the processing and/or consumption site, weather, topography, and the number of hunters involved (Bunn et al. 1988). Therefore, the relative abundances of skeletal elements from ungulate prey are analyzed to evaluate prey transport strategies of Neandertals at Pech IV. Sufficient samples for such evaluation were not available from all taxa, so this analysis focused on red deer, reindeer, and roe deer. Horse data were available (albeit limited) from Layers 6A and 6B. The small assemblages from Layers 3A, 3B, 4B, and 5B are excluded from the discussion.

It has been well established that the density of bone determines its survivorship in archeological assemblages (e.g., Brain 1969; Lyman 1984; Lam et al. 1998). Therefore, the degree of post-depositional bone attrition must be assessed before we can recognize and interpret economic choices of prehistoric humans from the skeletal element abundances in an assemblage.

For this study, the process began by plotting computed tomography (CT) bone density values from the relevant taxon (from Lam et al. 1999) against the relative abundance (i.e., % MAU) of skeletal elements from the Pech IV assemblages (excluding the neurocranium, carpals, small tarsals). Following Marean and Cleghorn (2003), elements were grouped into high-survival and low-survival sets. Comprising the ribs, vertebrae, pelves, tarsals, carpals, and phalanges of small ungulates, the low-survival set should reflect the degree of bone destruction by factors such as density-mediated attrition or carnivore consumption. High-survival elements include the long bones, mandible, and cranium.

Overall, the posterior axial skeleton (vertebrae, ribs, pelvis) is strongly underrepresented in all layers (Fig. 5.3). Significant and positive correlations are seen between bone density values and element abundance in Layers 4A, 4C, 5A, and 6A, suggesting that the bias is simply taphonomic. It is also possible that this pattern is the result of decisions by hunters to abandon the bulky rib and vertebral sections after removing the meat, although these elements were present in some layers in very low numbers, suggesting that they had been transported but deteriorated after deposition. The abundance of heads varies but mandibles are more frequent than crania except in Layer 8.

Whether the discrepancies in skeletal element abundances of some taxa are the result of transport decisions based on nutritional value can be explored through utility indices. Ratio MAU data from Pech IV are plotted against values for standardized food utility (S)FUI (Metcalfe and Jones 1988) and standardized marrow cavity volume (Binford 1978) of modern *Rangifer* (also used for red deer, as no data from *Cervus* exist). The food utility index averages values for meat, marrow, and overall nutritional benefit of skeletal elements, whereas the marrow index is based on the actual volume of each marrow-yielding long bone shaft; both indices are based on averaged data. Because the vertebrae and ribs are so poorly represented in the Pech IV assemblages, we limit our analysis to the "high survival set" of skeletal elements (Marean and Cleghorn 2003). These include elements with thick cortical bone but lacking fragile cancellous bone and comprise all of the long bone midshafts, mandibles, and cranium (Marean and Cleghorn 2003: 34). The increased likelihood to preserve in the fossil record means these skeletal parts represent the best opportunity for evaluating human subsistence strategies.

A significant and positive correlation is seen between overall food utility and skeletal abundance of reindeer in Layers 4A, 4C, and 5A; and red deer in Layer 6A (Fig. 5.4). Only the six long bones are included here, which certainly biases the results; the entire skeleton is normally considered in the S(FUI). These results at least suggest that no preferential transport of specific elements was practiced. However, the other utility indices suggest otherwise.

No statistical relationship exists between marrow cavity volume and ratio MAUs for the long bones and mandible in any layer. This is unexpected, as the overall frequency of these elements is high, particularly the tibia and non-meat bearing metatarsal (Fig. 5.4). This might be partly related to weaknesses in the return rate data on which these utility indices are based, which do not consider every variable involved in prey transport (Egeland and Byerly 2005). Subsequently, it is valuable to explore more closely the nutritional qualities of marrow specific to skeletal element and season (Binford 1978; Morin 2007).

In a reevaluation of Binford's marrow index, Morin (2007) confirmed that the overall quantity of unsaturated fatty acids was the most critical factor in whether an element was chosen for marrow processing by the Nunamiut, an Alaskan native group studied by Binford (1978). Recalculating the relevant data allowed Morin (2007: Table 4) to establish the Unsaturated Marrow Index (UMI) to replace the marrow cavity volume index. When plotted against the Pech IV ratio MAUs, the UMI shows significant correlations for the assemblages in Layer 4C and 6A perhaps relating to the higher rank of the femur in the UMI. However, neither index helps us understand the pattern in Layer 4C, where the frequencies of tibiae (MNE = 34) and metatarsals (MNE = 39) are extraordinarily high—by a factor of two or more—in comparison to equally robust humeri (MNE = 17) and radii (MNE = 11). These numbers suggest that the richest marrow bones in the reindeer were selectively transported to the cave in greater quantity than the upper limb elements for further processing. This might relate to the poor physical condition of some of the reindeer, in light of their having been hunted in spring when they tend to be undernourished (Burch 1972; Spiess 1979). Fat stores such as marrow become depleted in such situations, beginning with the upper limb elements (humerus and femur) (Speth 1983). The tibia and metatarsal retain fat longer (Binford 1978; Morin 2007), which may explain the higher abundance of these elements among the Layer 4C reindeer assemblage. Similar proximal-to-distal fat depletion processes affect other ungulates like bison (e.g., Speth 1983), so similar economic-based decisions may have influenced the greater abundances of tibiae (MNE = 29) and metatarsals (MNE = 20) from red deer in Layer 6B versus the humerus (MNE = 16).

It is important to consider the abundance of crania, as they are high-survival body parts; they are sometimes low in nutritional utility and bulky to transport, but these factors are highly dependent on taxon. For example, a bison skull is far heavier and more unwieldy than a red deer skull to transport. However, despite its bulk, an equid skull may warrant transport and extensive processing compared to other ungulates. Modern African hunter–gatherers value zebra crania because they are easily opened to access tissues (O'Connell et al. 1988; Lupo 1998), contain abundant tissue even after initial processing and removal of external flesh and tongue, especially in the large nasal cavity (Lupo 1998), and fat depletion of the skull is less than in other skeletal parts (Stiner 1994; Lupo 1998). Additional support for the economic utility of equid crania involves the higher amounts of fatty acids in their meat compared to other ungulates Levine (1998).

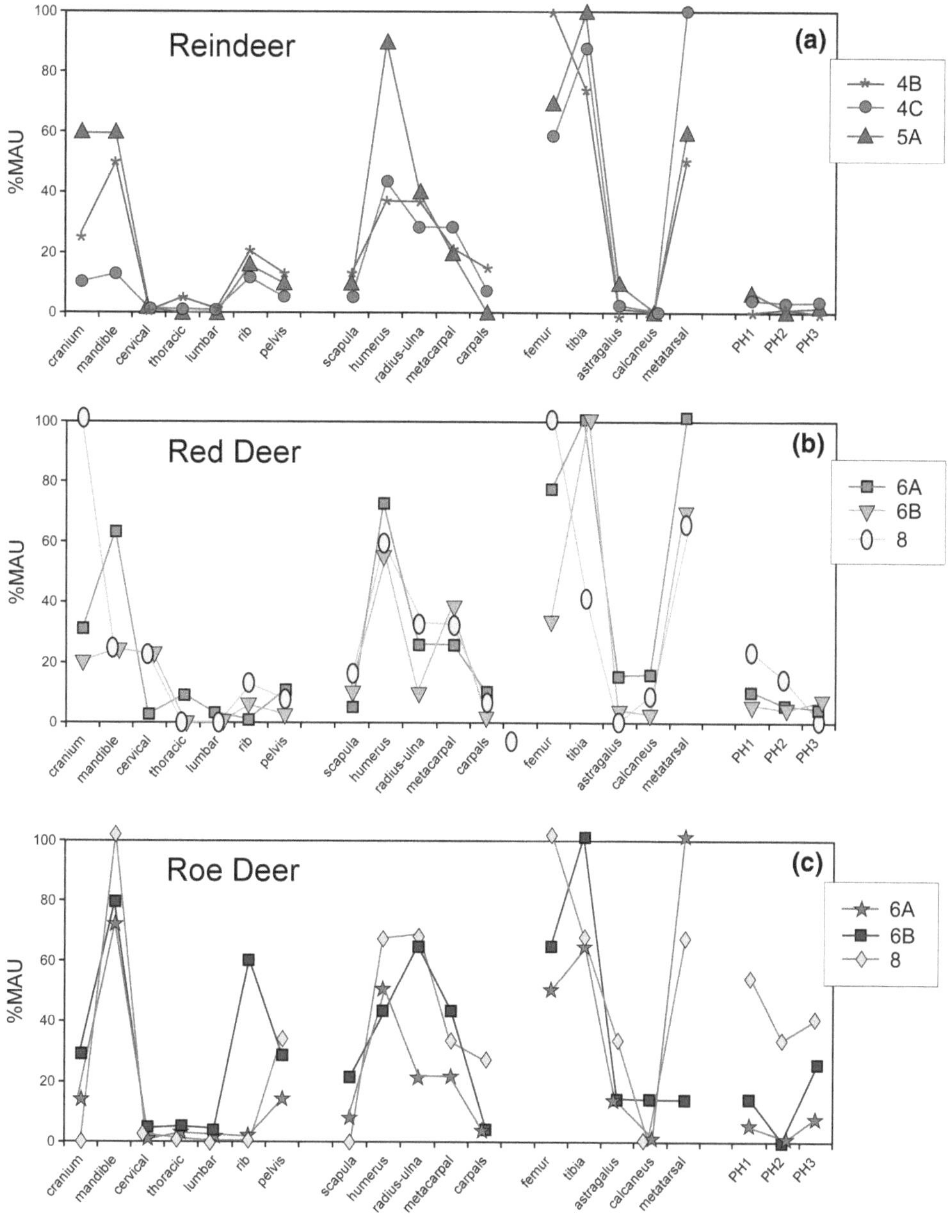


**Fig. 5.3** Skeletal element abundance expressed as ratio MAU (minimum animal units). Reindeer from Layers 4A, 4C, and 5A are plotted in (**a**), red deer from Layers 6A, 6B, and 8 in (**b**), and roe deer from Layers 6A, 6B, and 8 in (**c**)

Crania and mandibles, as well as the bony parts versus teeth of each element, are represented quite differently across the sequence. Overall, teeth from the maxilla and mandible are more numerous than bony parts of those elements, and MNIs for some taxa by layer were based on teeth. Lastly, some layers yielded few teeth of any kind, whereas others had a predominance of teeth for certain taxa. When quantifying teeth, only complete or nearly complete specimens were considered; fragments were counted but do not figure in the overall totals. Age was considered when quantifying MNIs; for example, two left and two right mandibular M3s would yield an MNI of two, but if two of the teeth are unworn and two are not, then the true number of individuals would be four.

Only Layers 4, 6, and 8 contained adequate samples of crania/mandibles and teeth to allow for discussion of possible subsistence decisions of Neandertals. The reindeer assemblages in Layer 4 show low frequencies of heads; mandibles are behind the 50.0 ratio MAU value in Layer 4A, although only four individuals are represented here, so this abundance might not be as significant as that in Layer 4C that contains 19 individuals.

Red deer mandibles are frequent in Layer 6A, suggesting that the small amount of marrow available this element was exploited along with the other marrow-rich elements in this layer. Crania are half as abundant. Crania and mandibles are about equally (and poorly) represented in Layer 6B, comprising only 25% of the MAU. Roe deer show more equal

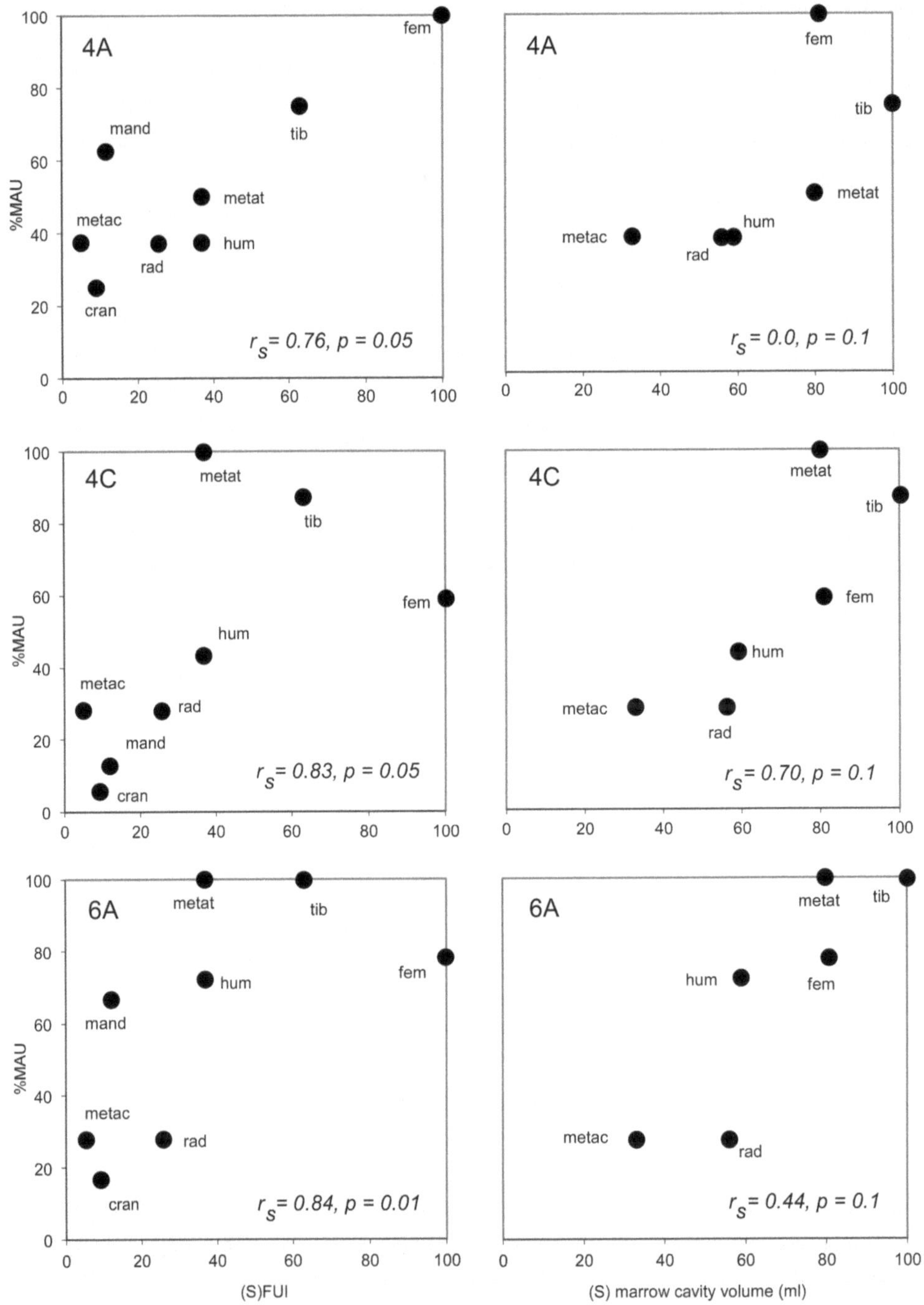


**Fig. 5.4** Relationship between ratio MAU and the standardized food utility index ([S]FUI) and standardized (S) marrow cavity volume for reindeer in Layers 4A and 4C and red deer in Layer 6A. Modified from Niven (2013: Fig. 9.2)

abundances of heads versus postcranial skeleton, with MNIs in 6A and 6B based on mandibular M3 (Fig. 5.3c). These layers offer the best opportunity to evaluate utilization of horse crania, as these elements are well-represented. At least three horses were found in both Layers 6A and 6B, with one juvenile in 6B. Postcranial element MNEs were nearly equal to crania but the presence of these large bulky skeletal parts may well be due to their nutritional yield, as discussed above.

Red deer and roe deer from Layer 8 are equally represented by heads and postcranial elements (Fig. 5.3b, c), suggesting that entire carcasses were transported to the cave for processing and consumption. In this case, the fire was presumably involved in these activities and may have facilitated different and/or more efficient processing for maximal nutritional yield. No horse cranial remains were recovered in this layer.

In summary, relative abundances of the high-survival set of skeletal elements (i.e., the long bones and heads) are consistently higher among all main prey taxa and across the Pech IV assemblages overall. Even the somewhat robust (but still considered low-survival; Marean and Cleghorn 2003) scapula and pelvis are poorly represented, indicating that these elements were rarely transported. Carpals, tarsals, and patellae are rare. Their underrepresentation is likely not taphonomic or a result of collection bias during excavation, as these bones are generally robust (an exception being the patella), diagnostic, and above the size cutoff for piece-plotting. Therefore, their abundance may relate to anthropogenic factors, namely that they were rarely articulated to their respective long bones because the limbs were disarticulated and defleshed before transport to the cave. This corroborates evidence for the layers in which we see the selective transport of the richest marrow bones. Layer 4C shows the most distinctive discrepancy between the relative abundance of certain skeletal elements, pointing to different carcass transport strategies during the cold-period occupations.

## Carcass Butchery and Utilization

Neandertals' utilization of ungulate prey is evaluated through frequencies of stone tool cutmarks, percussion marks, and fresh (i.e., green or spiral) breakage. The actual number of cutmark striations or percussion marks was not counted, but instead, the data presented here involve the NISP bearing one or more modifications. Data are summarized in %NISP (Table 5.5).

Distinctive patterns in cutmark abundance are seen across the faunal sequence. In general, the reindeer-dominated assemblages show the highest frequencies (Fig. 5.5a; Table 5.6). For example, cutmarks are more than twice as abundant on elements in Layers 4A, 4C, and 5A compared to the lower layers. Since articular ends of long bones are scarce in the assemblages overall, the majority of cutmarks are found on the shaft portions, indicating the removal of meat and tissue (e.g., Binford 1981; Nilssen 2000). The reasons behind these high frequencies of cutmarked reindeer elements are difficult to explain, particularly because cutmark production and abundance are not fully understood (e.g., Egeland 2003). If we compare cutmark frequencies from the Quina Mousterian Layer 4A to two other known Quina reindeer assemblages, the general pattern still shows high numbers at Pech IV; for example, Jonzac exhibits 20–25% cutmarks on the long bones (Niven et al. 2012). However, cutmark frequencies in Layer 9 of Les Pradelles range between 30 and 60% (Costamagno et al. 2006).

Differences in cutmark frequencies are noted between specific skeletal elements or between the meatier upper limbs and meat-poor lower limbs. The red deer assemblages in Layers 6 and 8 (Fig. 5.5b) show generally similar cutmark abundances. This pattern also applies to the reindeer in Layer 4C but not in 4A or 4B. The latter assemblages show different utilization of the meatier upper elements (humerus, femur) versus the lower limb elements with minimal muscle mass; perhaps these discrepancies reflect carcass condition (e.g., stiff versus supple) or alternatively, variations in morphology among each element that result in more or fewer cutmarks regardless of butchery effort (Egeland 2003).

The generally lower cutmark frequencies on the red deer-dominated assemblages in Layers 6 and 8 cannot be explained by bone surface preservation or other factors related to identification, even in the case of Layer 8 with its evidence for burning. When the cutmark frequencies between red deer and roe deer from 6A and 6B are evaluated in each of these three assemblages, little difference is evident. Frequencies are slightly higher among red deer in Layers 6A (red deer 29.9%, roe deer 25.6%) and 6B (red deer 27.7%, roe deer 23.9%) but opposite in Layer 8 (red deer 23.8%, roe deer 25.0%). Overall frequencies between these three assemblages are also consistent. Some discrepancies are seen between 6A and 6B, for both taxa, across the skeleton; for example, the neck portions from 6B are more cutmarked than in 6A (Fig. 5.6). A similar pattern is evident between taxa in 8. This may be related to the bias against these elements overall (i.e., if axial elements had been better preserved in 6A or for roe deer in 8, the pattern may have been similar). Cutmark frequencies are fairly consistent across the skeleton in both taxa. In summary, little variability is evident in cutmark frequencies between red deer and roe deer in any of the three assemblages in which they predominate.

Percussion impact marks are also abundant across the extensive cervid assemblages (Fig. 5.7). These distinctive modifications are produced when a butcher uses a hard hammer such as a cobblestone to break open the long bone for accessing the marrow. Some of these marks may be lost or obscured by fracture surfaces in the bone, which may be a factor in the low frequency of percussion impacts in the Layer 8 assemblage that was heavily fractured by burning and trampling. Lower frequencies in 6A and 6B could also be the result of subsequent breakage but they might also reflect various strategies of processing red deer long bones for marrow. Excluding the Layer 8 assemblage, which was affected by trampling that potentially biased the results, assemblages in Layer 6 show the smallest mean fragment size (50–52 mm) of freshly broken long bones across the sequence (Table 5.7). Considering that ethnoarcheological data on ungulate carcass processing from the Kua in Botswana revealed a distinctive pattern in which long bone fragmentation increased with time spent at the processing site (Bartram 1993: 86), we should reevaluate the importance of fragmentation along with bone surface modification data to better address questions on carcass utilization and site use (Egeland and Byerly 2005).

Fresh, spiral breakage of long bones is consistently high across the cervid assemblages but again, the

**Table 5.5** Comprehensive summary of bone modifications on red deer and reindeer across the six long bones expressed as NISP and % (modification)

| Level | NISP | Cut | %cut | Percussion | %perc | Green brk | %green |
|---|---|---|---|---|---|---|---|
| 3A | 52 | 25 | 48.1 | 6 | 12.5 | 43 | 81.1 |
| 3B | 121 | 52 | 43 | 8 | 7 | 112 | 92.6 |
| 4A | 118 | 60 | 50.8 | 21 | 17.8 | 95 | 80.5 |
| 4B | 69 | 34 | 49.3 | 9 | 13 | 57 | 82.6 |
| 4C | 581 | 378 | 65.1 | 114 | 19.6 | 522 | 90.5 |
| 5A | 221 | 160 | 72.4 | 42 | 19 | 184 | 83.3 |
| 5B | 107 | 45 | 42.1 | 11 | 10.3 | 70 | 65.4 |
| 6A | 510 | 108 | 21.2 | 62 | 12.2 | 392 | 76.9 |
| 6B | 508 | 137 | 26.9 | 46 | 9.1 | 377 | 74.2 |
| 8 | 293 | 61 | 20.8 | 7 | 2.4 | 230 | 78.5 |

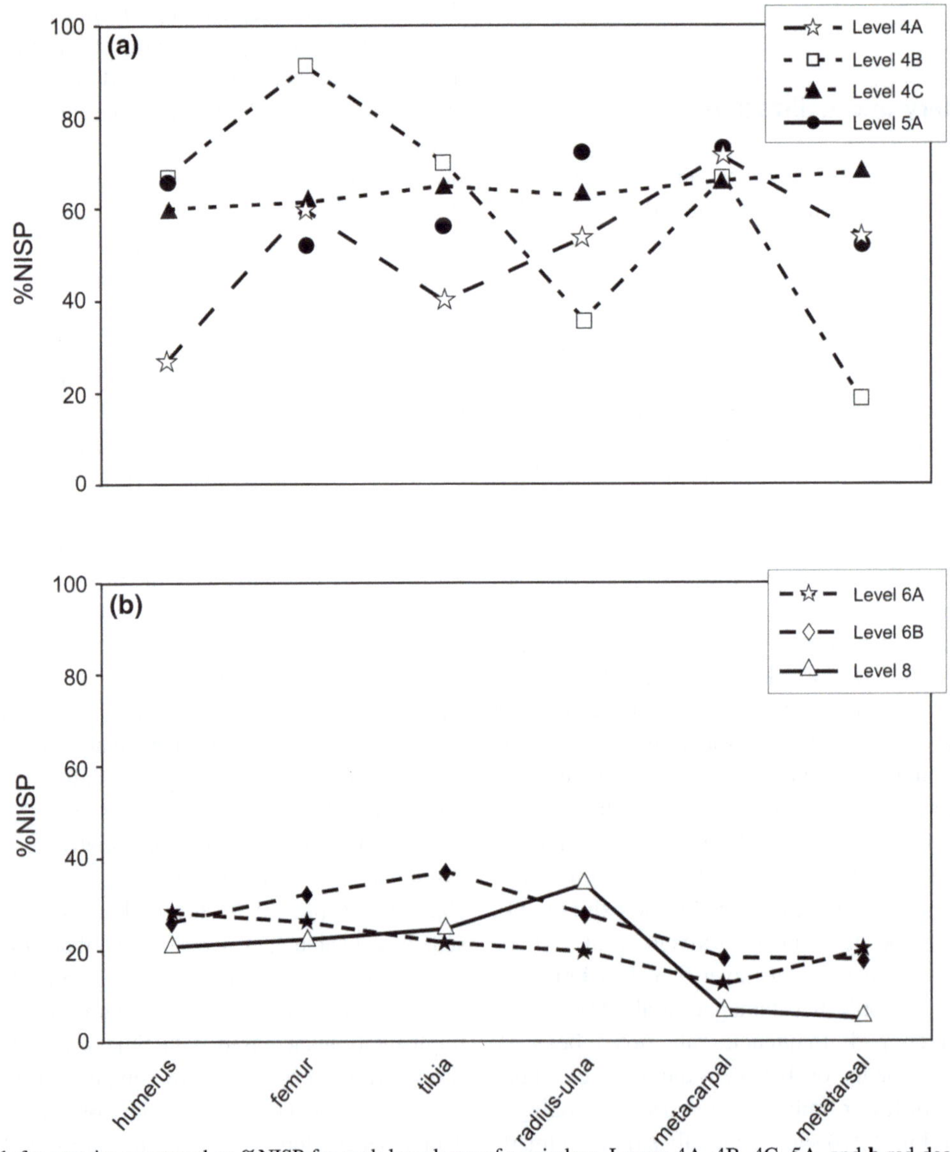


**Fig. 5.5** Cutmark frequencies expressed as %NISP for each long bone of **a** reindeer, Layers 4A, 4B, 4C, 5A, and **b** red deer, Layers 6A, 6B, 8

**Table 5.6** Summary of cutmark and percussion frequencies by skeletal element (long bones only) expressed as NISP and % (modification)

| Level | 4A | | | 4B | | | 4C | | | 5A | | | 6A | | | 6B | | | 8 | | |
|---|---|---|---|---|---|---|---|---|---|---|---|---|---|---|---|---|---|---|---|---|---|
| | NISP | % c | %p | NISP | % c | %p | NISP | % c | %p | NISP | % c | %p | NISP | % c | %p | NISP | % c | %p | NISP | %c | % p |
| Humerus | 11 | 27 | 55 | 9 | 67 | 22 | 50 | 60 | 26 | 44 | 66 | 18.2 | 64 | 28 | 20 | 81 | 27 | 15 | 51 | 22 | 0 |
| Femur | 20 | 60 | 30 | 11 | 91 | 27 | 59 | 61 | 27 | 29 | 52 | 13.8 | 66 | 26 | 12 | 65 | 32 | 9.2 | 52 | 23 | 3.8 |
| Tibia | 20 | 40 | 15 | 10 | 70 | 10 | 108 | 65 | 32 | 9 | 56 | 55.6 | 78 | 22 | 15 | 104 | 38 | 6.7 | 60 | 25 | 5 |
| Radius-ulna | 28 | 54 | 3.6 | 14 | 36 | 7.1 | 100 | 62 | 20 | 33 | 72 | 42.4 | 66 | 20 | 4.5 | 68 | 28 | 5.9 | 52 | 35 | 0 |
| Metacarpal | 7 | 71 | 57 | 3 | 67 | 33 | 42 | 67 | 31 | 37 | 73 | 32.4 | 39 | 13 | 28 | 85 | 19 | 7.1 | 27 | 7.4 | 0 |
| Metatarsal | 32 | 53 | 3.1 | 22 | 18 | 4.5 | 222 | 69 | 7.7 | 67 | 52 | 3 | 197 | 19 | 7.6 | 105 | 19 | 11 | 51 | 5.9 | 3.9 |

reindeer-dominated assemblages show a slightly different pattern. The frequency of green breakage is higher and long bone fragments are slightly longer. It is also notable that the frequencies of spiral breaks, cutmarks, and percussion marks increase and decrease more or less consistently in these assemblages (Fig. 5.8). The implication of these potential correlations is unclear, but the bone modification data together indicate that butchering and processing of reindeer in Layers 4 and 5 differed from how red deer were handled in Layers 6 and 8. The significantly higher cutmark frequencies in Layers 4A, 4C, and 5A may reflect more intensive processing of elements for their food yield, but the time investment in these activities was brief compared to the red deer-dominated layers if we presume that larger fragment size implies less time spent on processing (Bartram 1993).

The correlation between relative abundances of cutmarks and percussion impacts across the entire stratigraphic sequence was explored by plotting these data for all large cervids by layer. The highest frequencies of each modification served as the standard (100%) from which the other datasets were scaled. Figure 5.9 shows several correlations: overall frequencies of both cut and percussion marks are seen for all layers except 3B, 6A, and 8. This graph also emphasizes the abundance of butchery signatures in the reindeer assemblages, supporting our hypothesis that more intensive processing took place in these cold-period occupations. Although the true implications of cutmark and percussion impact frequencies remain unclear, it is remarkable that both signatures were often consistently similar in the Pech IV sequence.

Along the same lines, we explored possible correlations between cutmark frequencies and frequency of formal tools across the sequence. The highest frequencies of formal tools occur in 4A, 4C, and 5A (reindeer/cold), but similarly, tool abundances in the Asinipodian 6A–6B are not reflected in butchery traces (Fig. 5.10). Increased tool production in Layer 4 seems to correspond to the more intensive butchery of reindeer, but not necessarily with red deer. As previously discussed, we do see variable butchering strategies and intensity between the cold- and temperate period assemblages; the additional support for these arguments from the stone tool data are difficult to explain but do suggest that correlations between the lithics and fauna are worthy of further study.

## Summary

The long stratigraphic sequence at Pech IV provides a unique opportunity for evaluating Neandertal subsistence behavior and site use over a long period of time and within variable climate regimes. This section summarizes the data presented above, beginning with the earliest occupations, which allows us to reconstruct the ecological and anthropological changes reflected in the faunal record.

## Layer 8

Several lines of evidence point to these occupations taking place in generally temperate conditions with forested landscapes. Red deer were the predominant prey, with much smaller numbers of boar, roe deer, ibex, reindeer, and horse. Faunal data indicate that the cave was occupied repeatedly during most or all of the year, suggesting that at least red deer were available locally throughout the year. In conjunction with clear evidence for the use of fire in this layer and the production of tools on local raw materials, the faunal data point to longer or more frequent occupations.

The most distinctive aspect of Layer 8 is the extensive use of fire. Numerous lines of evidence indicate that animal bone was intentionally burned in the combustion features and a full report on this layer was published elsewhere (Dibble et al. 2009). However, a brief summary is useful here to understand the nature of the faunal assemblage that was significantly affected by burning. Of the piece-plotted bone and a sample of coarse water-screen bone subjected to special analysis, 54% was burned. The macroscopic evaluation showed that the majority was carbonized, i.e., subjected to low-temperature fires; microscopic FTIR analysis of burned bone showed similar results.

Our analysis of the burned bone shows several things. First, we attempted to evaluate whether the fire was used for cooking of red deer products. According to Speth and Tchernov (2007), evidence for cooking can be seen in different degrees of burning by anatomical part and in some cases, by a portion of the element. The Layer 8 red deer showed comparable percentages of burning across the skeleton, with slightly higher frequencies on the feet. Thus, the results are inconclusive for addressing questions of cooking activities. Second, we explored the question of whether the abundance of burned bone in the hearths was related to one or a combination of the following: that bone was being intentionally used as a fuel source; that it was burned incidentally (fires constructed on top of previously discarded bone); that bone being burned for site maintenance reasons; or whether it represents a combination of bone burned as fuel and as one aspect of site maintenance (i.e., removing bone refuse from living space; see Speth 2006). Plugging our data into a statistical model designed by Costamagno et al. (2009) to address this question specifically showed that Layer 8 reflects the last scenario. The abundance of lithic refuse in the hearths supports this argument as well. A more detailed discussion of these issues can be found in Dibble et al. (2009: 208–210).

The unique record from Layer 8 involving the controlled use of fire and use of bone as fuel adds to the list of other regional Mousterian sites with similar evidence, including Roc de Marsal (Aldeias et al. 2012; Goldberg et al. 2012; Sandgathe et al. 2011), Saint-Césaire (Morin 2004) and La Quina (Chase 1999).

## Layer 6

Minus the abundant evidence of fire present in Layer 8 (though still with high frequencies of burned flints), Layer 6 shows a similar pattern of occupation. The faunal composition is nearly identical and indicates similar temperate and

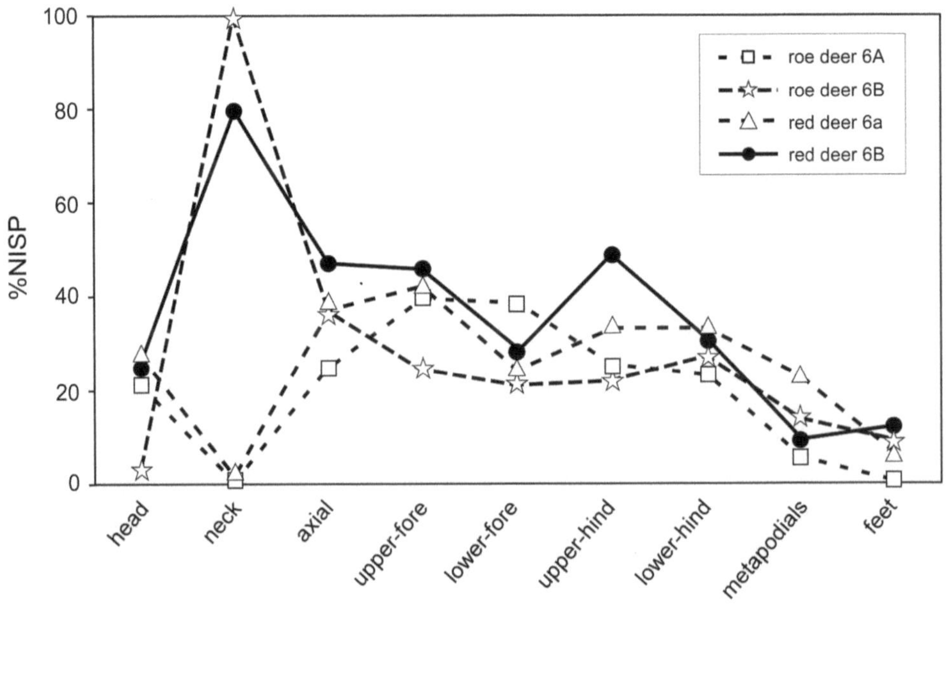


**Fig. 5.6** Cutmark frequencies expressed as %NISP for the entire skeleton of red deer and roe deer from Layers 6A and 6B

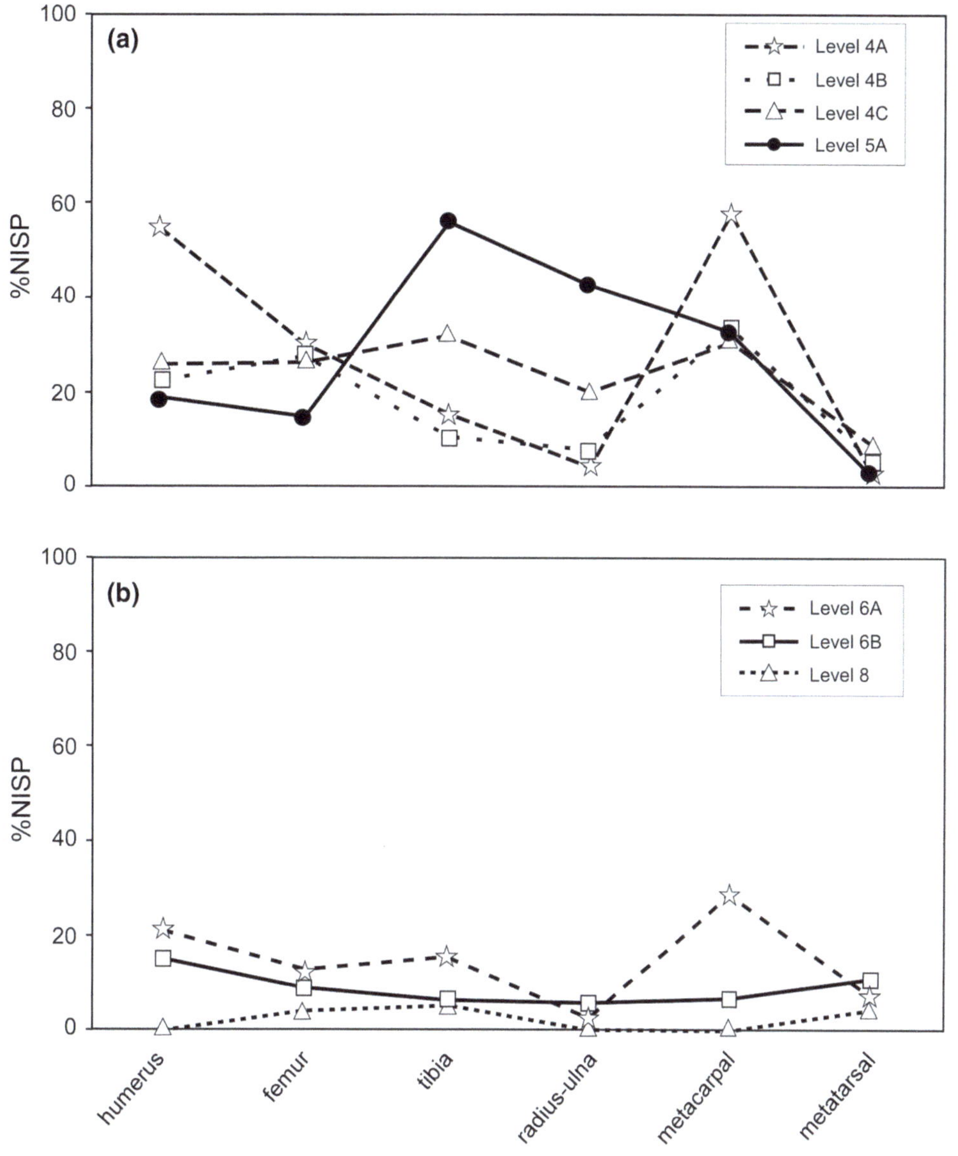


**Fig. 5.7** Percussion impact frequencies expressed as %NISP for each long bone of **a** reindeer, Layers 4A, 4B, 4C, 5A and **b** red deer, Layers 6A, 6B, 8

**Table 5.7** Summary of fragment length (in mm) of freshly broken long bones of large cervids through the Pech IV sequence

| Level | *n* | Length |
|---|---|---|
| 3A | 57 | 56.5 |
| 3B | 143 | 54.2 |
| 4A | 120 | 62.4 |
| 4B | 82 | 59.3 |
| 4C | 702 | 57.9 |
| 5A | 214 | 57.9 |
| 5B | 79 | 63.2 |
| 6A | 886 | 50.5 |
| 6B | 2193 | 52 |
| 8 | 785 | 44.5 |

Material includes unidentifiable long bone shaft fragments of red deer/reindeer. Note that Layer 8 was affected by trampling, which may have biased the results

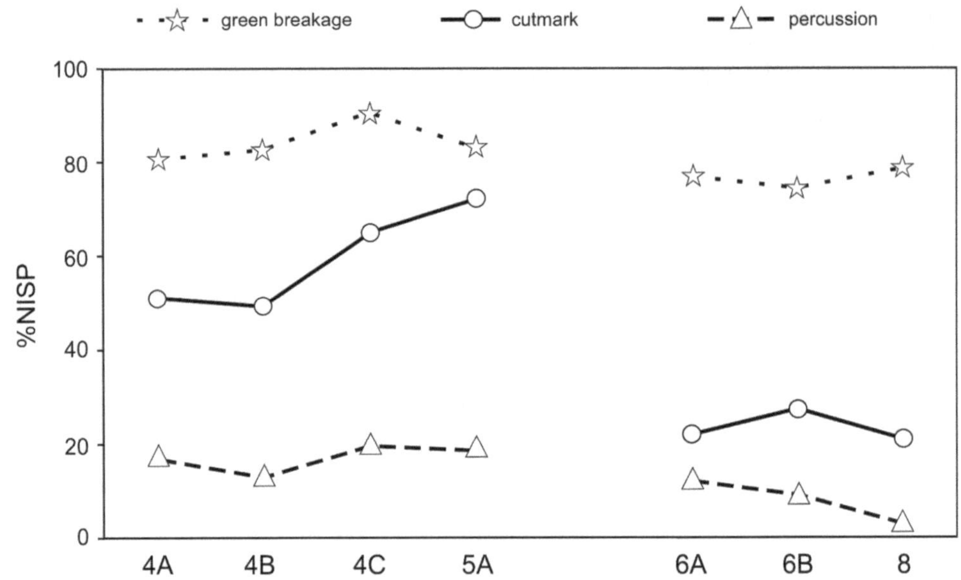


**Fig. 5.8** Bone modifications expressed as %NISP for reindeer (Levels 4A, 4B, 4C, 5A) and red deer (6A, 6B, 8)

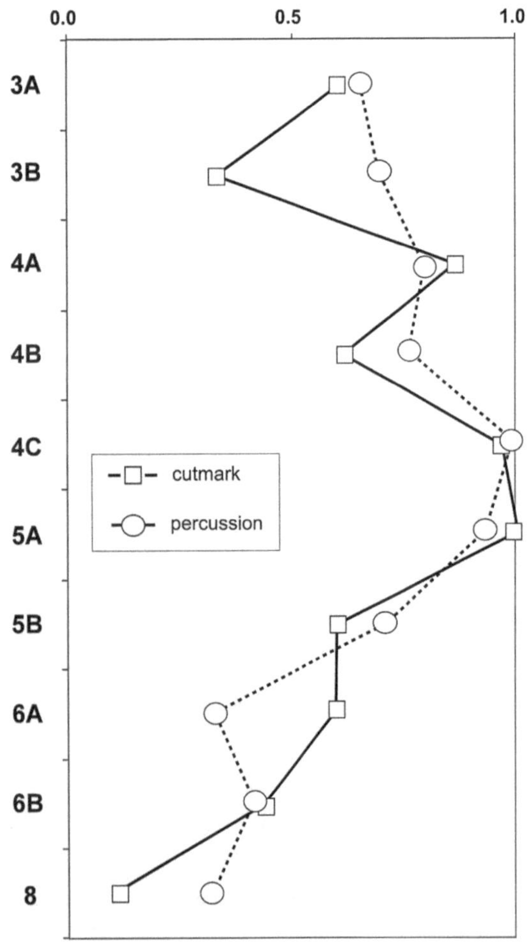


**Fig. 5.9** Standardized frequencies of cutmark and percussion impact marks on long bones of all large cervids through the Pech IV sequence

wooded conditions. The assemblages from 6A and 6B are the largest in the sequence, and the primary prey were red deer and roe deer. The horse makes more of an appearance in these occupations with three individuals represented in each layer.

Seasonality evidence from both layers points to occupations mainly in the winter, spring, and summer. Unlike Layer 8, this layer yielded a sizeable sample of cervid teeth that provided valuable insight into the age of prey: both the red deer and roe deer assemblages are composed of mainly prime adult animals.

The faunal evidence indicates extensive utilization of cervid prey in both sublayers. However, the frequency of butchery traces is significantly lower than what shows up later in Layer 4, despite the high production of tools in layer 6. These differences need further exploration but generally point to variable subsistence strategies and changing site use between the cold and temperate periods. Considering the multiple seasonal occupations, this layer resembles Layer 8 in terms of site use with the exception of the decreased frequency or intensity of fire use. Overall, the temperate layers of Pech IV show remarkable stability in the way the cave was used.

## Layer 5

This layer is remarkably poor in fauna reflecting a distinct change in subsistence and/or site use. Both reindeer and red deer are present; this is not a factor of misidentification, but instead, reflects the shift from warmer to colder conditions. Albeit in small numbers, reindeer predominate in 5A. High cutmark and percussion mark frequencies reflect a similar pattern to the later Layer 4C. The primary difference is in

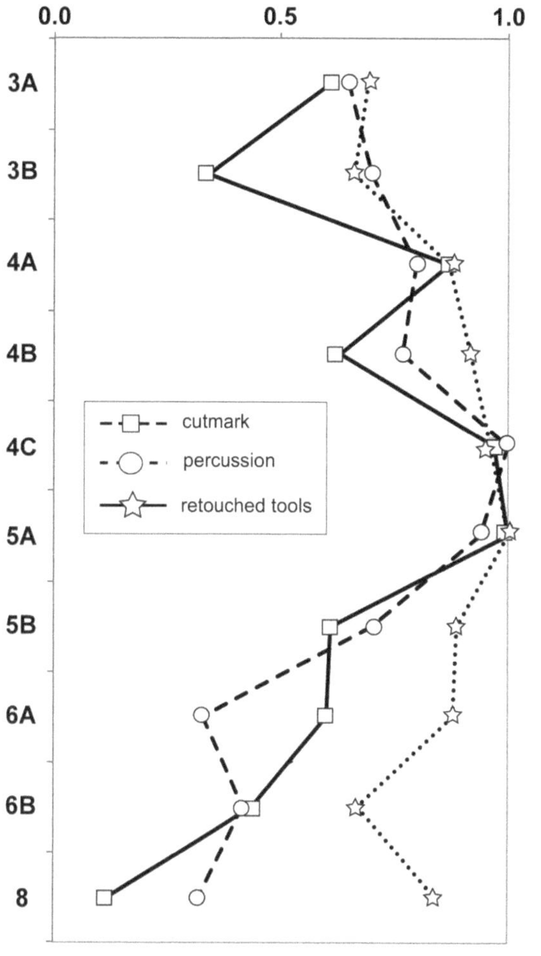


**Fig. 5.10** Standardized frequencies of cutmark and percussion impact marks on long bones of reindeer and red deer in comparison to standardized frequencies of formal stone tool abundance through the Pech IV sequence

overall prey abundance (low) and seasonality indications for winter and spring hunts.

## Layer 4

The cold period represented by this layer may have resulted in a different kind of occupation at the site by Neandertals. The faunal spectrum is remarkably limited; reindeer overwhelmingly dominate, but this was most likely due to climatic factors as opposed to human choice (Grayson and Delpech 2005). Both the occupations at the site and the hunting of *Rangifer* occurred in seasonally restricted timeframes, presumably when this prey species was available locally as it migrated through the area in the spring. The evidence for focused predation on this seasonally migrating ungulate from this layer of Pech IV adds to the number of cold-period Quina Mousterian sites in southwestern France showing this subsistence strategy [e.g., Les Pradelles (Costamagno et al. 2006); Jonzac (Niven et al. 2012)]. This does not necessarily mean that Neandertals followed the herds during the year, but it could be that they were knowledgeable about the seasonal abundance of a specific prey animal and took advantage of these opportunities. At other times of the year, Neandertals presumably hunted less migratory and more predictable animals such as horse and bison. Where Neandertals were based during the seasons they were not occupying Pech IV is unknown, but all of the evidence from Layer 4 indicates higher mobility among these groups. A higher percentage of exotic raw materials is seen in the Quina Mousterian Layer 4A, and the overall trend evident in the zooarcheological data is that reindeer were processed intensively in this layer as well as in Layers 4B and 4C, with the selective transport of the richest marrow bones (distal limb). Overall, the record in Layer 4 shows much less stability in site use compared to the earlier occupations in Layers 8–5. Differing records reflect changing adaptations by Neandertals to such factors as climate, ecological conditions, and prey availability.

## Layer 3

Poor preservation due to a variety of taphonomic factors resulted in low identifiability of bone to taxon or element in these assemblages. Low numbers of faunal remains point to less use of the cave for prey processing and consumption compared to other occupations. However, fetal horse bone indicates winter occupation, which was a common season for site use based on seasonal indicators in previous layers. The low faunal abundance hinders any broad conclusions for this layer, but the sparse data show that roe deer was present, indicating occupations during the intermittent warmer periods of OIS 3. Reindeer are also represented, showing that occupations in 3A and 3B occurred throughout various environmental regimes.

**Acknowledgements** Thanks to all of the Pech IV team members for their support, camaraderie, and insight. Numerous other individuals contributed to this study. Thanks to Jean-Christophe Castel, Charles Egeland, Sabine Gaudzinski-Windheuser, Olaf Jöris, Antje Justus, Lutz Kindler, Claire Letourneux, Natalie Munro, Isabelle Thèry-Parisot, Will Rendu, Daniel Richter, Uta Schwarz, Marie Soressi, Teresa Steele, Martin Street, and Elaine Turner. A portion of this research was conducted at the Archaeological Research Centre and Museum for Human Behavioural Evolution, Roman-German Central Museum, Neuwied/Mainz (Germany).

## References

Aldeias, V., Goldberg, P., Sandgathe, D., Berna, F., Dibble, H. L., McPherron, S. P., et al. (2012). Evidence for Neanderthal use of fire at Roc de Marsal (France). *Journal of Archaeological Science, 39,* 2414–2423.

Bartram, L. E. (1993). *An ethnoarchaeological analysis of Kua San (Botswana) bone food refuse*. Ph.D. Dissertation, University of Wisconsin.

Beauval, C. (2004). La faune des niveaux Mousteriens de "Chez-Pinaud" (Jonzac, Charente-Maritime, France). Première analyse. In J. Airvaux (Ed.), *Le site paléolithique de Chez-Pinaud à Jonzac, Charente-Maritime* (pp. 125–156). Préhistoire du Sud Ouest.

Bellomo, R. (1993). A methodological approach for identifying archaeological evidence for fire resulting from human activities. *Journal of Archaeological Science, 20,* 525–554.

Binford, L. R. (1978). *Nunamiut ethnoarchaeology*. New York: Academic Press.

Binford, L. R. (1981). *Bones: Ancient men and modern myths*. New York: Academic Press.

Binford, L. R. (1984). *Faunal remains from Klasies River Mouth*. New York: Academic Press.

Blasco, M. F. (1997). The pursuit of game: the Mousterian Cave Site of Gabasa 1 in the Spanish Pyrenees. *Journal of Anthropological Research, 53,* 177–217.

Brain, C. K. (1969). The contribution of Namib Desert Hottentots to an understanding of australopithecine bone accumulations. *Scientific Papers of the Namib Desert Research Station, 39,* 13–22.

Brain, C. K. (1981). *The hunters or the hunted? An introduction to African cave taphonomy*. Chicago: University of Chicago Press.

Britton, K., Grimes, V., Dau, J., & Richards, M. P. (2009). Reconstructing faunal migrations using intra-tooth sampling and strontium and oxygen isotope analyses: a case study of modern caribou (*Rangifer tarandus granti*). *Journal of Archaeological Science, 36,* 1163–1172.

Britton, K., Grimes, V., Niven, L., Steele, T. E., McPherron, S., Soressi, M., et al. (2011). Strontium isotope evidence for migration in late Pleistocene *Rangifer*: Implications for Neanderthal hunting strategies at the Middle Palaeolithic site of Jonzac, France. *Journal of Human Evolution, 61,* 176–185.

Bunn, H. T., Bartram, L. E., & Kroll, E. M. (1988). Variability in bone assemblage formation from Hadza hunting, scavenging, and carcass processing. *Journal of Anthropological Archaeology, 7,* 412–457.

Burch, E. S. (1972). The caribou/wild reindeer as a human resource. *American Antiquity, 37,* 339–368.

Burke, A. & Castanet, J. (1995). Histological observations of cementum growth in horse teeth and their application to archaeology. *Journal of Archaeological Science, 22,* 479–493.

Burke, A. (2000). The view from Starosele: Faunal exploitation at a Middle Palaeolithic site in Western Crimea. *International Journal of Osteoarchaeology, 10,* 325–335.

Carter, R. J. (2006). A method to estimate the ages at death of red deer (*Cervus elaphus*) and roe deer (*Capreolus capreolus*) from developing mandibular dentition and its application to Mesolithic NW Europe In D. Ruscillo (Ed.), *Recent advances in ageing and sexing animal bones* (pp. 40–61). Oxford: Oxbow Books.

Chase, P. G. (1986). *The hunters of Combe Grenal: Approaches to Middle Palaeolithic subsistence in Europe*. Oxford: British Archaeological Reports 286.

Chase, P. G. (1999). Bison in the context of complex utilization of faunal resources: A preliminary report on the Mousterian zooarchaeology of La Quina (Charente, France). In J.-P. Brugal, F. David, J. Enloe, & J. Jaubert (Eds.), *Le bison: Gibier et moyen de subsistence des hommes du Paléolithique aux Paléoindiens des Grandes Plains* (pp. 159–184). Antibes: Éditions APDCA.

Conard, N. J., & Prindiville, T. J. (2000). Middle Palaeolithic hunting economies in the Rhineland. *International Journal of Osteoarchaeology, 10,* 286–309.

Costamagno, S., Meignen, L., Beauval, C., Vandermeersch, B., & Maureille, B. (2006). Les Pradelles (Marillac-le-Franc, France): A Mousterian reindeer hunting camp? *Journal of Anthropological Archaeology, 25,* 466–484.

Costamagno, S., Théry-Parisot, I., Castel, J. C., & Brugal, J. P. (2009). Combustible ou non? Analyse multifactorielle et modèles explicatifs sur des ossements brûlées paléolithiques. In I. Théry-Parisot & A. Henry (Eds.), *Gestion des combustibles au Paléolithique et au Mésolithique: Nouveaux outils, nouvelles interpretations* (pp. 65–84). Oxford: BAR International Series 1914.

D'Errico, F., & Vanhaeren, M. (2002). Criteria for identifying red deer (*Cervus elaphus*) age and sex from their canines. Application to the study of Upper Palaeolithic and Mesolithic ornaments. *Journal of Archaeological Science, 29,* 211–232.

Dibble, H. L., Berna, F., Goldberg, P., McPherron, S. P., Mentzer, S., Niven, L., et al. (2009). A preliminary report on Pech de l'Azé IV, Layer 8 (Middle Paleolithic, France). *PaleoAnthropology,* 182–219.

Egeland, C. P. (2003). Carcass processing intensity and cutmark creation: An experimental approach. *Plains Anthropologist, 48,* 39–51.

Egeland, C. P., & Byerly, R. M. (2005). Application of return rates to large mammal butchery and transport among hunter-gatherers and its implications for Plio-Pleistocene hominid carcass foraging and site use. *Journal of Taphonomy, 3,* 135–157.

Enloe, J. G., & David, F. (1997). *Rangifer* herd behavior: Seasonality of hunting in the Magdalenian of the Paris Basin. In L. J. Jackson & P. T. Thacker (Eds.), *Caribou and reindeer hunters of the Northern Hemisphere* (pp. 52–68). Hampshire: Avebury.

Fiore, I., Gala, M., & Tagliacozzo, A. (2004). Ecology and subsistence strategies in the eastern Italian Alps during the Mousterian. *International Journal of Osteoarchaeology, 14,* 273–286.

Gaudzinski, S. (1995). Wallertheim revisited: A re-analysis of the fauna from the Middle Palaeolithic site of Wallertheim (Rheinhessen/Germany). *Journal of Archaeological Science, 22,* 51–66.

Goldberg P., Dibble, H., Berna, F., Sandgathe D., McPherron, S. J. P., Turq, A. (2012). New evidence on Neandertal use of fire: Examples from Roc de Marsal and Pech de l'Azé IV. *Quaternary International, 247,* 325–340.

Grayson, D. K., & Delpech, F. (2005). Pleistocene reindeer and global warming. *Conservation Biology, 19,* 557–562.

Habermehl, K.-H. (1975). *Die Altersbestimmung bei Haus- und Labortieren*. Berlin and Hamburg: Verlag Paul Parey.

Habermehl, K.-H. (1985). *Altersbestimmung bei Wild- und Pelztieren*. Hamburg and Berlin: Verlag Paul Parey.

Hanson, M., & Cain, C. R. (2007). Examining histology to identify burned bone. *Journal of Archaeological Science, 34,* 1902–1913.

Klevezal', G. A. (1996). *Recording structures of mammals: Determination of age and reconstruction of life history*. Rotterdam: A. A. Balkema.

Klevezal', G. A. & Kleinenberg, S. E. (1969). General description of the annual layers in the tissues of tooth and bone. In G. A. Klevezal' & S. E. Kleinenberg (Eds.), *Age determination of mammals from annual layers in teeth and bones* (pp. 6–28). USA: Israel Program for Scientific Translations, U.S. Dept. of Commerce, Clearinghouse for Federal Scientific and Technical Information.

Lam, Y. M., Chen, X., Marean, C. W., Frey, C. (1998). Bone density and long bone representation in archaeological faunas: Comparing results from CT and photon densitometry. *Journal of Archaeological Science, 25,* 559–570.

Lam, Y. M., Chen, X., & Pearson, O. M. (1999). Intertaxonomic variability in patterns of bone density and the differential representation of bovid, cervid, and equid elements in the archaeological record. *American Antiquity, 64,* 343–362.

Laquay, G. (1981). *Recherches sur les faunes du Würm I en Périgord.* Ph.D. Dissertation, Université de Bordeaux I.

Levine, M. A. (1998). Eating horses: The evolutionary significance of hippophagy. *Antiquity, 72,* 90–100.

Lieberman, D. E. & Meadow, R. H. (1992). The biology of cementum increments (with an archaeological application). *Mammal Review, 22,* 57–77.

Lupo, K. D. (1998). Experimentally derived extraction rates for marrow: Implications for body part exploitation strategies of Plio-Pleistocene hominid scavengers. *Journal of Archaeological Science, 25,* 75–657.

Lyman, R. L. (1984). Bone density and differential survivorship of fossil classes. *Journal of Anthropological Archaeology, 3,* 99–259.

Lyman, R. L. (1994). *Vertebrate taphonomy.* Cambridge: Cambridge University Press.

Marean, C. W., & Cleghorn, N. (2003). Large mammal skeletal element transport: Applying foraging theory in a complex taphonomic system. *Journal of Taphonomy, 1,* 15–42.

Mellars, P. (1996). *The Neanderthal legacy: An archaeological perspective from Western Europe.* Princeton: Princeton University Press.

Metcalfe, D., & Jones, K. T. (1988). A reconsideration of animal body-part utility indices. *American Antiquity, 53,* 486–504.

Miller, F. L. (1974). *Biology of the Kaminuriak population of barren-ground caribou part 2.* Ottawa: Canadian Wildlife Service.

Morin, E. (2004). *Late Pleistocene population interaction in western Europe and modern origins: New insights based on the faunal remains from Saint-Césaire, Southwestern France.* Ph.D. Dissertation, University of Michigan.

Morin, E. (2007). Fat composition and Nunamiut decision-making: A new look at the marrow and bone grease indices. *Journal of Archaeological Science, 34,* 69–82.

Morin, E., & Laroulandie, V. (2012). Presumed symbolic use of diurnal raptors by Neanderthals. *PLoS ONE, 7*(3), e32856.

Nicholson, R. A. (1993). A morphological investigation of burnt animal bone and an evaluation of its utility in archaeology. *Journal of Archaeological Science, 20,* 411–428.

Nilssen, P. J. (2000). *An actualistic butchery study in South Africa and its implications for reconstructing hominid strategies of carcass acquisition and butchery in the Upper Pleistocene and Plio-Pleistocene.* Ph.D. Dissertation, University of Cape Town.

Niven, L. (2013). A diachronic evaluation of Neanderthal cervid exploitation and site use at Pech de l'Azé IV, France. In J. Clark & J. Speth (Eds.), *Zooarchaeology and modern human origins: Human hunting behavior during the Later Pleistocene* (pp. 151–161). Berlin: Springer.

Niven, L., Steele, T. E., Rendu, W., Mallye, J.-B., McPherron, S. P., Soressi, M., et al. (2012). Neandertal mobility and large-game hunting: The exploitation of reindeer during the Quina Mousterian at Chez-Pinaud Jonzac (Charente-Maritime, France). *Journal of Human Evolution, 63,* 624–635.

Niven, L., Steele, T. E., Rendu, W., Mallye, J.-B., McPherron, S. P., Soressi, M., et al. (2012). Neanderthal subsistence and site use during the Quina Mousterian at Jonzac, France. *Journal of Human Evolution, 63,* 624–635.

O'Connell, J. F., Hawkes, K., & Blurton Jones, N. (1988). Hadza hunting, butchering, and bone transport and their archaeological implications. *Journal of Anthropological Research, 44,* 113–161.

Pellegrini, M., Donahue, R. E., Chenery, C., Evans, J., Lee-Thorp, J. A., Montgomery, J., et al. (2008). Faunal migration in Late-Glacial central Italy: Implications for human resource exploitation. *Rapid Communications in Mass Spectrometry, 22,* 1714–1726.

Peresani, M., Fiore, I., Gala, M., Romandini, M., & Tagliacozzo, A. (2011). Late Neanderthals and the intentional removal of feathers as evidenced from bird bone taphonomy at Fumane Cave 44 ky B.P., Italy. *Proceedings of the National Academy of Sciences, 108,* 3888–3893.

Pike-Tay, A., Valdés, V. C., & de Quirós, F. B. (1999). Seasonal variations of the Middle-Upper Paleolithic transition at El Castillo, Cueva Morín and El Pendo (Cantabria, Spain). *Journal of Human Evolution, 36,* 283–317.

Rabinovich, R., & Hovers, E. (2004). Faunal analysis from Amud Cave: Preliminary results and interpretations. *International Journal of Osteoarchaeology, 14,* 287–306.

Rendu, W. (2010). Hunting behavior and Neanderthal adaptability in the Late Pleistocene site of Pech-de-l'Aze I. *Journal of Archaeological Science, 37,* 1798–1810.

Rivals, F., Moncel, M.-H., & Patou-Mathis, M. (2009). Seasonality and intra-site variation of Neanderthal occupations in the Middle Palaeolithic locality of Payre (Ardèche, France) using dental wear analyses. *Journal of Archaeological Science, 36,* 1070–1078.

Sandgathe, D. M., Dibble, H. L., Goldberg, P., McPherron, S. P., Turq, A., Niven, L., et al. (2011). On the role of fire in Neandertal adaptations in Western Europe: Evidence from Pech de l'Azé IV and Roc de Marsal, France. *PaleoAnthropology,* 2011, 216–242.

Schiegl, S., Goldberg, P., Pfretzschner, H.-U., & Conard, N. J. (2003). Paleolithic burnt bone horizons from the Swabian Jura: Distinguishing between in situ fire places and dumping areas. *Geoarchaeology, 18,* 65–541.

Shahack-Gross, R., Bar-Yosef, O., & Weiner, S. (1997). Black-coloured bones in Hayonim Cave, Israel: Differentiating between burning and oxide staining. *Journal of Archaeological Science, 24,* 46–439.

Shipman, P., Foster, G., & Schoeninger, M. (1984). Burnt bones and teeth: An experimental study of color, morphology, crystal structure and shrinkage. *Journal of Archaeological Science, 11,* 307–325.

Soressi, M., Rendu, W., Texier, J. P., Claud, E., & Daulny, L. (2008). Pech-de-l'Azé I (Dordogne, France): Nouveau regard sur un gisement moustérien de tradition acheuléenne connu depuis le XIXe siècle. In J. Jaubert, J. G. Bordes, & I. Ortega, *Les sociétés paléolithiques d'un grand Sud-Ouest: Nouveaux gisements, nouvelles méthodes, nouveaux résultats* (pp. 95–132). Mémoire de la Société Préhistorique française XLVII.

Speth, J. D. (1983). *Bison kills and bone counts: Decision making by ancient hunters.* Chicago: University of Chicago Press.

Speth, J. D. (2006). Housekeeping, Neanderthal-style: Hearth placement and midden formation in Kebara Cave (Israel). In E. Hovers & S. L. Kuhn (Eds.), *Transitions before the transition: Evolution and stability in the Middle Paleolithic and Middle Stone Age* (pp. 171–188). New York: Springer.

Speth, J. D., & Clark, J. L. (2006). Hunting and overhunting in the Levantine Late Middle Palaeolithic. *Before Farming, 3,* 1–42.

Speth, J. D., & Tchernov, E. (2007). The Middle Paleolithic occupations at Kebara Cave: A faunal perspective. In O. Bar-Yosef & L. Meignen (Eds.), *Kebara Cave Mt Carmel, Israel: The Middle and Upper Paleolithic archaeology part I* (pp. 165–260). Cambridge: Peabody Museum of Archaeology and Ethnology.

Spiess, A. E. (1979). *Reindeer and caribou hunters: An archaeological study.* San Francisco: Academic Press.

Stiner, M. C. (1990). The use of mortality patterns in archaeological studies of hominid predatory adaptations. *Journal of Anthropological Archaeology, 9,* 305–351.

Stiner, M. C. (1994). *Honor among thieves: A Zooarchaeological study of Neanderthal Ecology*. Princeton: Princeton University Press.

Stiner, M. C. (2005). *The faunas of Hayonim Cave, Israel: A 200,000-year record of Paleolithic diet, demography and society*. American School of Prehistoric Research Bulletin 48, Peabody Museum, Harvard University.

Stiner, M. C., Kuhn, S. L., Weiner, S., & Bar-Yosef, O. (1995). Differential burning, recrystallization, and fragmentation of archaeological bone. *Journal of Archaeological Science, 22,* 223–237.

Stuart, A. J. (1982). *Pleistocene vertebrates of the British Isles*. London: Longman.

# 6 The Lithic Assemblages

Shannon J.P. McPherron, Harold L. Dibble, Dennis M. Sandgathe, Paul Goldberg, Sam C. Lin, and Alain Turq

## Introduction

Prior to excavating Pech IV, we studied Bordes' collection from his 8 years of excavation at the site. After dealing with various problems in the data set (discussed in Chap. 1), just over 57,000 lithic artifacts with good contextual information were studied and published prior to the start of excavations (McPherron and Dibble 2000). The new excavations recovered nearly 20,000 more lithic objects, and a basic description of those materials is presented here. As described in earlier chapters, one of the primary goals of the lithic analysis of the new collection was to establish that our levels could be linked to the existing collection, meaning that the new observations made on the site's geology, site formation, dating, and fauna could be applied to what we already knew of the site's lithics. As discussed in Chap. 2, it has been possible to correlate the interpretation of Bordes' stratigraphy with our own, thus making it possible to combine both collections into a unified stratigraphic sequence, using a new set of labels ("Units" designated with Roman numerals) for the combined strata. Basic descriptions of these combined lithic assemblages will also be presented in this chapter, along with discussions of the lithic raw materials present in the collections (prepared by A. Turq) and analyses focused on lithic transport (prepared by S. Lin). This is followed by an exploration of changes in blank production techniques and blank selection criteria. Finally, we comment on some specific issues raised by the assemblages.

The depositional sequence of Pech IV includes eight major stratigraphic units, with five of these composed of two or more subunits, for a total of 13 independent assemblages. Layers 8 through 3 (from bottom to top) are of Pleistocene age (see Chap. 3), dating between approximately 100 and 45 ka BP. The overlying Layers 1 and 2 represent Holocene deposits, much of which is redeposited Pleistocene sediments upslope from the site (see Chap. 2), and contain limited archaeological material. These layers are only briefly discussed here.

S.J.P. McPherron (✉) · H.L. Dibble
Department of Human Evolution, Max Planck Institute for Evolutionary Anthropology, Leipzig, Germany
e-mail: mcpherron@eva.mpg.de

H.L. Dibble
e-mail: hdibble@sas.upenn.edu

H.L. Dibble
Department of Anthropology, University of Pennsylvania, Philadelphia, PA, USA

H.L. Dibble
School of Human Evolution and Social Change, Institute for Human Origins, Arizona State University, Tempe, AZ, USA

D.M. Sandgathe
Department of Archaeology and Human Evolution Studies Program, Simon Fraser University, Burnaby, BC, Canada
e-mail: dms@sfu.ca

D.M. Sandgathe
Museum of Archaeology and Anthropology, University of Pennsylvania, Philadelphia, USA

P. Goldberg
Department of Archaeology, Boston University, Boston, USA
e-mail: paulberg@bu.edu

P. Goldberg · S.C. Lin
School of Earth and Environmental Sciences, Centre for Archaeological Science, University of Wollongong, Wollongong, NSW 2522, Australia
e-mail: samlin@uow.edu.au

P. Goldberg
Institute for Archaeological Sciences, Eberhard Karls Universitat Tübingen, Tübingen, Germany

A. Turq
Musée National de Préhistoire, Les Eyzies-de-Tayac-Sireuil, France
e-mail: alain.turq@orange.fr

A. Turq
PACEA, UMR 5199, Talence, France


H.L. Dibble et al. (eds.), *The Middle Paleolithic Site of Pech de l'Azé IV*,
Cave and Karst Systems of the World, https://doi.org/10.1007/978-3-319-57524-7_6

## Methods

What follows is a description of the lithic recording system used for the Pech IV collection. It serves as a reference for the analysis presented here and for the database which we have made available. This recording system is a continuation of the one developed for Combe-Capelle Bas (Dibble et al. 1995), Fontéchevade (Chase et al. 2009) and our previous study of Bordes' collection from Pech IV (McPherron and Dibble 2000), though with some additional observations having been made. By and large, however, the range of variables examined and the codes or measurements used are very similar among the datasets.

All observations were made using self-authored data entry software, E4 (www.oldstoneage.com\software), which saves the input data into a Microsoft Access database, and this, in turn, is integrated into the main database of the site. Additionally, E4 tracks already analyzed artifacts and the list of artifacts to be studied so as to eliminate missing and duplicate records. Artifact IDs (Unit-ID) were read from the artifact bag labels using barcode scanners directly cabled to the computer. Size measurements were made with digital calipers with 0.05 mm precision and cabled directly to the computer. Weights were measured on electronic scales with 1 g precision but were not cabled directly to the computer.

Most of the data entry was done by Dibble and McPherron; however, many other individuals contributed to the data set. All of these individuals were trained by Dibble and McPherron, and nearly all of the data entry was done as a group with Dibble and/or McPherron present. All raw material observations were done by Turq.

In addition to the piece by piece attribute analysis detailed here, we also looked at the totality of lithics of each layer laid out on tables so that the entire assemblage could be viewed at the same time. All authors participated in this stage of analysis, and their observations are incorporated into the following presentation.

## Dataclass

Dataclass is a fundamental variable in our recording system that structures both the data entry that follows (which attributes are recorded for particular kinds of pieces) and subsequent analyses. It corresponds roughly to blank type in other recording systems. Valid entries include: complete flake, proximal flake, distal flake, medial flake, complete tool, proximal tool, distal tool, medial tool, core tool, core, core fragment, biface, biface fragment, uniface, uniface fragment, shatter, manuport, hammerstone, hammerstone fragment, and *rognon* (unworked flint nodules). For these purposes, tools represent retouched pieces in Bordes' essential count (Bordes 1961; Debénath and Dibble 1994), although unretouched flakes that are included in Bordes' type list (e.g. Levallois flakes) are classified as flakes. Completeness is not 100% strict in that if a very small portion of a piece is missing and it will not affect the size measurements and will not substantially affect the weight measurement it may be classified as complete. However, completeness refers only to breakage, not to the removal of mass due to retouching. To be proximal, the point of percussion or at least 50% of the platform should be present (cf. Hiscock 2002). Core tools are tools made on non-flake blanks (e.g., choppers). The distinction between medial flakes and shatter can be difficult to make, but if the piece is judged to have an interior (ventral) surface, it is considered a medial flake.

## Technique

This field records the technology of blank production for flakes and for cores. Valid options include normal (i.e., non-diagnostic), Clactonian, Levallois, discoidal, blade, burin, tranchet, biface-retouch, retouch-flake, Kombewa, splitpebble, other, and N/A. If the technology cannot be identified the piece is classified as normal. Blade technique is determined solely on the presence of parallel flake scars on the flake surface and not on the ratio of the length to the width (but see Form below). Retouch flakes are flakes removed during the retouching a tool. Normally these will have an exterior surface of prior retouch originating at the proximal end and a plain platform (representing the interior surface of the flake being retouched). Clactonian flakes have a large, plain platform, a large interior platform angle (>105°), and a large, diffuse bulb of percussion. They are also typically wider than they are long. Levallois flakes are flakes coming from the Levallois method. While this can be difficult to determine objectively and as a consequence varies among different researchers (Dibble 1995; see also Dibble and Bar-Yosef 1995), we look for relatively thin flakes with little or no cortex and a complex flake scar pattern. Discoidal flakes come from discoidal cores and typically show flake scars from two adjacent directions (e.g., pseudo-Levallois points). Burin is for burin spalls, not the piece that results from such a removal. Biface-retouch flakes have three of the following four characteristics: a steeply curved profile, a complex scar morphology, a lip on the interior surface just below the platform, or a small platform that may appear to be facetted or ground and has a small exterior platform angle. Kombewa is reserved for flakes that have been removed from the interior of another flake. Evidence for use of the Kombewa technique consists of two internal surfaces or two platforms. Splitpebble is used for tools made on split pebbles. A tranchet flake is one that has removed the edge of another flake or flake tool. If the flake likely comes from a particular technology, for instance, bifacial thinning, but it is not entirely clear (or if one of two possibilities are likely), then N/A is used.

## Form

This field is used to describe the blank type and is used for flakes, flake tools, and bifaces. Valid entries include normal, angular, naturally backed, burin, débordant, pustule, overshot, biface edge, LSF, tranchet, point, blade, expanding, broad, shatter, scraper, core tool, lame-à-crête, and other. Angular pieces are defined as those with a steep ridge or two exterior surfaces that meet at an angle of less than 90°. Points are flakes with margins that converge toward the distal end. Naturally backed pieces have a stripe of cortex along one edge, and this cortical surface is roughly perpendicular to the surface of the flake. Conversely, débordant is used for flakes that remove part of a core edge, and thus the perpendicular lateral surface is non-cortical; most pseudo-Levallois points are thus débordant. Burin is for burin spalls. Overshot flakes have a termination that includes part of the distal core from which they were removed. Biface edge flakes look as though they have been removed from the edge of a biface meaning usually that they have a bifacial crest. Lateral struck flakes (LSF) are based on the definition of the La Cotte de St. Brelade publication (Callow and Cornford 1986). They are similar to biface edge flakes except that they appear to have removed a scraper edge. Blades are flakes that are more than twice as long as they are wide whether or not they have parallel flake scars on their surface. Expanding flakes are narrow near the platform and expand towards the distal end. Broad flakes are wider than they are long. Shatter is a piece that shows no obvious flake characteristics (in particular, no interior surface). This category includes pieces that are sometimes referred to as chunks. Pustule is a type added for Pech IV to describe blanks that appear to be small knobs of flint struck from larger nodules.

For bifaces, the form is a way of describing a piece that is bifacially worked but which can resemble either a classic handaxe a core, ébauche (rough-out), or scraper. Additionally, if it is a classic handaxe and has a cortical base, this is recorded here as well. The definition of a rough-out is highly subjective but it is defined here as a handaxe form with relatively few removals such that it is unclear whether further reduction would have produced a handaxe or something else.

## Tool, Biface and Core Types

We followed Bordes' typology as described by Bordes (1961) and Debénath and Dibble (1994). Truncated-facetted pieces are given Type 64, and scrapers on the platform are given Type 65. Types 46–49 are all given Type 48. Space was allocated for two types per piece. When two types are present on the same piece, atypical types are listed second. If the two types are still equally weighed, we followed Bordes' rule and gave the least represented type in the collection the first position. When three or more types are present on the same piece, it can be typed as miscellaneous (Type 63); however, every effort was made to avoid using this type.

Cores are typed using these categories: inform, Levallois, discoidal, Mousterian disc, single surface, globular, chopper, chopping-tool, prismatic, pyramidal, tested, Kombewa and other. Inform cores lack any kind of patterning or organization. Levallois cores are those that show the Levallois-like preparation of a flaking surface and with a central, primary Levallois flake removal (cf. Bordes 1961). Cores that do not show this primary removal were classified as single-surface cores (typically centripetal). Discoidal cores are similar but do not show a distinction between a flaking surface and a preparation surface. These are often bi-conic. Mousterian discs are typically small, flat, thin, and circular cores (cf. Bordes 1961). Globular cores are similar to inform cores, though in this case, they are rounded chunks of raw material with flakes removed with no clear organization. Choppers have a series of removals across one edge. Chopping-tools have a bifacial series of removals across one edge. Prismatic and pyramidal cores follow classic definitions (Bordes 1961) and are rare in the Pech IV collection. The term "tested" was used for chunks of raw material with only one or two removals. Kombewa refers to flakes with a removal on the interior surface that removes the bulb of percussion (thereby creating a Kombewa flake). Other types of flaked flakes typically fell into the category of truncated-facetted.

Biface types follow Bordes (1961) and include triangular, triangular-elongated, subtriangular, cordiform, cordiform-elongated, subcordiform, ovate, limande, disc, amygdaloide, lanceolate, Micoquian, ficron, Abbevillian, cleaver, cleaver on a flake, lageniform, losanform, naviform, nucleiform, biface-a-dos (i.e., backed), proximal distal, medial, partial, and divers. Broken bifaces are given proximal, distal, medial, or partial when it is unclear what portion is preserved.

## Platforms

Platform surfaces are described as plain, dihedral, faceted, chapeau de gendarme, punctiform, cortical, and removed. A platform is removed when it is missing because of retouch, as opposed to breakage.

Exterior platform angle is the angle between the platform surface and the exterior flake surface. It is measured with a goniometer with one-degree precision. On curving flakes, the angle is taken to the point along the exterior approximately equal to the platform thickness. If the platform surface is not flat or difficult to isolate, or if the exterior of the flake has an irregular surface, the measurement is not recorded.

Platform width and thickness are recorded with digital calipers on all pieces with preserved platforms with clearly defined limits. Cortical platforms on primary flakes, for instance, are often problematic for measurement because the limits are not easily identified. The width is recorded from one lateral edge of the platform to the opposite edge, where the interior and exterior surfaces meet. Platform thickness is the depth of the platform behind the point of percussion.

## Cortex

The percentage of cortex is noted on all pieces including shatter. Cortex is assessed using the following intervals: 0, >0–10, >10–40, >40–60, >60–90, >90–<100, and 100%. On flake blanks, this percentage is relative to the exterior surface. On all other pieces, including shatter, cores, and bifaces, the percentage is relative to the total, volumetric surface area. For analysis, the intervals can be converted to a ratio scale using the midpoint of the interval. In a measurement methods study using 3-D scanning of cortical flakes, this interval estimate method was found to be sufficiently accurate to estimate the amount of cortex in an assemblage (Lin et al. 2010).

## Taphonomic Variables

Edge damage, alteration, and burning were noted for all complete pieces, cores, and bifaces. Valid entries for edge damage are none, 1-side, 2-side (meaning the interior or exterior surfaces or both), and rolled. For more discussion on how this variable was recorded, see Chap. 2. The alteration field records any post-depositional modifications to the surface of the lithic. Valid entries for alteration include unaltered, speckled, light patina, heavy patina, white, double patina, burned, other, and N/A. Thus patina, when present, is divided into three stages: light, heavy and white. Double patina refers to pieces that have two stages of patina due to material having been removed after the original piece had been patinated. Speckling is possibly a stage prior to light patina where patina is visible in small dots across the piece. Burned pieces should show signs of thermal alteration including crazing, pot-lid fracture, and color changes (though much less emphasis was placed on this latter variable). A discussion of the burned pieces and the results can also be found in Chap. 2.

## Size

The length, width, and thickness were recorded for all complete flakes and tools, cores, and bifaces. Length and width were also recorded for all pieces with cortex. Weight was recorded on all pieces. We followed the Jelinek (1977; see also Dogandžić et al. 2015) system of measurement for complete flakes and tools. Length is recorded from the point of percussion to the most distal point, width perpendicular to this axis at its midpoint, and thickness at the intersection of these two axes. On broken pieces, the length and width are the maximum length and the maximum width perpendicular to the maximum length. On cores and core tools, length is the maximum dimension and width and thickness are maximum dimensions perpendicular to length. Handaxes are measured following the systems outlined by Bordes (1961), Roe (1969), and McPherron (2003) and include the width and thickness 1/5 and 4/5 the length of the handaxe, the width at three-fourth the length of the handaxe, the width and thickness at the midpoint, the width at the tip midpoint, and the distances from the base to the maximum width and to the maximum thickness. The percent of edge with modifications is also recorded on all handaxes using the categories 100, 75–100, 50–75, 25–50, and 0–25%.

## Assemblage Indices

In the tables that follow, several indices are used to summarize various technological and typological characteristics. These are defined as follows (see also Debénath and Dibble 1994; Dibble et al. 1995; Chase et al. 2009). The typological Levallois index (ILty) is the number of Levallois types 1–4 relative to the total number of tools (1–65). The scraper index (IR) is the number of scraper types 9–29 relative to the total number of tools. The unifacial Acheulian index (IAu) is the number of backed knives (36–37) relative to the total number of tools. The Group II index (Mousterian Group) is the number of types 5–29 relative to the total number of tools. It differs from the scraper index in that it also includes types 5–8. The Group III index (Upper Paleolithic Group) is the number of types 30–37 and 40 relative to the total number of tools. The Group IV index (Denticulate Group) is the number of denticulates (43) relative to the total number of tools. The Levallois index (IL) is the number of Levallois blanks (retouched or not) relative to the total number of blanks. The faceting index (IF) is the number of faceted and dihedral platforms relative to the total number of identifiable platforms. The strict faceting index (Ifs) counts only faceted platforms. The blade index (ILam) is the number of blanks with a blade form (length twice the width) relative to the number of blanks. The truncated-faceted index (TF) is the number of truncated-faceted artifacts (64) relative to the number of retouched tools (essential count). The Kombewa index is the number of Kombewa blanks relative to the total number of blanks with platforms. The core edge index is the number of core edge blanks relative to the total number of blanks with platforms. The naturally backed index is the number of naturally backed blanks, which is more loosely defined than a Type 38, relative to the total number of blanks. The naturally backed index, the Kombewa index, and the Levallois index are calculated on blanks with platforms only.

## Lithic Descriptions

### Dibble-McPherron Excavation

The samples from the new excavations are summarized in Table 6.1. Layer 7 is one of the smallest in terms of liters excavated, but it is by the far the richest in terms of number of artifacts recovered. However, this layer has undergone substantial post-depositional alteration due to cryoturbation (see Chap. 2) and so, for this reason, while the basic numbers are presented here, Layer 7 is not considered in inter-layer summaries and comparisons. At the other extreme, Layers 4B and 4A have very few artifacts. In the case of Layer 4A, this is despite having excavated the second highest number of liters from this deposit. These two layers are characterized mostly by thin bands of artifactual material separated by sterile sediments, whereas the other Pech IV layers are rich in artifacts throughout, that is, relatively homogenous. However, because of their low sample sizes, some of these assemblages with low *N* are excluded from some inter-level summaries and comparisons.

Tables 6.2, 6.3, 6.4, 6.5, 6.6, 6.7, 6.8, 6.9, 6.10, 6.11, 6.12, 6.13, 6.14, 6.15, 6.16, 6.17, 6.18, 6.19, 6.20, 6.21, 6.22, 6.23, 6.24, 6.25, 6.26, 6.27, 6.28, 6.29, 6.30, 6.31, 6.32, 6.33, 6.34, 6.35, 6.36, 6.37, 6.38, 6.39, 6.40, 6.41, 6.42, 6.43, 6.44, 6.45, 6.46, and 6.47 present the basic data for Layers 8-3A. Included here are Bordes type counts, basic counts of tool and flake categories, some typological and technological indices (as defined above), and some size measurements for complete flakes, scrapers, notched tools, other retouched tools, and cores. Table 6.46 gives a summary of some of the indices, and the core types are presented in Table 6.47.

One of the interesting aspects of Pech IV is that there is considerable variability in the stone tool assemblages. Starting at the base of the sequence, Layer 8 is one of the richer layers. Of the retouched component, scrapers dominate, and of these, simple scrapers are most frequent. Of the notched pieces, notches are more common than denticulates. There are some truncated-facetted artifacts, and there are some small Levallois cores and flakes similar to what is found in the overlying Layer 6. Overall, however, this layer is characterized by large artifacts including large scrapers. Levallois blanks are common, and there are relatively high percentage of prepared platforms. Likewise, the core types are heavily dominated by single-surface types.

Layer 7 is an assemblage that was heavily altered post-depositionally. As a result, breakage and edge damage rates are quite high making it more difficult to identify the technology of blank production and retouch. Not surprisingly then, the frequency of artifacts with abrupt and alternating retouch is 81% of the real counts. Similarly, the frequency of notched tools is quite high in Layer 7, and the scraper percentage is correspondingly lower. The frequency of Levallois blanks is much lower than in Layer 8, but the frequency of facetted platforms remains high. So too does

**Table 6.1** Basic counts and density for stone artifacts (complete and broken flakes, tools and cores)

| Level | Tools | Flakes | Cores | *N* | Liters | Density |
|---|---|---|---|---|---|---|
| 3A | 132 | 1267 | 50 | 1449 | 1148 | 1.26 |
| 3B | 228 | 2293 | 90 | 2611 | 791 | 3.3 |
| 4A | 26 | 174 | 4 | 204 | 2443 | 0.08 |
| 4B | 7 | 34 | 2 | 43 | 742 | 0.06 |
| 4C | 78 | 486 | 20 | 584 | 735 | 0.79 |
| 5A | 101 | 1199 | 47 | 1347 | 679 | 1.98 |
| 5B | 56 | 530 | 43 | 629 | 1435 | 0.44 |
| 6A | 202 | 1825 | 179 | 2206 | 2163 | 1.02 |
| 6B | 162 | 1723 | 264 | 2149 | 2527 | 0.85 |
| 7 | 233 | 3088 | 259 | 3580 | 791 | 4.53 |
| 8 | 154 | 2086 | 43 | 2283 | 1862 | 1.23 |
| Totals | 1379 | 14,705 | 1001 | 17,085 | 15,316 | 1.12 |

**Table 6.2** Type counts for Layer 8

| Type | Name | Real | Percent | Essential | Ess. perc. |
|---|---|---|---|---|---|
| 1 | Levallois flake | 120 | 26.7 | 0 | 0 |
| 2 | Atypical Levallois flake | 40 | 8.9 | 0 | 0 |
| 3 | Levallois point | 3 | 0.7 | 0 | 0 |
| 4 | Retouched Levallois point | 1 | 0.2 | 1 | 0.6 |
| 5 | Pseudo-Levallois point | 12 | 2.7 | 0 | 0 |
| 6 | Mousterian point | 5 | 1.1 | 5 | 3.2 |
| 9 | Single straight scraper | 20 | 4.4 | 20 | 12.7 |
| 10 | Single convex scraper | 44 | 9.8 | 44 | 28 |
| 11 | Single concave scraper | 5 | 1.1 | 5 | 3.2 |
| 13 | Double straight-convex scraper | 3 | 0.7 | 3 | 1.9 |
| 14 | Double straight-concave scraper | 1 | 0.2 | 1 | 0.6 |
| 15 | Double convex scraper | 7 | 1.6 | 7 | 4.5 |
| 17 | Double convex–concave scraper | 2 | 0.4 | 2 | 1.3 |
| 18 | Straight convergent scraper | 3 | 0.7 | 3 | 1.9 |
| 19 | Convex convergent scraper | 11 | 2.4 | 11 | 7 |
| 23 | Convex transverse scraper | 3 | 0.7 | 3 | 1.9 |
| 24 | Concave transverse scraper | 2 | 0.4 | 2 | 1.3 |
| 25 | Scraper on interior surface | 1 | 0.2 | 1 | 0.6 |
| 31 | Atypical endscraper | 1 | 0.2 | 1 | 0.6 |
| 32 | Typical burin | 5 | 1.1 | 5 | 3.2 |
| 33 | Atypical burin | 1 | 0.2 | 1 | 0.6 |
| 36 | Typical backed knife | 1 | 0.2 | 1 | 0.6 |
| 38 | Naturally backed knife | 46 | 10.2 | 0 | 0 |
| 40 | Truncation | 2 | 0.4 | 2 | 1.3 |
| 42 | Notch | 15 | 3.3 | 15 | 9.6 |
| 43 | Denticulate | 10 | 2.2 | 10 | 6.4 |
| 45 | Flake with irregular retouch on | 1 | 0.2 | 0 | 0 |
| 48 | Flake with abrupt and alter | 71 | 15.8 | 0 | 0 |
| 54 | End-notched flake | 1 | 0.2 | 1 | 0.6 |
| 64 | Truncated-facetted piece | 13 | 2.9 | 13 | 8.3 |
|  | Totals | 450 | 99.8 | 157 | 99.9 |

**Table 6.3** Artifact counts for Layer 8

|  | *N* |
|---|---|
| Complete and proximal flakes | 1286 |
| Other flake fragments | 800 |
| Complete and proximal tools | 96 |
| Other tool fragments | 58 |
| Handaxes | 0 |
| Handaxe fragments | 0 |
| Shatter | 203 |
| Complete cores | 27 |
| Core fragments | 16 |

**Table 6.4** Typological and technological indices for Layer 8

| Typological indices | | | Technological indices | |
|---|---|---|---|---|
| | Real | Essential | IL | 12.16 |
| ILty | 36.44 | | IF | 30.41 |
| IR | 22.67 | 64.97 | IFs | 17.79 |
| IAu | 0.22 | 0.01 | Ilam | 19.14 |
| II | 23.78 | 68.15 | TF | 8.28 |
| III | 0.02 | 5.1 | Kombewa | 0 |
| IV | 5.56 | 15.92 | Core edge | 1.09 |
| | | | Nat. backed | 7.96 |

**Table 6.5** Basic size measurements for complete flakes, scrapers, notched tools, other retouched tools, and cores from Level 8

| | | Length | Width | Thickness | Weight |
|---|---|---|---|---|---|
| Flakes | *N* | 722 | 723 | 723 | 720 |
| | Mean | 35.36 | 23.43 | 5.83 | 6.24 |
| | S.D. | 10.95 | 6.67 | 2.65 | 5.87 |
| Scrapers | *N* | 57 | 57 | 57 | 57 |
| | Mean | 50.99 | 30.32 | 7.74 | 15.3 |
| | S.D. | 12.06 | 6.27 | 2.69 | 9.87 |
| Notches | *N* | 11 | 11 | 11 | 11 |
| | Mean | 37.62 | 25.47 | 9.04 | 10.45 |
| | S.D. | 10.21 | 5.58 | 3.48 | 7.46 |
| Other tools | *N* | 11 | 11 | 11 | 11 |
| | Mean | 40.83 | 24.47 | 7.94 | 9.91 |
| | S.D. | 16.29 | 8.23 | 2.65 | 7.13 |
| Cores | *N* | 27 | 27 | 27 | 26 |
| | Mean | 41.89 | 31.83 | 13.69 | 24 |
| | S.D. | 12.57 | 9.88 | 4.61 | 27.34 |

the frequency of single-surface cores. Thus, the lower percentage of Levallois flakes could be simply a result of the difficulty of recognizing this technology in a highly altered assemblage, whereas some aspects of Layer 7 connect it with the preceding Layer 8, other aspects anticipate the distinctive small flakes character of Layer 6. In particular, Layer 7 shows an increase in truncated-faceted tools and Kombewa blanks. Kombewa cores are also quite common, and the mean core weight is very low (like that of Layer 6B). Thus, taking into consideration the bias introduced into the assemblage by the post-depositional alterations, Layer 7 appears to represent something intermediate between the scraper rich Layer 8 below and the small flake production of Layer 6 above.

Layer 6 is what we (Dibble and McPherron 2006, 2007) and what Bordes (1975) described as an assemblage emphasizing small flake production. Bordes gave it the name Asinipodian and considered it a new facies of the Mousterian. Its distinctive character comes from very small Levallois cores (some less than 25 mm), a high incidence of truncated-faceted pieces, a high incidence of Kombewa flakes and cores, and a low level of tool production (see more below). In these measures, Layer 6B is the peak with the TF index reaching nearly 35%. Both Layers 6A and 6B have the highest percentages of Levallois blanks and Layer 6B has the highest percentage of facetted platforms. Of the retouched tools, notches are more common than scrapers in Layer 6B with the high frequency of truncated-facetted

**Table 6.6** Type counts for Layer 7

| Type | Name | Real | Percent | Essential | Ess. perc. |
|---|---|---|---|---|---|
| 1 | Levallois flake | 59 | 3 | 0 | 0 |
| 2 | Atypical Levallois flake | 20 | 1 | 0 | 0 |
| 3 | Levallois point | 1 | 0.1 | 0 | 0 |
| 4 | Retouched Levallois point | 2 | 0.1 | 2 | 0.9 |
| 5 | Pseudo-Levallois point | 24 | 1.2 | 0 | 0 |
| 9 | Single straight scraper | 19 | 1 | 19 | 8.2 |
| 10 | Single convex scraper | 18 | 0.9 | 18 | 7.8 |
| 11 | Single concave scraper | 4 | 0.2 | 4 | 1.7 |
| 12 | Double straight scraper | 2 | 0.1 | 2 | 0.9 |
| 14 | Double straight-concave scraper | 1 | 0.1 | 1 | 0.4 |
| 15 | Double convex scraper | 2 | 0.1 | 2 | 0.9 |
| 17 | Double convex–concave scraper | 2 | 0.1 | 2 | 0.9 |
| 18 | Straight convergent scraper | 2 | 0.1 | 2 | 0.9 |
| 19 | Convex convergent scraper | 5 | 0.3 | 5 | 2.2 |
| 22 | Straight transverse scraper | 2 | 0.1 | 2 | 0.9 |
| 23 | Convex transverse scraper | 3 | 0.2 | 3 | 1.3 |
| 25 | Scraper on interior surface | 3 | 0.2 | 3 | 1.3 |
| 26 | Abrupt scraper | 2 | 0.1 | 2 | 0.9 |
| 30 | Typical endscraper | 4 | 0.2 | 4 | 1.7 |
| 31 | Atypical endscraper | 1 | 0.1 | 1 | 0.4 |
| 32 | Typical burin | 4 | 0.2 | 4 | 1.7 |
| 33 | Atypical burin | 2 | 0.1 | 2 | 0.9 |
| 34 | Typical percoir | 3 | 0.2 | 3 | 1.3 |
| 37 | Atypical backed knife | 1 | 0.1 | 1 | 0.4 |
| 38 | Naturally backed knife | 38 | 1.9 | 0 | 0 |
| 40 | Truncation | 4 | 0.2 | 4 | 1.7 |
| 42 | Notch | 49 | 2.5 | 49 | 21.1 |
| 43 | Denticulate | 52 | 2.6 | 52 | 22.4 |
| 44 | Alternate retouched bec | 2 | 0.1 | 2 | 0.9 |
| 45 | Flake with irregular retouch on | 2 | 0.1 | 0 | 0 |
| 48 | Flake with abrupt and alter | 1590 | 80.9 | 0 | 0 |
| 51 | Tayac point | 2 | 0.1 | 2 | 0.9 |
| 54 | End-notched flake | 2 | 0.1 | 2 | 0.9 |
| 57 | Stemmed point | 1 | 0.1 | 1 | 0.4 |
| 64 | Truncated-facetted piece | 38 | 1.9 | 38 | 16.4 |
| | Totals | 1966 | 100.3 | 232 | 100.3 |

pieces, but in Layer 6A scrapers and notched tools are equally represented. The core types of both layers are dominated by single-surface types followed by Levallois and Kombewa cores. The cores of Layer 6B are, along with Layer 7, by far the smallest in the site (mean core weight of approximately 18 g and mean core length of roughly 39 mm). Layer 6B is striking among the Pech IV assemblages for the lack of large lithics. Layer 6A, however,

**Table 6.7** Artifact counts for Layer 7

| | *N* |
|---|---|
| Complete and proximal flakes | 1840 |
| Other flake fragments | 1248 |
| Complete and proximal tools | 138 |
| Other tool fragments | 95 |
| Handaxes | 1 |
| Handaxe fragments | 1 |
| Shatter | 341 |
| Complete cores | 170 |
| Core fragments | 89 |

**Table 6.8** Typological and technological indices for Layer 7

| Typological indices | | | Technological indices | |
|---|---|---|---|---|
| | Real | Essential | IL | 4.15 |
| ILty | 4.17 | | IF | 29.15 |
| IR | 3.31 | 28.02 | IFs | 18.08 |
| IAu | 0.05 | 0 | Ilam | 13.25 |
| II | 3.31 | 28.02 | TF | 16.38 |
| III | 0.01 | 6.47 | Kombewa | 0.25 |
| IV | 5.14 | 43.53 | Core edge | 1.72 |
| | | | Nat. backed | 4.8 |

**Table 6.9** Basic size measurements for complete flakes, scrapers, notched tools, other retouched tools, and cores from Level 7

| | | Length | Width | Thickness | Weight |
|---|---|---|---|---|---|
| Flakes | *N* | 749 | 748 | 748 | 744 |
| | Mean | 33.64 | 23.89 | 8.42 | 9.03 |
| | S.D. | 9.98 | 6.16 | 3.16 | 7.99 |
| Scrapers | *N* | 21 | 21 | 21 | 21 |
| | Mean | 46.06 | 29.47 | 8.04 | 14.14 |
| | S.D. | 12.54 | 5.81 | 2.42 | 6.66 |
| Notches | *N* | 39 | 39 | 39 | 38 |
| | Mean | 38.04 | 27.91 | 9.31 | 13.84 |
| | S.D. | 10 | 6.9 | 2.51 | 9.28 |
| Other tools | *N* | 25 | 25 | 25 | 25 |
| | Mean | 36.13 | 25.94 | 10.02 | 12.88 |
| | S.D. | 9.62 | 5.4 | 2.96 | 9.28 |
| Cores | *N* | 169 | 169 | 169 | 168 |
| | Mean | 37.81 | 28.96 | 14.78 | 18.86 |
| | S.D. | 6.89 | 5.88 | 5.48 | 11.78 |

shows some return to the patterns seen in Layer 8 and again in the overlying Layer 5.

Layer 5 represents another shift in the lithic assemblages. In terms of overall assemblage size and density, Layer 5B is both small and has a rather low artifact density. Layer 5A, by contrast, has both a larger overall assemblage and one of the higher artifact densities. In these assemblages, Levallois blank production remains important, and there are some

**Table 6.10** Type counts for Layer 6B

| Type | Name | Real | Percent | Essential | Ess. perc. |
|---|---|---|---|---|---|
| 1 | Levallois flake | 126 | 20.8 | 0 | 0 |
| 2 | Atypical Levallois flake | 62 | 10.2 | 0 | 0 |
| 3 | Levallois point | 2 | 0.3 | 0 | 0 |
| 5 | Pseudo-Levallois point | 23 | 3.8 | 0 | 0 |
| 6 | Mousterian point | 1 | 0.2 | 1 | 0.6 |
| 9 | Single straight scraper | 5 | 0.8 | 5 | 3.1 |
| 10 | Single convex scraper | 9 | 1.5 | 9 | 5.5 |
| 25 | Scraper on interior surface | 3 | 0.5 | 3 | 1.8 |
| 26 | Abrupt scraper | 2 | 0.3 | 2 | 1.2 |
| 29 | Alternate scraper | 1 | 0.2 | 1 | 0.6 |
| 30 | Typical endscraper | 3 | 0.5 | 3 | 1.8 |
| 32 | Typical burin | 1 | 0.2 | 1 | 0.6 |
| 33 | Atypical burin | 1 | 0.2 | 1 | 0.6 |
| 34 | Typical percoir | 2 | 0.3 | 2 | 1.2 |
| 35 | Atypical percoir | 1 | 0.2 | 1 | 0.6 |
| 37 | Atypical backed knife | 2 | 0.3 | 2 | 1.2 |
| 38 | Naturally backed knife | 42 | 6.9 | 0 | 0 |
| 39 | Raclette | 1 | 0.2 | 1 | 0.6 |
| 40 | Truncation | 2 | 0.3 | 2 | 1.2 |
| 42 | Notch | 29 | 4.8 | 29 | 17.8 |
| 43 | Denticulate | 27 | 4.4 | 27 | 16.6 |
| 44 | Alternate retouched bec | 2 | 0.3 | 2 | 1.2 |
| 45 | Flake with irregular retouch on | 1 | 0.2 | 0 | 0 |
| 48 | Flake with abrupt and alter | 188 | 31 | 0 | 0 |
| 54 | End-notched flake | 6 | 1 | 6 | 3.7 |
| 56 | Rabot | 1 | 0.2 | 1 | 0.6 |
| 58 | Stemmed tool | 1 | 0.2 | 1 | 0.6 |
| 62 | Miscellaneous | 6 | 1 | 6 | 3.7 |
| 64 | Truncated-facetted piece | 57 | 9.4 | 57 | 35 |
| | Totals | 607 | 100.2 | 163 | 99.8 |

**Table 6.11** Artifact counts for Layer 6B

| | *N* |
|---|---|
| Complete and proximal flakes | 1165 |
| Other flake fragments | 558 |
| Complete and proximal tools | 84 |
| Other tool fragments | 78 |
| Handaxes | 1 |
| Handaxe fragments | 0 |
| Shatter | 267 |
| Complete cores | 189 |
| Core fragments | 75 |

**Table 6.12** Typological and technological indices for Layer 6B

| Typological indices | | | Technological indices | |
|---|---|---|---|---|
| | Real | Essential | IL | 15.77 |
| ILty | 31.3 | | IF | 35.27 |
| IR | 3.29 | 12.27 | IFs | 23.45 |
| IAu | 0.33 | 0.01 | Ilam | 9.8 |
| II | 3.46 | 12.88 | TF | 34.97 |
| III | 0.02 | 6.13 | Kombewa | 0.08 |
| IV | 9.23 | 34.36 | Core edge | 2.96 |
| | | | Nat. backed | 8.01 |

**Table 6.13** Basic size measurements for complete flakes, scrapers, notched tools, other retouched tools, and cores from Level 6B

| | | Length | Width | Thickness | Weight |
|---|---|---|---|---|---|
| Flakes | *N* | 655 | 655 | 656 | 654 |
| | Mean | 33.33 | 24.43 | 6.92 | 7.38 |
| | S.D. | 9.31 | 6.8 | 2.86 | 7.02 |
| Scrapers | *N* | 6 | 6 | 6 | 6 |
| | Mean | 48.07 | 32.81 | 8.77 | 19.33 |
| | S.D. | 15.05 | 2.86 | 4.42 | 11.76 |
| Notches | *N* | 28 | 28 | 28 | 28 |
| | Mean | 35.08 | 25.54 | 8.64 | 10 |
| | S.D. | 9.29 | 7.68 | 2.28 | 5.61 |
| Other tools | *N* | 29 | 30 | 30 | 30 |
| | Mean | 39.68 | 26.4 | 10.59 | 14.17 |
| | S.D. | 9.08 | 6.83 | 3.65 | 8.96 |
| Cores | *N* | 189 | 189 | 189 | 189 |
| | Mean | 38.6 | 29.77 | 13.96 | 18.33 |
| | S.D. | 9.07 | 6.43 | 5.11 | 14.07 |

Kombewa flakes; however, truncated-facetted artifacts, while still present, drop to the frequency seen in Layer 8 and scrapers become dominant (again back to Layer 8 frequencies). The scrapers show light levels of retouch. In Layer 5A, the notched artifacts also seem to be on edge damaged flakes. As will be seen later, there is also again an emphasis on tool production (i.e., more blanks being made into retouched tools). The cores, when they can be identified, are single surface, faceted platforms remain frequent, but also core size increases.

The stone tool assemblages change again in Layer 4, though not at the outset. In terms of assemblage size and density, the Layer 4 assemblages are the smallest. Particularly striking are Layers 4B and 4A which are characterized by very low artifact densities. In fact, these layers contain thin lenses of archaeological material within nearly sterile deposits, something not seen in the rest of the sequence. All three layers (4C, 4B, and 4A) are heavily dominated by scrapers (>75%). In Layer 4A, some of the scrapers show Quina retouch. At the same time, the frequency of Levallois, faceted platforms, truncated-faceted, and Kombewa drop to the lowest levels seen in the sequence thus far. Single scrapers remain the most frequent scraper type, but of potential interest is the presence of transverse scrapers in

**Table 6.14** Type counts for Layer 6A

| Type | Name | Real | Percent | Essential | Ess. perc. |
|---|---|---|---|---|---|
| 1 | Levallois flake | 171 | 28.1 | 0 | 0 |
| 2 | Atypical Levallois flake | 70 | 11.5 | 0 | 0 |
| 5 | Pseudo-Levallois point | 12 | 2 | 0 | 0 |
| 9 | Single straight scraper | 16 | 2.6 | 16 | 7.9 |
| 10 | Single convex scraper | 35 | 5.8 | 35 | 17.2 |
| 12 | Double straight scraper | 1 | 0.2 | 1 | 0.5 |
| 15 | Double convex scraper | 1 | 0.2 | 1 | 0.5 |
| 19 | Convex convergent scraper | 1 | 0.2 | 1 | 0.5 |
| 20 | Concave convergent scraper | 1 | 0.2 | 1 | 0.5 |
| 21 | Dejete scraper | 1 | 0.2 | 1 | 0.5 |
| 23 | Convex transverse scraper | 1 | 0.2 | 1 | 0.5 |
| 25 | Scraper on interior surface | 5 | 0.8 | 5 | 2.5 |
| 26 | Abrupt scraper | 1 | 0.2 | 1 | 0.5 |
| 27 | Scraper with thinned back | 1 | 0.2 | 1 | 0.5 |
| 30 | Typical endscraper | 2 | 0.3 | 2 | 1 |
| 31 | Atypical endscraper | 2 | 0.3 | 2 | 1 |
| 32 | Typical burin | 1 | 0.2 | 1 | 0.5 |
| 33 | Atypical burin | 1 | 0.2 | 1 | 0.5 |
| 35 | Atypical percoir | 1 | 0.2 | 1 | 0.5 |
| 37 | Atypical backed knife | 2 | 0.3 | 2 | 1 |
| 38 | Naturally backed knife | 52 | 8.6 | 0 | 0 |
| 40 | Truncation | 3 | 0.5 | 3 | 1.5 |
| 42 | Notch | 42 | 6.9 | 42 | 20.7 |
| 43 | Denticulate | 29 | 4.8 | 29 | 14.3 |
| 45 | Flake with irregular retouch on | 3 | 0.5 | 0 | 0 |
| 48 | Flake with abrupt and alter | 97 | 16 | 0 | 0 |
| 54 | End-notched flake | 3 | 0.5 | 3 | 1.5 |
| 62 | Miscellaneous | 2 | 0.3 | 2 | 1 |
| 64 | Truncated-facetted piece | 51 | 8.4 | 51 | 25.1 |
|  | Totals | 608 | 100.4 | 203 | 100.2 |

**Table 6.15** Artifact counts for Layer 6A

|  | *N* |
|---|---|
| Complete and proximal flakes | 1238 |
| Other flake fragments | 587 |
| Complete and proximal tools | 134 |
| Other tool fragments | 68 |
| Handaxes | 0 |
| Handaxe fragments | 1 |
| Shatter | 213 |
| Complete cores | 115 |
| Core fragments | 64 |

**Table 6.16** Typological and technological indices for Layer 6A

| Typological indices | | | Technological indices | |
|---|---|---|---|---|
| | Real | Essential | IL | 16.55 |
| ILty | 39.64 | | IF | 27.95 |
| IR | 10.53 | 31.53 | IFs | 18.43 |
| IAu | 0.33 | 0.01 | Ilam | 17.25 |
| II | 10.53 | 31.53 | TF | 25.12 |
| III | 0.01 | 4.43 | Kombewa | 0.36 |
| IV | 11.68 | 34.98 | Core edge | 2.04 |
| | | | Nat. backed | 10.5 |

**Table 6.17** Basic size measurements for complete flakes, scrapers, notched tools, other retouched tools, and cores from Level 6A

| | | Length | Width | Thickness | Weight |
|---|---|---|---|---|---|
| Flakes | *N* | 772 | 773 | 772 | 774 |
| | Mean | 38.44 | 26.06 | 6.93 | 9.75 |
| | S.D. | 13.12 | 8.19 | 3.3 | 10.99 |
| Scrapers | *N* | 42 | 42 | 42 | 42 |
| | Mean | 58.99 | 35.64 | 9.51 | 28.9 |
| | S.D. | 16.99 | 12.38 | 4.12 | 29.78 |
| Notches | *N* | 28 | 28 | 28 | 28 |
| | Mean | 49.38 | 31.99 | 9.45 | 21.11 |
| | S.D. | 17.22 | 9.82 | 3.22 | 17.03 |
| Other tools | *N* | 30 | 30 | 30 | 30 |
| | Mean | 39.29 | 29.77 | 8.62 | 12.23 |
| | S.D. | 10.99 | 8.53 | 2.64 | 8.11 |
| Cores | *N* | 115 | 115 | 115 | 113 |
| | Mean | 40.74 | 31.27 | 16.1 | 29.96 |
| | S.D. | 12.75 | 7.97 | 8.08 | 42.69 |

Layer 4A. The scrapers from this layer are also quite thick. As is discussed more below, there seems to be a shift in Layer 4 from heavily Levallois blank production in Layer 5 to more of a Quina-type blank production in Layer 4A. Additionally, in Layers 4A and 4C there is a notable presence of flakes coming from notch production, and the notches are large, Clactonian notches.

Finally, Layer 3 represents again a shift in the lithic assemblages. In terms of quantity and density, Layer 3B is second only to Layer 7 (the post-depositionally altered assemblage). In both these measures, Layer 3A is more average for the sequence. In terms of retouched tool types, notched tools become dominant and reach their highest relative proportion in the whole sequence. Some of the denticulates show a very regular, saw-tooth type pattern of complex notches and are made on thin flakes. In Layer 3A, Upper Paleolithic type tools reach their highest value in the sequence, and of these backed knives (Type 36) are most common. The backed knives are on large flakes. In terms of blank production, Levallois is poorly represented, especially

**Table 6.18** Type counts for Layer 5B

| Type | Name | Real | Percent | Essential | Ess. perc. |
|---|---|---|---|---|---|
| 1 | Levallois flake | 24 | 9.7 | 0 | 0 |
| 2 | Atypical Levallois flake | 9 | 3.6 | 0 | 0 |
| 3 | Levallois point | 2 | 0.8 | 0 | 0 |
| 5 | Pseudo-Levallois point | 2 | 0.8 | 0 | 0 |
| 9 | Single straight scraper | 10 | 4 | 10 | 17.5 |
| 10 | Single convex scraper | 11 | 4.5 | 11 | 19.3 |
| 11 | Single concave scraper | 1 | 0.4 | 1 | 1.8 |
| 13 | Double straight-convex scraper | 1 | 0.4 | 1 | 1.8 |
| 15 | Double convex scraper | 2 | 0.8 | 2 | 3.5 |
| 19 | Convex convergent scraper | 1 | 0.4 | 1 | 1.8 |
| 21 | Dejete scraper | 1 | 0.4 | 1 | 1.8 |
| 22 | Straight transverse scraper | 1 | 0.4 | 1 | 1.8 |
| 25 | Scraper on interior surface | 1 | 0.4 | 1 | 1.8 |
| 26 | Abrupt scraper | 1 | 0.4 | 1 | 1.8 |
| 27 | Scraper with thinned back | 1 | 0.4 | 1 | 1.8 |
| 31 | Atypical endscraper | 2 | 0.8 | 2 | 3.5 |
| 32 | Typical burin | 2 | 0.8 | 2 | 3.5 |
| 34 | Typical percoir | 1 | 0.4 | 1 | 1.8 |
| 38 | Naturally backed knife | 19 | 7.7 | 0 | 0 |
| 40 | Truncation | 2 | 0.8 | 2 | 3.5 |
| 42 | Notch | 7 | 2.8 | 7 | 12.3 |
| 43 | Denticulate | 4 | 1.6 | 4 | 7 |
| 48 | Flake with abrupt and alter | 134 | 54.3 | 0 | 0 |
| 54 | End-notched flake | 2 | 0.8 | 2 | 3.5 |
| 62 | Miscellaneous | 2 | 0.8 | 2 | 3.5 |
| 64 | Truncated-facetted piece | 4 | 1.6 | 4 | 7 |
| | Totals | 247 | 99.8 | 57 | 100.3 |

**Table 6.19** Artifact counts for Layer 5B

| | *N* |
|---|---|
| Complete and proximal flakes | 323 |
| Other flake fragments | 207 |
| Complete and proximal tools | 41 |
| Other tool fragments | 15 |
| Handaxes | 0 |
| Handaxe fragments | 0 |
| Shatter | 65 |
| Complete cores | 26 |
| Core fragments | 17 |

**Table 6.20** Typological and technological indices for Layer 5B

| Typological indices | | | Technological indices | |
|---|---|---|---|---|
| | Real | Essential | IL | 11.81 |
| ILty | 14.17 | | IF | 26.78 |
| IR | 12.55 | 54.39 | IFs | 18.98 |
| IAu | 0 | 0 | Ilam | 19.05 |
| II | 12.55 | 54.39 | TF | 7.02 |
| III | 0.02 | 8.77 | Kombewa | 0.27 |
| IV | 4.45 | 19.3 | Core edge | 2.2 |
| | | | Nat. backed | 11.26 |

**Table 6.21** Basic size measurements for complete flakes, scrapers, notched tools, other retouched tools, and cores from Level 5B

| | | Length | Width | Thickness | Weight |
|---|---|---|---|---|---|
| Flakes | *N* | 189 | 189 | 187 | 190 |
| | Mean | 37.03 | 24.76 | 7.39 | 9.43 |
| | S.D. | 10.66 | 6.41 | 3.17 | 8.53 |
| Scrapers | *N* | 23 | 23 | 23 | 23 |
| | Mean | 55.89 | 32.75 | 9.69 | 25.96 |
| | S.D. | 14.71 | 9.6 | 3.76 | 18.34 |
| Notches | *N* | 6 | 6 | 6 | 6 |
| | Mean | 35.92 | 27.37 | 8.78 | 10 |
| | S.D. | 10.9 | 9.35 | 3.17 | 5.97 |
| Other tools | *N* | 8 | 8 | 8 | 8 |
| | Mean | 41.86 | 28.9 | 8.72 | 13.38 |
| | S.D. | 9.9 | 4.77 | 1.7 | 5.18 |
| Cores | *N* | 26 | 26 | 26 | 26 |
| | Mean | 47.81 | 34.43 | 18.93 | 36.08 |
| | S.D. | 8.79 | 7.47 | 8.49 | 23.49 |

in Layer 3A where it is the lowest in the sequence. Likewise, the frequency of faceted platforms is also low. In terms of core types, single-surface cores are the most frequent with a scattering of other types. Layer 3A, and to a lesser extent Layer 3B, have a very high percentage of pseudo-Levallois points (Type 5). This, the high percentage of core edge flakes, and a high frequency of what we called globular core shapes suggest a shift toward discoidal flake production at the top of the sequence.

## Raw Material Types and Sources (Prepared by A. Turq)

All studies of raw materials variability begin with an estimation of the resources available in the region which means beginning with the geology (Demars 1982; Morala 1983; Geneste 1985; Seronie-Vivien 1987; Turq 2000). The Sarlat region where Pech IV is located corresponds to a “synclinorium”-oriented southeast–northwest with Upper

**Table 6.22** Type counts for Layer 5A

| Type | Name | Real | Percent | Essential | Ess. perc. |
|---|---|---|---|---|---|
| 1 | Levallois flake | 63 | 23.6 | 0 | 0 |
| 2 | Atypical Levallois flake | 27 | 10.1 | 0 | 0 |
| 3 | Levallois point | 3 | 1.1 | 0 | 0 |
| 5 | Pseudo-Levallois point | 9 | 3.4 | 0 | 0 |
| 6 | Mousterian point | 1 | 0.4 | 1 | 1 |
| 7 | Elongated mousterian point | 2 | 0.7 | 2 | 2 |
| 9 | Single straight scraper | 12 | 4.5 | 12 | 11.9 |
| 10 | Single convex scraper | 31 | 11.6 | 31 | 30.7 |
| 11 | Single concave scraper | 1 | 0.4 | 1 | 1 |
| 12 | Double straight scraper | 1 | 0.4 | 1 | 1 |
| 13 | Double straight-convex scraper | 2 | 0.7 | 2 | 2 |
| 15 | Double convex scraper | 1 | 0.4 | 1 | 1 |
| 17 | Double convex–concave scraper | 1 | 0.4 | 1 | 1 |
| 18 | Straight convergent scraper | 1 | 0.4 | 1 | 1 |
| 19 | Convex convergent scraper | 6 | 2.2 | 6 | 5.9 |
| 21 | Dejete scraper | 2 | 0.7 | 2 | 2 |
| 22 | Straight transverse scraper | 3 | 1.1 | 3 | 3 |
| 23 | Convex transverse scraper | 1 | 0.4 | 1 | 1 |
| 25 | Scraper on interior surface | 1 | 0.4 | 1 | 1 |
| 30 | Typical endscraper | 1 | 0.4 | 1 | 1 |
| 32 | Typical burin | 1 | 0.4 | 1 | 1 |
| 33 | Atypical burin | 1 | 0.4 | 1 | 1 |
| 34 | Typical percoir | 2 | 0.7 | 2 | 2 |
| 35 | Atypical percoir | 1 | 0.4 | 1 | 1 |
| 36 | Typical backed knife | 1 | 0.4 | 1 | 1 |
| 37 | Atypical backed knife | 1 | 0.4 | 1 | 1 |
| 38 | Naturally backed knife | 44 | 16.5 | 0 | 0 |
| 39 | Raclette | 1 | 0.4 | 1 | 1 |
| 42 | Notch | 13 | 4.9 | 13 | 12.9 |
| 43 | Denticulate | 3 | 1.1 | 3 | 3 |
| 45 | Flake with irregular retouch on | 2 | 0.7 | 0 | 0 |
| 48 | Flake with abrupt and alter | 18 | 6.7 | 0 | 0 |
| 54 | End-notched flake | 3 | 1.1 | 3 | 3 |
| 64 | Truncated-facetted piece | 7 | 2.6 | 7 | 6.9 |
| | Totals | 267 | 100 | 101 | 100.3 |

Cretaceous formations (Turonian, Coniacian, Santonian, and the base of the Campanian) (Fig. 6.1). Within this structure, there are several anticlinal ridges, one of which contains Pech IV (Turq et al. 1999), which contributed to the development of the karstic system that provided shelter during Paleolithic times and which permitted access to layers

**Table 6.23** Artifact counts for Layer 5A

| | *N* |
|---|---|
| Complete and proximal flakes | 854 |
| Other flake fragments | 345 |
| Complete and proximal tools | 63 |
| Other tool fragments | 38 |
| Handaxes | 1 |
| Handaxe fragments | 0 |
| Shatter | 76 |
| Complete cores | 24 |
| Core fragments | 23 |

**Table 6.24** Typological and technological indices for Layer 5A

| Typological indices | | | Technological indices | |
|---|---|---|---|---|
| | Real | Essential | IL | 11.12 |
| ILty | 34.83 | | IF | 33.09 |
| IR | 23.6 | 62.38 | IFs | 22.38 |
| IAu | 0.75 | 0.02 | Ilam | 21.26 |
| II | 24.72 | 65.35 | TF | 6.93 |
| III | 0.03 | 7.92 | Kombewa | 0.11 |
| IV | 5.99 | 15.84 | Core edge | 2.51 |
| | | | Nat. backed | 12.76 |

**Table 6.25** Basic size measurements for complete flakes, scrapers, notched tools, other retouched tools, and cores from Level 5A

| | | Length | Width | Thickness | Weight |
|---|---|---|---|---|---|
| Flakes | *N* | 603 | 602 | 603 | 598 |
| | Mean | 36.33 | 23.52 | 6.21 | 7.35 |
| | S.D. | 11.2 | 7.28 | 3.23 | 15.45 |
| Scrapers | *N* | 34 | 34 | 34 | 34 |
| | Mean | 61.75 | 33.84 | 8.87 | 23.56 |
| | S.D. | 12.36 | 6.56 | 2.77 | 11.65 |
| Notches | *N* | 9 | 9 | 9 | 9 |
| | Mean | 42.65 | 30.44 | 8.7 | 13.78 |
| | S.D. | 8.03 | 5.1 | 3.94 | 9 |
| Other tools | *N* | 6 | 6 | 6 | 6 |
| | Mean | 43.87 | 30.75 | 9.48 | 18.33 |
| | S.D. | 9.54 | 9.84 | 5.07 | 14.71 |
| Cores | *N* | 24 | 24 | 24 | 24 |
| | Mean | 49.64 | 35.67 | 19.84 | 43.42 |
| | S.D. | 12.42 | 8.54 | 8.48 | 36.84 |

of flint at the base of the Coniacian. This anticlinal ridge, which is seen only along the Farge stream, measures only 2 km in length.

In the surrounding area, Coniacian flint appears in limestones a few meters from the Pech de l'Azé sites on the hillslopes and along the valley in front of the site and more

**Table 6.26** Type counts for Layer 4C

| Type | Name | Real | Percent | Essential | Ess. perc. |
|---|---|---|---|---|---|
| 1 | Levallois flake | 15 | 11.6 | 0 | 0 |
| 2 | Atypical Levallois flake | 10 | 7.8 | 0 | 0 |
| 3 | Levallois point | 2 | 1.6 | 0 | 0 |
| 5 | Pseudo-Levallois point | 6 | 4.7 | 0 | 0 |
| 6 | Mousterian point | 1 | 0.8 | 1 | 1.3 |
| 9 | Single straight scraper | 8 | 6.2 | 8 | 10.1 |
| 10 | Single convex scraper | 26 | 20.2 | 26 | 32.9 |
| 11 | Single concave scraper | 2 | 1.6 | 2 | 2.5 |
| 13 | Double straight-convex scraper | 2 | 1.6 | 2 | 2.5 |
| 15 | Double convex scraper | 4 | 3.1 | 4 | 5.1 |
| 19 | Convex convergent scraper | 5 | 3.9 | 5 | 6.3 |
| 21 | Dejete scraper | 6 | 4.7 | 6 | 7.6 |
| 22 | Straight transverse scraper | 2 | 1.6 | 2 | 2.5 |
| 23 | Convex transverse scraper | 3 | 2.3 | 3 | 3.8 |
| 25 | Scraper on interior surface | 1 | 0.8 | 1 | 1.3 |
| 27 | Scraper with thinned back | 1 | 0.8 | 1 | 1.3 |
| 32 | Typical burin | 2 | 1.6 | 2 | 2.5 |
| 38 | Naturally backed knife | 11 | 8.5 | 0 | 0 |
| 42 | Notch | 2 | 1.6 | 2 | 2.5 |
| 43 | Denticulate | 8 | 6.2 | 8 | 10.1 |
| 44 | Alternate retouched bec | 1 | 0.8 | 1 | 1.3 |
| 48 | Flake with abrupt and alter | 6 | 4.7 | 0 | 0 |
| 51 | Tayac point | 1 | 0.8 | 1 | 1.3 |
| 54 | End-notched flake | 2 | 1.6 | 2 | 2.5 |
| 61 | Chopping-tool | 1 | 0.8 | 1 | 1.3 |
| 64 | Truncated-facetted piece | 1 | 0.8 | 1 | 1.3 |
| | Totals | 129 | 100.7 | 79 | 100 |

**Table 6.27** Artifact counts for Layer 4C

| | *N* |
|---|---|
| Complete and proximal flakes | 354 |
| Other flake fragments | 132 |
| Complete and proximal tools | 53 |
| Other tool fragments | 25 |
| Handaxes | 0 |
| Handaxe fragments | 0 |
| Shatter | 41 |
| Complete cores | 12 |
| Core fragments | 8 |

generally throughout the heart of the anticline. In situ Coniacian flint can be found today, to the north in the Enéa valley (7 km toward Saint-Nathalène), to the south–southeast in the Dordogne Valley, slightly upstream from Domme (5 km), and on the north side of the Dordogne River at Vitrac (4 km). In secondary positions, Coniacian flint is

**Table 6.28** Typological and technological indices for Layer 4C

| Typological indices | | | Technological indices | |
|---|---|---|---|---|
| | Real | Essential | IL | 7.62 |
| ILty | 20.93 | | IF | 22.13 |
| IR | 46.51 | 75.95 | IFs | 11.75 |
| IAu | 0 | 0 | Ilam | 12.45 |
| II | 47.29 | 77.22 | TF | 1.27 |
| III | 0.02 | 2.53 | Kombewa | 0 |
| IV | 7.75 | 12.66 | Core edge | 3.44 |
| | | | Nat. backed | 7.62 |

**Table 6.29** Basic size measurements for complete flakes, scrapers, notched tools, other retouched tools, and cores from Level 4C

| | | Length | Width | Thickness | Weight |
|---|---|---|---|---|---|
| Flakes | *N* | 242 | 241 | 242 | 241 |
| | Mean | 33.67 | 24.94 | 6.28 | 7.27 |
| | S.D. | 11.81 | 8.04 | 3.17 | 9.4 |
| Scrapers | *N* | 32 | 32 | 32 | 32 |
| | Mean | 56.17 | 36.03 | 10 | 30.28 |
| | S.D. | 15.99 | 11.3 | 4.12 | 25.6 |
| Notches | *N* | 6 | 6 | 6 | 6 |
| | Mean | 45.9 | 26.19 | 10.33 | 20.67 |
| | S.D. | 16.3 | 4.79 | 5.04 | 14.33 |
| Other tools | *N* | 2 | 2 | 2 | 2 |
| | Mean | 52.07 | 43.63 | 17.46 | 38.5 |
| | S.D. | 7.88 | 3.14 | 1.28 | 0.71 |
| Cores | *N* | 12 | 12 | 12 | 12 |
| | Mean | 51.99 | 41.5 | 18.67 | 58.08 |
| | S.D. | 16.47 | 12.89 | 9.89 | 59 |

**Table 6.30** Type counts for Layer 4B

| Type | Name | Real | Percent | Essential | Ess. perc. |
|---|---|---|---|---|---|
| 1 | Levallois flake | 1 | 9.1 | 0 | 0 |
| 2 | Atypical Levallois flake | 1 | 9.1 | 0 | 0 |
| 9 | Single straight scraper | 1 | 9.1 | 1 | 14.3 |
| 10 | Single convex scraper | 3 | 27.3 | 3 | 42.9 |
| 19 | Convex convergent scraper | 1 | 9.1 | 1 | 14.3 |
| 21 | Dejete scraper | 1 | 9.1 | 1 | 14.3 |
| 38 | Naturally backed knife | 1 | 9.1 | 0 | 0 |
| 48 | Flake with abrupt and alter | 1 | 9.1 | 0 | 0 |
| 54 | End-notched flake | 1 | 9.1 | 1 | 14.3 |
| | Totals | 11 | 100.1 | 7 | 100.1 |

present in the *altérites* a few hundred meters west of Pech IV, but more frequently in the river deposits of the Farge, the Enéa, and the Dordogne, including upstream of the confluence of the Dordogne and Enéa.

Less abundant than Coniacian flint, other kinds of flint exist at the base of the upper part of the Santonian in a thin, chalky white limestone (Capdeville 1988: 14). This flint can be found in secondary positions in the *altérites* that cover flat

**Table 6.31** Artifact counts for Layer 4B

| | *N* |
|---|---|
| Complete and proximal flakes | 27 |
| Other flake fragments | 7 |
| Complete and proximal tools | 4 |
| Other tool fragments | 3 |
| Handaxes | 0 |
| Handaxe fragments | 0 |
| Shatter | 6 |
| Complete cores | 2 |
| Core fragments | 0 |

**Table 6.32** Typological and technological indices for Layer 4B

| Typological indices | | | Technological indices | |
|---|---|---|---|---|
| | Real | Essential | IL | 9.68 |
| ILty | 18.18 | | IF | 11.11 |
| IR | 54.55 | 85.71 | IFs | 7.41 |
| IAu | 0 | 0 | Ilam | 18.18 |
| II | 54.55 | 85.71 | TF | 0 |
| III | 0 | 0 | Kombewa | 0 |
| IV | 0 | 0 | Core edge | 0 |
| | | | Nat. backed | 9.68 |

**Table 6.33** Basic size measurements for complete flakes, scrapers, notched tools, other retouched tools, and cores from Level 4B

| | | Length | Width | Thickness | Weight |
|---|---|---|---|---|---|
| Flakes | *N* | 22 | 22 | 22 | 22 |
| | Mean | 36.63 | 26.15 | 7.74 | 10 |
| | S.D. | 13.19 | 6.79 | 4.32 | 9.05 |
| Scrapers | *N* | 4 | 4 | 4 | 4 |
| | Mean | 52.2 | 35.24 | 12.29 | 32 |
| | S.D. | 14.46 | 3.01 | 4.78 | 15.98 |
| Cores | *N* | 2 | 2 | 2 | 2 |
| | Mean | 84.48 | 66.8 | 50.07 | 483.5 |
| | S.D. | 43.63 | 40.92 | 35.18 | 564.98 |

areas, or within the alluvial deposits of the Enéa. Campanian flint occurs only to the north of Sarlat on highlands and can be locally silicified.

Chalcedony can have two origins in the Pech IV region: geodes forming in the base of the Cretaceous and silicification (*meulérisation*) of Tertiary limestone. The former exists in discrete locations in the Enéa Valley. The latter exists on the south side of the Dordogne, at the plateau Plaine de Bor, and nearby but further upslope in the forest, in the *altérites* and slope deposits. This formation, drained by

**Table 6.34** Type counts for Layer 4A

| Type | Name | Real | Percent | Essential | Ess. perc. |
|---|---|---|---|---|---|
| 1 | Levallois flake | 10 | 19.6 | 0 | 0 |
| 2 | Atypical Levallois flake | 3 | 5.9 | 0 | 0 |
| 5 | Pseudo-Levallois point | 3 | 5.9 | 0 | 0 |
| 9 | Single straight scraper | 4 | 7.8 | 4 | 15.4 |
| 10 | Single convex scraper | 9 | 17.6 | 9 | 34.6 |
| 15 | Double convex scraper | 1 | 2 | 1 | 3.8 |
| 23 | Convex transverse scraper | 5 | 9.8 | 5 | 19.2 |
| 27 | Scraper with thinned back | 1 | 2 | 1 | 3.8 |
| 30 | Typical endscraper | 1 | 2 | 1 | 3.8 |
| 38 | Naturally backed knife | 5 | 9.8 | 0 | 0 |
| 42 | Notch | 1 | 2 | 1 | 3.8 |
| 43 | Denticulate | 2 | 3.9 | 2 | 7.7 |
| 48 | Flake with abrupt and alter | 4 | 7.8 | 0 | 0 |
| 54 | End-notched flake | 1 | 2 | 1 | 3.8 |
| 64 | Truncated-facetted piece | 1 | 2 | 1 | 3.8 |
| | Totals | 51 | 100.1 | 26 | 99.7 |

**Table 6.35** Artifact counts for Layer 4A

| | *N* |
|---|---|
| Complete and proximal flakes | 114 |
| Other flake fragments | 60 |
| Complete and proximal tools | 18 |
| Other tool fragments | 8 |
| Handaxes | 0 |
| Handaxe fragments | 0 |
| Shatter | 27 |
| Complete cores | 3 |
| Core fragments | 1 |

**Table 6.36** Typological and technological indices for Layer 4A

| Typological indices | | | Technological indices | |
|---|---|---|---|---|
| | Real | Essential | IL | 7.58 |
| ILty | 25.49 | | IF | 13.21 |
| IR | 39.22 | 76.92 | IFs | 6.6 |
| IAu | 0 | 0 | Ilam | 11.9 |
| II | 39.22 | 76.92 | TF | 3.85 |
| III | 0.02 | 3.85 | Kombewa | 0 |
| IV | 5.88 | 11.54 | Core edge | 1.52 |
| | | | Nat. backed | 9.09 |

**Table 6.37** Basic size measurements for complete flakes, scrapers, notched tools, other retouched tools, and cores from Level 4A

| | | Length | Width | Thickness | Weight |
|---|---|---|---|---|---|
| Flakes | *N* | 84 | 84 | 84 | 84 |
| | Mean | 34.71 | 25.74 | 6.77 | 9.45 |
| | S.D. | 11.56 | 8.94 | 4.12 | 13.16 |
| Scrapers | *N* | 10 | 10 | 10 | 10 |
| | Mean | 52.11 | 40.49 | 14.45 | 47 |
| | S.D. | 11.31 | 11.36 | 5.89 | 28.4 |
| Notches | *N* | 2 | 2 | 2 | 3 |
| | Mean | 21.38 | 21.63 | 3.88 | 4.67 |
| | S.D. | 0.16 | 2.71 | 0.79 | 4.73 |
| Other tools | *N* | 1 | 1 | 1 | 1 |
| | Mean | 87.89 | 35.86 | 16.83 | 67 |
| | S.D. | NA | NA | NA | NA |
| Cores | *N* | 3 | 3 | 3 | 3 |
| | Mean | 47.63 | 38.56 | 24.4 | 39 |
| | S.D. | 8.91 | 4.91 | 16.91 | 12.12 |

**Table 6.38** Type counts for Layer 3B

| Type | Name | Real | Percent | Essential | Ess. perc. |
|---|---|---|---|---|---|
| 1 | Levallois flake | 98 | 18 | 0 | 0 |
| 2 | Atypical Levallois flake | 38 | 7 | 0 | 0 |
| 3 | Levallois point | 2 | 0.4 | 0 | 0 |
| 5 | Pseudo-Levallois point | 21 | 3.9 | 0 | 0 |
| 9 | Single straight scraper | 5 | 0.9 | 5 | 2.2 |
| 10 | Single convex scraper | 16 | 2.9 | 16 | 7 |
| 11 | Single concave scraper | 4 | 0.7 | 4 | 1.7 |
| 13 | Double straight-convex scraper | 3 | 0.6 | 3 | 1.3 |
| 15 | Double convex scraper | 1 | 0.2 | 1 | 0.4 |
| 19 | Convex convergent scraper | 2 | 0.4 | 2 | 0.9 |
| 22 | Straight transverse scraper | 2 | 0.4 | 2 | 0.9 |
| 23 | Convex transverse scraper | 1 | 0.2 | 1 | 0.4 |
| 25 | Scraper on interior surface | 1 | 0.2 | 1 | 0.4 |
| 26 | Abrupt scraper | 2 | 0.4 | 2 | 0.9 |
| 30 | Typical endscraper | 2 | 0.4 | 2 | 0.9 |
| 32 | Typical burin | 5 | 0.9 | 5 | 2.2 |
| 34 | Typical percoir | 3 | 0.6 | 3 | 1.3 |
| 35 | Atypical percoir | 3 | 0.6 | 3 | 1.3 |
| 36 | Typical backed knife | 1 | 0.2 | 1 | 0.4 |
| 37 | Atypical backed knife | 1 | 0.2 | 1 | 0.4 |
| 38 | Naturally backed knife | 60 | 11 | 0 | 0 |
| 39 | Raclette | 1 | 0.2 | 1 | 0.4 |
| 40 | Truncation | 4 | 0.7 | 4 | 1.7 |
| 42 | Notch | 73 | 13.4 | 73 | 31.7 |
| 43 | Denticulate | 72 | 13.2 | 72 | 31.3 |
| 44 | Alternate retouched bec | 1 | 0.2 | 1 | 0.4 |
| 48 | Flake with abrupt and alter | 96 | 17.6 | 0 | 0 |

(continued)

**Table 6.38** (continued)

| Type | Name | Real | Percent | Essential | Ess. perc. |
|---|---|---|---|---|---|
| 50 | Bifacially retouched flake | 1 | 0.2 | 1 | 0.4 |
| 51 | Tayac point | 2 | 0.4 | 2 | 0.9 |
| 54 | End-notched flake | 12 | 2.2 | 12 | 5.2 |
| 56 | Rabot | 1 | 0.2 | 1 | 0.4 |
| 59 | Chopper | 1 | 0.2 | 1 | 0.4 |
| 62 | Miscellaneous | 1 | 0.2 | 1 | 0.4 |
| 64 | Truncated-facetted piece | 9 | 1.7 | 9 | 3.9 |
| | Totals | 545 | 100.5 | 230 | 99.7 |

**Table 6.39** Artifact counts for Layer 3B

| | *N* |
|---|---|
| Complete and proximal flakes | 1710 |
| Other flake fragments | 583 |
| Complete and proximal tools | 150 |
| Other tool fragments | 78 |
| Handaxes | 1 |
| Handaxe fragments | 0 |
| Shatter | 239 |
| Complete cores | 52 |
| Core fragments | 38 |

**Table 6.40** Typological and technological indices for Layer 3B

| Typological indices | | | Technological indices | |
|---|---|---|---|---|
| | Real | Essential | IL | 6.88 |
| ILty | 25.32 | | IF | 17.73 |
| IR | 6.79 | 16.09 | IFs | 11.01 |
| IAu | 0.37 | 0.01 | Ilam | 14.29 |
| II | 6.79 | 16.09 | TF | 3.91 |
| III | 0.03 | 6.52 | Kombewa | 0.11 |
| IV | 26.61 | 63.04 | Core edge | 2.74 |
| | | | Nat. backed | 10.81 |

the hydrological system of the Germaine and the Céou, is largely fed by Tertiary silicification of the Dordogne alluvial deposits where they mix with those of the Massif Central.

Raw material sources in Quaternary deposits are particularly rich in terms of quantity, quality, and in their being resupplied after each flood (a mixing of alluvial materials and transported materials, as well as material eroded from the stream banks). The raw material resources differ between the valleys of the Dordogne and Enéa. The latter transports and contains only material of local origin (limestone, flints, and sandstones) coming from the slopes along the drainages, whereas the Dordogne system includes local and upstream

**Table 6.41** Basic size measurements for complete flakes, scrapers, notched tools, other retouched tools, and cores from Level 3B

| | | Length | Width | Thickness | Weight |
|---|---|---|---|---|---|
| Flakes | *N* | 1226 | 1226 | 1219 | 1224 |
| | Mean | 32.69 | 23.33 | 6.91 | 6.87 |
| | S.D. | 9.35 | 6.79 | 3.23 | 6.71 |
| Scrapers | *N* | 13 | 13 | 13 | 13 |
| | Mean | 37.66 | 25.2 | 8.6 | 11.69 |
| | S.D. | 11.82 | 5.79 | 3.54 | 9.35 |
| Notches | *N* | 78 | 78 | 78 | 77 |
| | Mean | 37.74 | 26.22 | 9.57 | 12.18 |
| | S.D. | 12.64 | 6.73 | 3.36 | 10.06 |
| Other tools | *N* | 17 | 17 | 17 | 17 |
| | Mean | 37.51 | 25.97 | 9.33 | 12.71 |
| | S.D. | 12.47 | 5.83 | 3.75 | 8.51 |
| Cores | *N* | 50 | 49 | 50 | 50 |
| | Mean | 45.76 | 33.99 | 17.07 | 32.14 |
| | S.D. | 12.09 | 10.57 | 6.46 | 26.02 |

**Table 6.42** Type counts for Layer 3A

| Type | Name | Real | Percent | Essential | Ess. perc. |
|---|---|---|---|---|---|
| 1 | Levallois flake | 17 | 6.2 | 0 | 0 |
| 2 | Atypical Levallois flake | 9 | 3.3 | 0 | 0 |
| 3 | Levallois point | 2 | 0.7 | 0 | 0 |
| 5 | Pseudo-Levallois point | 42 | 15.3 | 0 | 0 |
| 9 | Single straight scraper | 7 | 2.6 | 7 | 5.3 |
| 10 | Single convex scraper | 4 | 1.5 | 4 | 3 |
| 11 | Single concave scraper | 1 | 0.4 | 1 | 0.8 |
| 23 | Convex transverse scraper | 1 | 0.4 | 1 | 0.8 |
| 25 | Scraper on interior surface | 1 | 0.4 | 1 | 0.8 |
| 30 | Typical endscraper | 1 | 0.4 | 1 | 0.8 |
| 32 | Typical burin | 3 | 1.1 | 3 | 2.3 |
| 34 | Typical percoir | 2 | 0.7 | 2 | 1.5 |
| 36 | Typical backed knife | 5 | 1.8 | 5 | 3.8 |
| 37 | Atypical backed knife | 1 | 0.4 | 1 | 0.8 |
| 38 | Naturally backed knife | 37 | 13.5 | 0 | 0 |
| 40 | Truncation | 1 | 0.4 | 1 | 0.8 |
| 42 | Notch | 44 | 16.1 | 44 | 33.3 |
| 43 | Denticulate | 45 | 16.4 | 45 | 34.1 |
| 44 | Alternate retouched bec | 1 | 0.4 | 1 | 0.8 |
| 48 | Flake with abrupt and alter | 35 | 12.8 | 0 | 0 |
| 54 | End-notched flake | 2 | 0.7 | 2 | 1.5 |
| 62 | Miscellaneous | 5 | 1.8 | 5 | 3.8 |
| 64 | Truncated-facetted piece | 7 | 2.6 | 7 | 5.3 |
| 65 | Scraper on the platform | 1 | 0.4 | 1 | 0.8 |
| | Totals | 274 | 100.3 | 132 | 100.3 |

**Table 6.43** Artifact counts for Layer 3A

| | *N* |
|---|---|
| Complete and proximal flakes | 997 |
| Other flake fragments | 270 |
| Complete and proximal tools | 104 |
| Other tool fragments | 28 |
| Handaxes | 0 |
| Handaxe fragments | 0 |
| Shatter | 153 |
| Complete cores | 29 |
| Core fragments | 21 |

**Table 6.44** Typological and technological indices for Layer 3A

| Typological indices | | | Technological indices | |
|---|---|---|---|---|
| | Real | Essential | IL | 2.36 |
| ILty | 10.22 | | IF | 12.67 |
| IR | 5.11 | 10.61 | IFs | 6.49 |
| IAu | 2.19 | 0.05 | Ilam | 13.37 |
| II | 5.11 | 10.61 | TF | 5.3 |
| III | 0.04 | 9.09 | Kombewa | 0 |
| IV | 32.48 | 67.42 | Core edge | 6.81 |
| | | | Nat. backed | 12.08 |

**Table 6.45** Basic size measurements for complete flakes, scrapers, notched tools, other retouched tools, and cores from Level 3A

| | | Length | Width | Thickness | Weight |
|---|---|---|---|---|---|
| Flakes | *N* | 750 | 748 | 742 | 749 |
| | Mean | 32.31 | 23.27 | 7.77 | 7.65 |
| | S.D. | 9.02 | 7.16 | 3.35 | 8.03 |
| Scrapers | *N* | 7 | 7 | 7 | 7 |
| | Mean | 45.82 | 29.25 | 9.31 | 13.57 |
| | S.D. | 16.73 | 11.76 | 3.4 | 11.76 |
| Notches | *N* | 47 | 47 | 47 | 47 |
| | Mean | 39.55 | 27.49 | 11.07 | 15.06 |
| | S.D. | 12.32 | 7.96 | 3.64 | 10.43 |
| Other tools | *N* | 19 | 19 | 19 | 19 |
| | Mean | 40.5 | 29.45 | 11.53 | 17.16 |
| | S.D. | 12.54 | 10.26 | 2.98 | 12.44 |
| Cores | *N* | 29 | 29 | 29 | 29 |
| | Mean | 42.61 | 32.66 | 19.47 | 34.45 |
| | S.D. | 12.83 | 7.81 | 6.6 | 26.17 |

**Table 6.46** Summary technological and typological data

| | Group II | Group III | Group IV | Il | Ifs | Tf | Kombewa | Core edge | Nat. backed | Type 5s |
|---|---|---|---|---|---|---|---|---|---|---|
| 3A | 10.61 | 9.09 | 67.42 | 2.36 | 6.49 | 5.3 | 0 | 6.81 | 12.08 | 15 |
| 3B | 16.09 | 6.52 | 63.04 | 6.88 | 11.01 | 3.91 | 0.11 | 2.74 | 10.81 | 4 |
| 4A | 76.92 | 3.85 | 11.54 | 7.58 | 6.6 | 3.85 | 0 | 1.52 | 9.09 | 6 |
| 4B | 85.71 | 0 | 0 | 9.68 | 7.41 | 0 | 0 | 0 | 9.68 | 0 |
| 4C | 77.22 | 2.53 | 12.66 | 7.62 | 11.75 | 1.27 | 0 | 3.44 | 7.62 | 5 |
| 5A | 65.35 | 7.92 | 15.84 | 11.12 | 22.38 | 6.93 | 0.11 | 2.51 | 12.76 | 3 |
| 5B | 54.39 | 8.77 | 19.3 | 11.81 | 18.98 | 7.02 | 0.27 | 2.2 | 11.26 | 1 |
| 6A | 31.53 | 4.43 | 34.98 | 16.55 | 18.43 | 25.12 | 0.36 | 2.04 | 10.5 | 2 |
| 6B | 12.88 | 6.13 | 34.36 | 15.77 | 23.45 | 34.97 | 0.08 | 2.96 | 8.01 | 4 |
| 7 | 28.02 | 6.47 | 43.53 | 4.15 | 18.08 | 16.38 | 0.25 | 1.72 | 4.8 | 1 |
| 8 | 68.15 | 5.1 | 15.92 | 12.16 | 17.79 | 8.28 | 0 | 1.09 | 7.96 | 3 |

**Table 6.47** Core types (completes and broken) by layer

| Level | Biface | Chopper | Chopping-tool | Disc-core | Globular | Inform | Kombewa | Levallois | Moust-disk | Other | Prismatic | Pyramidal | Sing-surf | Tested |
|---|---|---|---|---|---|---|---|---|---|---|---|---|---|---|
| 3A | 0 | 0 | 0 | 0 | 9 | 8 | 3 | 1 | 0 | 0 | 1 | 1 | 11 | 2 |
| 3B | 0 | 0 | 0 | 2 | 0 | 8 | 6 | 1 | 0 | 2 | 1 | 1 | 54 | 1 |
| 4A | 0 | 0 | 0 | 0 | 0 | 0 | 0 | 0 | 0 | 0 | 0 | 0 | 3 | 0 |
| 4B | 0 | 0 | 0 | 0 | 0 | 0 | 0 | 0 | 0 | 1 | 0 | 0 | 0 | 1 |
| 4C | 0 | 1 | 0 | 0 | 2 | 2 | 2 | 1 | 0 | 2 | 0 | 0 | 3 | 0 |
| 5A | 0 | 0 | 0 | 0 | 0 | 5 | 4 | 0 | 0 | 2 | 0 | 0 | 20 | 0 |
| 5B | 0 | 0 | 0 | 0 | 2 | 11 | 4 | 2 | 0 | 5 | 0 | 0 | 14 | 0 |
| 6A | 0 | 0 | 0 | 1 | 4 | 10 | 18 | 12 | 0 | 5 | 1 | 2 | 101 | 3 |
| 6B | 2 | 0 | 1 | 1 | 2 | 28 | 16 | 24 | 2 | 25 | 1 | 0 | 136 | 0 |
| 7 | 0 | 1 | 2 | 5 | 11 | 33 | 29 | 10 | 1 | 21 | 0 | 1 | 99 | 3 |
| 8 | 0 | 0 | 1 | 0 | 0 | 5 | 3 | 2 | 1 | 3 | 0 | 0 | 19 | 1 |

materials (metamorphic and plutonic rocks, quartz, and quartzites) and other silicified material, such as jasper of the Hettangian (Jurassic) and Tertiary flints.

As is typical for European Middle Paleolithic sites in general, the majority of the lithic assemblage is composed of locally available raw materials (Table 6.48). There are sources of reasonable quality flints available within the region of the Pech de l'Azé complex of sites, although these tend to occur in relatively small nodules. On average, 95% of the Pech IV assemblage is on local raw material. This remains quite consistent for all the layers except for Layer 8 (Unit V). This layer has by far the highest percentage of exotic raw materials. These exotic materials appear to have been coming into the site in flake or tool form rather than being produced there from cores that were carried to the site. This could reflect different use of the site during this period of initial occupation, or a difference in overall settlement patterns. Whatever the reason, it may also be related to the intense use of fire during this period.

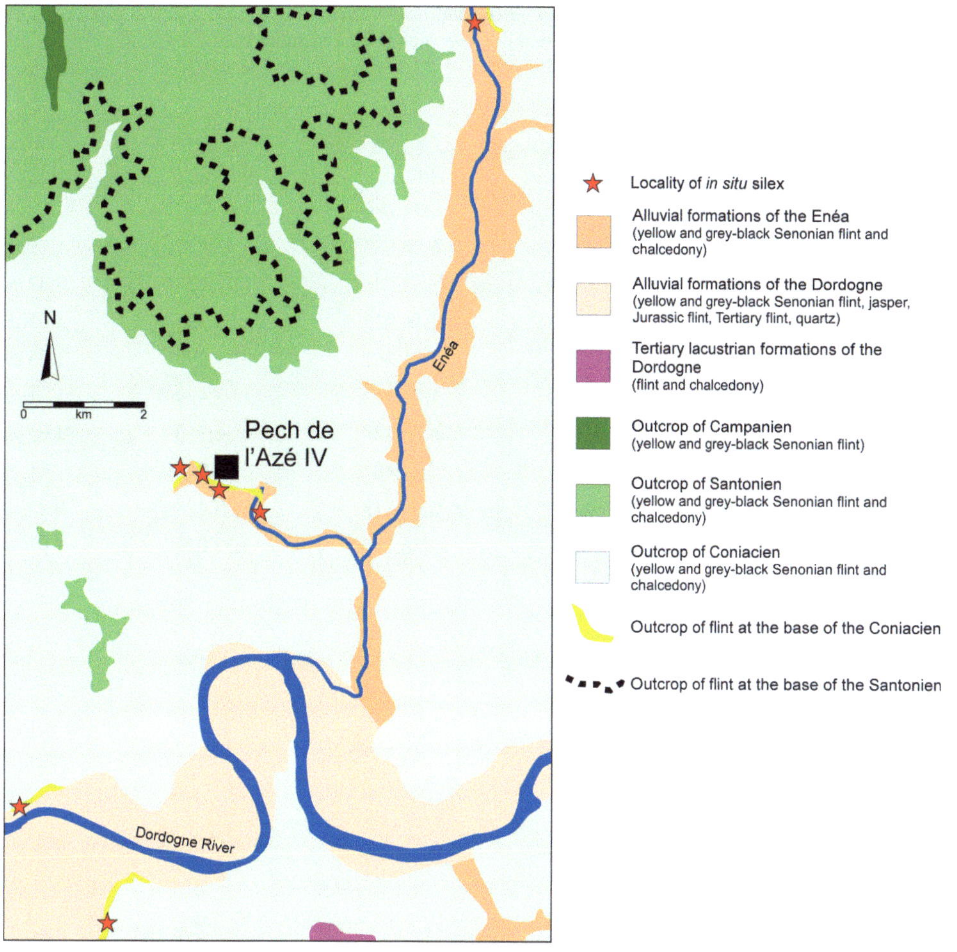


**Fig. 6.1** Location of flint sources [missing—see also Turq et al. (2011)]

**Table 6.48** Raw materials in the Pech IV collection

| Layer | *N* | Local flint | Tertiary flint | Other flint | Jasper | Quartz | Bergerac | Unid |
|---|---|---|---|---|---|---|---|---|
| 3A | 1635 | 96.6 | 1.5 | 0.0 | 0.1 | 0.8 | 0.1 | 1.0 |
| 3B | 2896 | 97.8 | 1.1 | 0.0 | 0.0 | 0.0 | 0.0 | 1.0 |
| 4A | 242 | 90.5 | 7.0 | 0.0 | 0.0 | 0.0 | 0.0 | 2.5 |
| 4B | 52 | 98.1 | 1.9 | 0.0 | 0.0 | 0.0 | 0.0 | 0.0 |
| 4C | 635 | 95.6 | 2.2 | 0.0 | 0.0 | 0.0 | 0.0 | 2.2 |
| 5A | 1447 | 96.6 | 1.4 | 0.0 | 0.2 | 0.0 | 0.0 | 1.8 |
| 5B | 847 | 95.0 | 1.5 | 0.0 | 0.1 | 1.1 | 0.0 | 2.2 |
| 6A | 2466 | 95.7 | 1.9 | 0.0 | 0.0 | 0.4 | 0.1 | 1.8 |
| 6B | 2455 | 95.5 | 1.7 | 0.0 | 0.2 | 0.3 | 0.0 | 2.3 |
| 7 | 3943 | 100.0 | 0.0 | 0.0 | 0.0 | 0.0 | 0.0 | 0.0 |
| 8 | 2503 | 86.2 | 0.5 | 0.0 | 0.0 | 0.0 | 0.0 | 13.2 |
| Totals | 19,121 | 95.7 | 1.2 | 0.0 | 0.1 | 0.2 | 0.0 | 2.8 |

**Table 6.49** Artifact counts, measured cortex, estimated cortex, the Cortex ratio, and the probability of that ratio departing from 1 based on resampling data from experiments reducing nodules less than 2 kg and more than 2 kg

| | | | | | | | Monte carlo *p* (from assemblage | |
|---|---|---|---|---|---|---|---|---|
| Layer | Artifacts | Cores | No. of expected nodules | Observed cortex $mm^2$ | Expected cortex $mm^2$ | Cortex ratio ±75% | <2 kg nodules | 2 kg nodules |
| 3A | 1547 | 29 | 9.6 | 4,54,295 | 3,54,296 | 1.28 ± 0.14 | 0.015 | 0.091 |
| 3B | 2762 | 51 | 14.8 | 6,50,882 | 5,45,835 | 1.19 ± 0.13 | 0.022 | 0.1 |
| 4A | 216 | 3 | 2 | 62,985 | 72,222 | 0.87 ± 0.09 | 0.43 | 0.59 |
| 4B-C | 647 | 14 | 5.6 | 2,16,923 | 2,07,359 | 1.05 ± 0.12 | 0.87 | 0.92 |
| 5A | 1366 | 24 | 8.1 | 3,71,261 | 2,99,270 | 1.25 ± 0.13 | 0.034 | 0.16 |
| 5B | 622 | 25 | 5.2 | 1,93,922 | 1,93,424 | 1 ± 0.11 | 0.85 | 0.89 |
| 6A | 2309 | 112 | 17.8 | 6,60,788 | 6,56,330 | 1.01 ± 0.11 | 0.94 | 0.96 |
| 6B | 2304 | 184 | 14.2 | 5,76,487 | 5,22,843 | 1.1 ± 0.12 | 0.27 | 0.42 |
| 7 | 2784 | 169 | 18.7 | 5,72,112 | 6,90,488 | 0.83 ± 0.09 | 0.0019 | 0.076 |
| 8 | 2076 | 22 | 9 | 3,36,748 | 3,32,025 | 1.01 ± 0.1 | 0.92 | 0.93 |

**Table 6.50** Stratigraphic correlations at Pech IV

| Bordes level | Dibble/McPherron layer | Combined assemblage units |
|---|---|---|
| A–E | 1, 2 | N/A |
| $F_1$ | 3A | I-A |
| $F_2$ | | |
| $F_3$ | | |
| $F_4$ | | |
| | 3B | I-B |
| G | 4A | II-A |
| $H_1/H_2$ | 4B | II-B |
| $I_1$ | | |
| $I_2$ | 4C | II-C |
| $J_1$ | 5A | III-A |
| $J_2$ | 5B | III-B |
| $J_3$ | | |
| $J_3A$ | 6A | IV-A |
| $J_3B$ | | |
| $J_3C$ | 6B | IV-B |
| X | 7 | N/A |
| Y | 8 | V |
| Z | | |

During the renew excavations, a new stratigraphy was defined. This new stratigraphy can be correlated with Bordes' to produce a system of unified layers and subdivisions within these. Thus when this chapter makes reference to Unit I, this includes artifacts excavated by Bordes and by ourselves

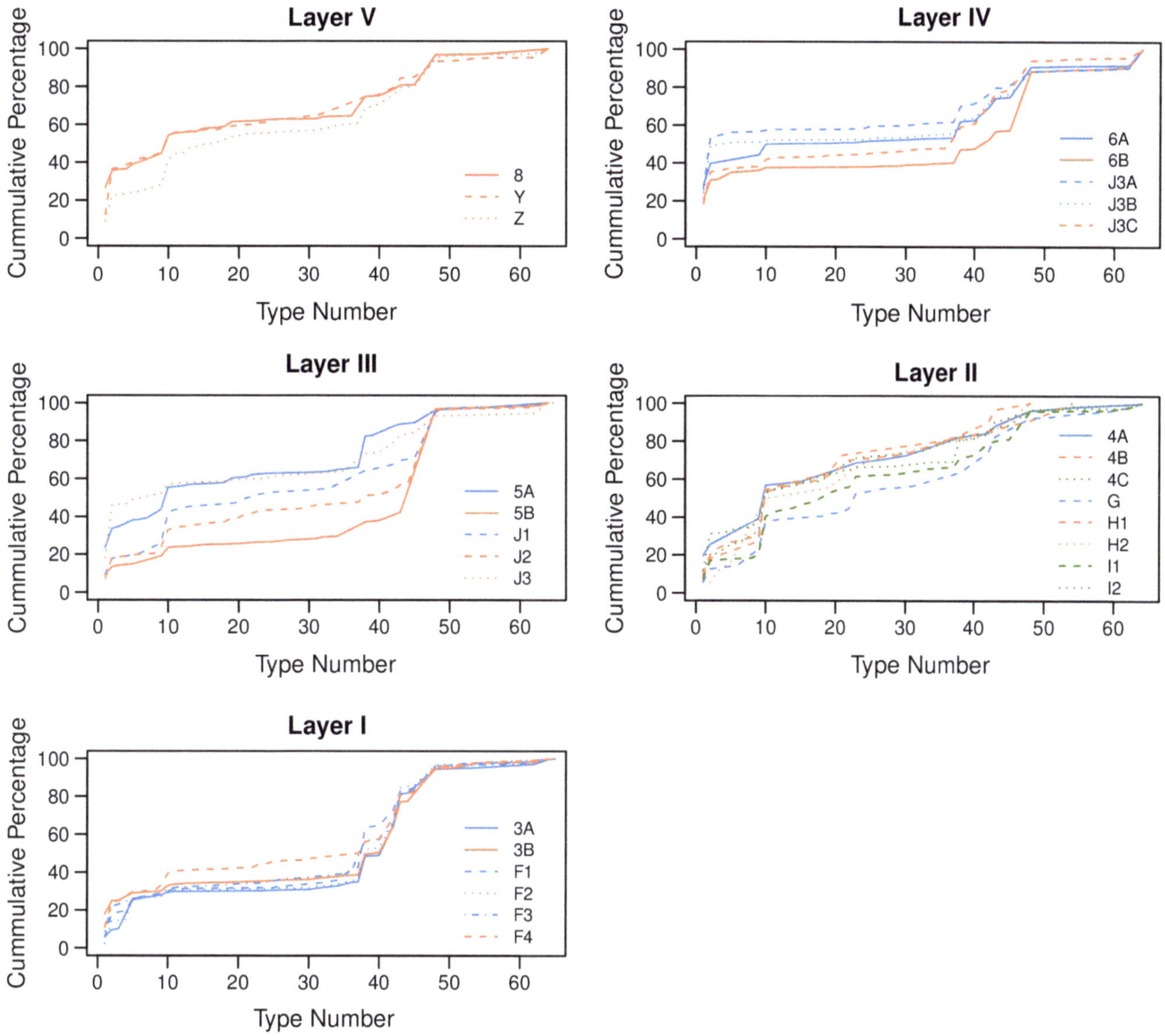


**Fig. 6.2** Cumulative diagrams for Bordes excavation and our excavation organized by the combined layers. *Line types* and *colors* represent corresponding layers

## Raw Material Transport as Measured by Cortex (Prepared by S. Lin)

One of the issues with looking at raw material movement in the Pech IV assemblages is that much of it is of a local type that cannot be easily distinguished. Thus, when we say above that most of the raw materials are local, this still does not rule out the possibility that these local types are moving in and out of the site in differing forms (e.g., as tools, as flakes or as cores). To better address this issue, we applied the Cortex Ratio developed by Dibble and colleagues (Dibble et al. 2005). The Cortex Ratio is a comparison of the actual amount of cortex in an assemblage compared to the expected amount of cortex.

**Table 6.51** Type counts for Unit V

| Type | Name | Real | Percent | Essential | Ess. perc. |
|---|---|---|---|---|---|
| 1 | Levallois flake | 213 | 16.3 | 0 | 0 |
| 2 | Atypical Levallois flake | 203 | 15.6 | 0 | 0 |
| 3 | Levallois point | 7 | 0.5 | 0 | 0 |
| 4 | Retouched Levallois point | 1 | 0.1 | 1 | 0.2 |
| 5 | Pseudo-Levallois point | 31 | 2.4 | 0 | 0 |
| 6 | Mousterian point | 16 | 1.2 | 16 | 2.8 |
| 7 | Elongated mousterian point | 1 | 0.1 | 1 | 0.2 |
| 9 | Single straight scraper | 47 | 3.6 | 47 | 8.3 |
| 10 | Single convex scraper | 138 | 10.6 | 138 | 24.4 |
| 11 | Single concave scraper | 26 | 2 | 26 | 4.6 |
| 12 | Double straight scraper | 2 | 0.2 | 2 | 0.4 |
| 13 | Double straight-convex scraper | 15 | 1.1 | 15 | 2.7 |
| 14 | Double straight-concave scraper | 4 | 0.3 | 4 | 0.7 |
| 15 | Double convex scraper | 16 | 1.2 | 16 | 2.8 |
| 16 | Double concave scraper | 1 | 0.1 | 1 | 0.2 |
| 17 | Double convex–concave scraper | 7 | 0.5 | 7 | 1.2 |
| 18 | Straight convergent scraper | 3 | 0.2 | 3 | 0.5 |
| 19 | Convex convergent scraper | 30 | 2.3 | 30 | 5.3 |
| 20 | Concave convergent scraper | 2 | 0.2 | 2 | 0.4 |
| 21 | Dejete scraper | 7 | 0.5 | 7 | 1.2 |
| 22 | Straight transverse scraper | 2 | 0.2 | 2 | 0.4 |
| 23 | Convex transverse scraper | 7 | 0.5 | 7 | 1.2 |
| 24 | Concave transverse scraper | 4 | 0.3 | 4 | 0.7 |
| 25 | Scraper on interior surface | 9 | 0.7 | 9 | 1.6 |
| 26 | Abrupt scraper | 3 | 0.2 | 3 | 0.5 |
| 27 | Scraper with thinned back | 1 | 0.1 | 1 | 0.2 |
| 28 | Scraper with bifacial retouch | 1 | 0.1 | 1 | 0.2 |
| 29 | Alternate scraper | 3 | 0.2 | 3 | 0.5 |
| 30 | Typical endscraper | 4 | 0.3 | 4 | 0.7 |
| 31 | Atypical endscraper | 3 | 0.2 | 3 | 0.5 |
| 32 | Typical burin | 14 | 1.1 | 14 | 2.5 |
| 33 | Atypical burin | 10 | 0.8 | 10 | 1.8 |
| 34 | Typical percoir | 5 | 0.4 | 5 | 0.9 |
| 35 | Atypical percoir | 1 | 0.1 | 1 | 0.2 |
| 36 | Typical backed knife | 2 | 0.2 | 2 | 0.4 |
| 37 | Atypical backed knife | 2 | 0.2 | 2 | 0.4 |
| 38 | Naturally backed knife | 112 | 8.6 | 0 | 0 |
| 39 | Raclette | 3 | 0.2 | 3 | 0.5 |
| 40 | Truncation | 13 | 1 | 13 | 2.3 |
| 42 | Notch | 49 | 3.8 | 49 | 8.7 |
| 43 | Denticulate | 49 | 3.8 | 49 | 8.7 |
| 44 | Alternate retouched bec | 1 | 0.1 | 1 | 0.2 |
| 45 | Flake with irregular retouch on | 5 | 0.4 | 0 | 0 |
| 48 | Flake with abrupt and alter | 169 | 13 | 0 | 0 |

(continued)

**Table 6.51** (continued)

| Type | Name | Real | Percent | Essential | Ess. perc. |
|---|---|---|---|---|---|
| 50 | Bifacially retouched flake | 5 | 0.4 | 5 | 0.9 |
| 51 | Tayac point | 1 | 0.1 | 1 | 0.2 |
| 54 | End-notched flake | 10 | 0.8 | 10 | 1.8 |
| 61 | Chopping-tool | 2 | 0.2 | 2 | 0.4 |
| 62 | Miscellaneous | 2 | 0.2 | 2 | 0.4 |
| 64 | Truncated-facetted piece | 43 | 3.3 | 43 | 7.6 |
| | Totals | 1305 | 100.5 | 565 | 100.3 |

**Table 6.52** Artifact counts for Unit V

| | *N* |
|---|---|
| Complete and proximal flakes | 3214 |
| Other flake fragments | 1782 |
| Complete and proximal tools | 360 |
| Other tool fragments | 204 |
| Handaxes | 1 |
| Handaxe fragments | 1 |
| Shatter | 848 |
| Complete cores | 126 |
| Core fragments | 60 |

**Table 6.53** Typological and technological indices for Unit V

| Typological indices | | | Technological indices | |
|---|---|---|---|---|
| | Real | Essential | IL | 9.37 |
| ILty | 32.49 | | IF | 32.25 |
| IR | 25.13 | 58.05 | IFs | 21.63 |
| IAu | 0.31 | 0.01 | Ilam | 19.06 |
| II | 26.44 | 61.06 | TF | 7.61 |
| III | 0.03 | 7.26 | Kombewa | 0.22 |
| IV | 7.51 | 17.35 | Core edge | 1.23 |
| | | | Nat. backed | 5.85 |

The actual amount is computed by taking the amount of cortex (using the intervals described in the methods) on each artifact scaled by the size of that artifact and totaled over the assemblage. The expected cortex is the amount of cortex that should be present in the assemblage if unworked nodules were brought to the site, knapped, and nothing was subsequently transported off site. Cortex ratios of less or more than one will then indicate that cortex is missing or over-represented, respectively. Estimating the expected level of cortex requires making several assumptions about the size and shape of nodules imported into the site.

**Table 6.54** Basic size measurements for complete flakes, scrapers, notched tools, other retouched tools, and cores from Level V

| | | Length | Width | Thickness | Weight |
|---|---|---|---|---|---|
| Flakes | *N* | 1837 | 1735 | 1737 | 1798 |
| | Mean | 36.61 | 24.72 | 6.34 | 7.73 |
| | S.D. | 11.37 | 7.77 | 2.89 | 9 |
| Scrapers | *N* | 181 | 183 | 181 | 181 |
| | Mean | 50.86 | 30.12 | 7.86 | 17.06 |
| | S.D. | 14.12 | 7.54 | 2.91 | 13.13 |
| Notches | *N* | 52 | 54 | 54 | 57 |
| | Mean | 43.73 | 29.03 | 9.52 | 15.53 |
| | S.D. | 10.72 | 7.53 | 3.81 | 10.91 |
| Other tools | *N* | 52 | 53 | 53 | 53 |
| | Mean | 44.69 | 29.54 | 9.74 | 17.89 |
| | S.D. | 15.34 | 9.23 | 3.88 | 14.03 |
| Cores | *N* | 125 | 125 | 125 | 123 |
| | Mean | 46.09 | 36.19 | 16.03 | 32.89 |
| | S.D. | 11.42 | 9.54 | 6.32 | 24.48 |

**Table 6.55** Type counts for Unit IV-B

| Type | Name | Real | Percent | Essential | Ess. perc. |
|---|---|---|---|---|---|
| 1 | Levallois flake | 233 | 19.6 | 0 | 0 |
| 2 | Atypical Levallois flake | 161 | 13.5 | 0 | 0 |
| 3 | Levallois point | 2 | 0.2 | 0 | 0 |
| 5 | Pseudo-Levallois point | 34 | 2.9 | 0 | 0 |
| 6 | Mousterian point | 3 | 0.3 | 3 | 0.8 |
| 7 | Elongated mousterian point | 1 | 0.1 | 1 | 0.3 |
| 9 | Single straight scraper | 8 | 0.7 | 8 | 2.2 |
| 10 | Single convex scraper | 30 | 2.5 | 30 | 8.1 |
| 11 | Single concave scraper | 6 | 0.5 | 6 | 1.6 |
| 13 | Double straight-convex scraper | 1 | 0.1 | 1 | 0.3 |
| 15 | Double convex scraper | 1 | 0.1 | 1 | 0.3 |
| 18 | Straight convergent scraper | 1 | 0.1 | 1 | 0.3 |
| 19 | Convex convergent scraper | 5 | 0.4 | 5 | 1.4 |
| 21 | Dejete scraper | 1 | 0.1 | 1 | 0.3 |
| 23 | Convex transverse scraper | 1 | 0.1 | 1 | 0.3 |
| 24 | Concave transverse scraper | 1 | 0.1 | 1 | 0.3 |
| 25 | Scraper on interior surface | 6 | 0.5 | 6 | 1.6 |
| 26 | Abrupt scraper | 3 | 0.3 | 3 | 0.8 |
| 29 | Alternate scraper | 1 | 0.1 | 1 | 0.3 |
| 30 | Typical endscraper | 9 | 0.8 | 9 | 2.4 |
| 31 | Atypical endscraper | 4 | 0.3 | 4 | 1.1 |
| 32 | Typical burin | 3 | 0.3 | 3 | 0.8 |
| 33 | Atypical burin | 2 | 0.2 | 2 | 0.5 |
| 34 | Typical percoir | 2 | 0.2 | 2 | 0.5 |
| 35 | Atypical percoir | 1 | 0.1 | 1 | 0.3 |
| 36 | Typical backed knife | 2 | 0.2 | 2 | 0.5 |
| 37 | Atypical backed knife | 2 | 0.2 | 2 | 0.5 |

(continued)

**Table 6.55** (continued)

| Type | Name | Real | Percent | Essential | Ess. perc. |
|---|---|---|---|---|---|
| 38 | Naturally backed knife | 108 | 9.1 | 0 | 0 |
| 39 | Raclette | 1 | 0.1 | 1 | 0.3 |
| 40 | Truncation | 12 | 1 | 12 | 3.3 |
| 42 | Notch | 86 | 7.2 | 86 | 23.3 |
| 43 | Denticulate | 66 | 5.5 | 66 | 17.9 |
| 44 | Alternate retouched bec | 4 | 0.3 | 4 | 1.1 |
| 45 | Flake with irregular retouch on | 8 | 0.7 | 0 | 0 |
| 48 | Flake with abrupt and alter | 276 | 23.2 | 0 | 0 |
| 50 | Bifacially retouched flake | 1 | 0.1 | 1 | 0.3 |
| 54 | End-notched flake | 11 | 0.9 | 11 | 3 |
| 56 | Rabot | 2 | 0.2 | 2 | 0.5 |
| 58 | Stemmed tool | 2 | 0.2 | 2 | 0.5 |
| 62 | Miscellaneous | 8 | 0.7 | 8 | 2.2 |
| 64 | Truncated-facetted piece | 82 | 6.9 | 82 | 22.2 |
| | Totals | 1191 | 100.6 | 369 | 100.1 |

**Table 6.56** Artifact counts for Unit IV-B

| | *N* |
|---|---|
| Complete and proximal flakes | 2702 |
| Other flake fragments | 1367 |
| Complete and proximal tools | 230 |
| Other tool fragments | 145 |
| Handaxes | 1 |
| Handaxe fragments | 0 |
| Shatter | 672 |
| Complete cores | 287 |
| Core fragments | 156 |

**Table 6.57** Typological and technological indices for Unit IV-B

| Typological indices | | | Technological indices | |
|---|---|---|---|---|
| | Real | Essential | IL | 10.33 |
| ILty | 33.25 | | IF | 33.41 |
| IR | 5.46 | 17.62 | IFs | 24.03 |
| IAu | 0.34 | 0.01 | Ilam | 12.29 |
| II | 5.79 | 18.7 | TF | 22.22 |
| III | 0.02 | 6.78 | Kombewa | 0.2 |
| IV | 12.76 | 41.19 | Core edge | 2.01 |
| | | | Nat. backed | 5.87 |

**Table 6.58** Basic size measurements for complete flakes, scrapers, notched tools, other retouched tools, and cores from Level IV-B

| | | Length | Width | Thickness | Weight |
|---|---|---|---|---|---|
| Flakes | *N* | 1392 | 1289 | 1290 | 1406 |
| | Mean | 34.22 | 24.52 | 6.93 | 7.16 |
| | S.D. | 9.43 | 6.51 | 3.01 | 6.36 |
| Scrapers | *N* | 29 | 29 | 29 | 30 |
| | Mean | 46.57 | 31.05 | 8.7 | 17.37 |
| | S.D. | 11.91 | 6.4 | 3.24 | 11.23 |
| Notches | *N* | 72 | 71 | 71 | 74 |
| | Mean | 36.48 | 26.41 | 8.41 | 10.57 |
| | S.D. | 10.1 | 6.64 | 2.41 | 6.29 |
| Other tools | *N* | 59 | 60 | 60 | 59 |
| | Mean | 41.2 | 27.99 | 10.12 | 14.2 |
| | S.D. | 10.04 | 6.85 | 3.4 | 7.9 |
| Cores | *N* | 284 | 282 | 282 | 285 |
| | Mean | 39.28 | 30.56 | 13.92 | 19.2 |
| | S.D. | 9.52 | 6.95 | 4.98 | 13.9 |

**Table 6.59** Type counts for Unit IV-A

| Type | Name | Real | Percent | Essential | Ess. perc. |
|---|---|---|---|---|---|
| 1 | Levallois flake | 792 | 25.7 | 0 | 0 |
| 2 | Atypical Levallois flake | 713 | 23.1 | 0 | 0 |
| 3 | Levallois point | 5 | 0.2 | 0 | 0 |
| 5 | Pseudo-Levallois point | 50 | 1.6 | 0 | 0 |
| 6 | Mousterian point | 1 | 0 | 1 | 0.1 |
| 9 | Single straight scraper | 22 | 0.7 | 22 | 2.4 |
| 10 | Single convex scraper | 55 | 1.8 | 55 | 6 |
| 11 | Single concave scraper | 9 | 0.3 | 9 | 1 |
| 12 | Double straight scraper | 1 | 0 | 1 | 0.1 |
| 15 | Double convex scraper | 2 | 0.1 | 2 | 0.2 |
| 19 | Convex convergent scraper | 2 | 0.1 | 2 | 0.2 |
| 20 | Concave convergent scraper | 1 | 0 | 1 | 0.1 |
| 21 | Dejete scraper | 2 | 0.1 | 2 | 0.2 |
| 22 | Straight transverse scraper | 2 | 0.1 | 2 | 0.2 |
| 23 | Convex transverse scraper | 4 | 0.1 | 4 | 0.4 |
| 24 | Concave transverse scraper | 2 | 0.1 | 2 | 0.2 |
| 25 | Scraper on interior surface | 28 | 0.9 | 28 | 3.1 |
| 26 | Abrupt scraper | 5 | 0.2 | 5 | 0.5 |
| 27 | Scraper with thinned back | 1 | 0 | 1 | 0.1 |
| 28 | Scraper with bifacial retouch | 1 | 0 | 1 | 0.1 |
| 29 | Alternate scraper | 1 | 0 | 1 | 0.1 |
| 30 | Typical endscraper | 6 | 0.2 | 6 | 0.7 |
| 31 | Atypical endscraper | 11 | 0.4 | 11 | 1.2 |
| 32 | Typical burin | 12 | 0.4 | 12 | 1.3 |
| 33 | Atypical burin | 9 | 0.3 | 9 | 1 |
| 34 | Typical percoir | 7 | 0.2 | 7 | 0.8 |
| 35 | Atypical percoir | 7 | 0.2 | 7 | 0.8 |

(continued)

**Table 6.59** (continued)

| Type | Name | Real | Percent | Essential | Ess. perc. |
|---|---|---|---|---|---|
| 36 | Typical backed knife | 4 | 0.1 | 4 | 0.4 |
| 37 | Atypical backed knife | 5 | 0.2 | 5 | 0.5 |
| 38 | Naturally backed knife | 238 | 7.7 | 0 | 0 |
| 39 | Raclette | 16 | 0.5 | 16 | 1.8 |
| 40 | Truncation | 19 | 0.6 | 19 | 2.1 |
| 42 | Notch | 210 | 6.8 | 210 | 23.1 |
| 43 | Denticulate | 117 | 3.8 | 117 | 12.9 |
| 44 | Alternate retouched bec | 4 | 0.1 | 4 | 0.4 |
| 45 | Flake with irregular retouch on | 33 | 1.1 | 0 | 0 |
| 48 | Flake with abrupt and alter | 346 | 11.2 | 0 | 0 |
| 50 | Bifacially retouched flake | 4 | 0.1 | 4 | 0.4 |
| 54 | End-notched flake | 22 | 0.7 | 22 | 2.4 |
| 58 | Stemmed tool | 1 | 0 | 1 | 0.1 |
| 62 | Miscellaneous | 22 | 0.7 | 22 | 2.4 |
| 64 | Truncated-facetted piece | 295 | 9.6 | 295 | 32.4 |
| | Totals | 3087 | 100 | 910 | 99.7 |

**Table 6.60** Artifact counts for Unit IV-A

| | *N* |
|---|---|
| Complete and proximal flakes | 7729 |
| Other flake fragments | 3352 |
| Complete and proximal tools | 529 |
| Other tool fragments | 402 |
| Handaxes | 0 |
| Handaxe fragments | 2 |
| Shatter | 2104 |
| Complete cores | 693 |
| Core fragments | 380 |

**Table 6.61** Typological and technological indices for Unit IV-A

| Typological indices | | | Technological indices | |
|---|---|---|---|---|
| | Real | Essential | IL | 10.61 |
| ILty | 48.91 | | IF | 32.52 |
| IR | 4.47 | 15.16 | IFs | 23.71 |
| IAu | 0.29 | 0.01 | Ilam | 15.55 |
| II | 4.5 | 15.27 | TF | 32.42 |
| III | 0.02 | 6.7 | Kombewa | 0.25 |
| IV | 10.59 | 35.93 | Core edge | 1.15 |
| | | | Nat. backed | 5.03 |

**Table 6.62** Basic size measurements for complete flakes, scrapers, notched tools, other retouched tools, and cores from Level IV-A

| | | Length | Width | Thickness | Weight |
|---|---|---|---|---|---|
| Flakes | *N* | 4053 | 3544 | 3568 | 4040 |
| | Mean | 36.78 | 25.83 | 6.88 | 8.26 |
| | S.D. | 11.3 | 7.69 | 3.09 | 8.98 |
| Scrapers | *N* | 74 | 74 | 74 | 75 |
| | Mean | 52.73 | 34.11 | 9.43 | 24.07 |
| | S.D. | 16.45 | 11.6 | 3.97 | 26.58 |
| Notches | *N* | 128 | 123 | 124 | 131 |
| | Mean | 43.89 | 29.37 | 8.93 | 14.59 |
| | S.D. | 13.04 | 8.53 | 3.36 | 10.67 |
| Other tools | *N* | 136 | 133 | 134 | 136 |
| | Mean | 39.43 | 29.51 | 8.99 | 12.99 |
| | S.D. | 9.93 | 7.06 | 2.9 | 7.31 |
| Cores | *N* | 664 | 643 | 641 | 662 |
| | Mean | 39.89 | 31.68 | 14.02 | 23.02 |
| | S.D. | 10.21 | 7.32 | 5.99 | 28.21 |

**Table 6.63** Type counts for Unit III-B

| Type | Name | Real | Percent | Essential | Ess. perc. |
|---|---|---|---|---|---|
| 1 | Levallois flake | 186 | 12.8 | 0 | 0 |
| 2 | Atypical Levallois flake | 260 | 17.9 | 0 | 0 |
| 3 | Levallois point | 5 | 0.3 | 0 | 0 |
| 4 | Retouched Levallois point | 1 | 0.1 | 1 | 0.2 |
| 5 | Pseudo-Levallois point | 24 | 1.7 | 0 | 0 |
| 6 | Mousterian point | 2 | 0.1 | 2 | 0.4 |
| 8 | Limace | 2 | 0.1 | 2 | 0.4 |
| 9 | Single straight scraper | 49 | 3.4 | 49 | 10.4 |
| 10 | Single convex scraper | 91 | 6.3 | 91 | 19.4 |
| 11 | Single concave scraper | 13 | 0.9 | 13 | 2.8 |
| 13 | Double straight-convex scraper | 6 | 0.4 | 6 | 1.3 |
| 15 | Double convex scraper | 12 | 0.8 | 12 | 2.6 |
| 17 | Double convex–concave scraper | 3 | 0.2 | 3 | 0.6 |
| 18 | Straight convergent scraper | 2 | 0.1 | 2 | 0.4 |
| 19 | Convex convergent scraper | 18 | 1.2 | 18 | 3.8 |
| 20 | Concave convergent scraper | 1 | 0.1 | 1 | 0.2 |
| 21 | Dejete scraper | 14 | 1 | 14 | 3 |
| 22 | Straight transverse scraper | 3 | 0.2 | 3 | 0.6 |
| 23 | Convex transverse scraper | 9 | 0.6 | 9 | 1.9 |
| 24 | Concave transverse scraper | 1 | 0.1 | 1 | 0.2 |
| 25 | Scraper on interior surface | 10 | 0.7 | 10 | 2.1 |
| 26 | Abrupt scraper | 6 | 0.4 | 6 | 1.3 |
| 27 | Scraper with thinned back | 2 | 0.1 | 2 | 0.4 |
| 28 | Scraper with bifacial retouch | 1 | 0.1 | 1 | 0.2 |
| 29 | Alternate scraper | 4 | 0.3 | 4 | 0.9 |
| 30 | Typical endscraper | 9 | 0.6 | 9 | 1.9 |
| 31 | Atypical endscraper | 6 | 0.4 | 6 | 1.3 |

(continued)

**Table 6.63** (continued)

| Type | Name | Real | Percent | Essential | Ess. perc. |
|---|---|---|---|---|---|
| 32 | Typical burin | 5 | 0.3 | 5 | 1.1 |
| 33 | Atypical burin | 4 | 0.3 | 4 | 0.9 |
| 34 | Typical percoir | 4 | 0.3 | 4 | 0.9 |
| 35 | Atypical percoir | 3 | 0.2 | 3 | 0.6 |
| 36 | Typical backed knife | 2 | 0.1 | 2 | 0.4 |
| 37 | Atypical backed knife | 2 | 0.1 | 2 | 0.4 |
| 38 | Naturally backed knife | 99 | 6.8 | 0 | 0 |
| 39 | Raclette | 4 | 0.3 | 4 | 0.9 |
| 40 | Truncation | 6 | 0.4 | 6 | 1.3 |
| 42 | Notch | 55 | 3.8 | 55 | 11.7 |
| 43 | Denticulate | 44 | 3 | 44 | 9.4 |
| 44 | Alternate retouched bec | 1 | 0.1 | 1 | 0.2 |
| 45 | Flake with irregular retouch on | 12 | 0.8 | 0 | 0 |
| 48 | Flake with abrupt and alter | 394 | 27.2 | 0 | 0 |
| 54 | End-notched flake | 9 | 0.6 | 9 | 1.9 |
| 58 | Stemmed tool | 1 | 0.1 | 1 | 0.2 |
| 62 | Miscellaneous | 10 | 0.7 | 10 | 2.1 |
| 64 | Truncated-facetted piece | 53 | 3.7 | 53 | 11.3 |
| 65 | Scraper on the platform | 1 | 0.1 | 1 | 0.2 |
| | Totals | 1449 | 99.8 | 469 | 99.8 |

**Table 6.64** Artifact counts for Unit III-B

| | *N* |
|---|---|
| Complete and proximal flakes | 2221 |
| Other flake fragments | 957 |
| Complete and proximal tools | 359 |
| Other tool fragments | 118 |
| Handaxes | 1 |
| Handaxe fragments | 0 |
| Shatter | 519 |
| Complete cores | 163 |
| Core fragments | 72 |

**Table 6.65** Typological and technological indices for Unit III-B

| Typological indices | | | Technological indices | |
|---|---|---|---|---|
| | Real | Essential | IL | 11.98 |
| ILty | 31.19 | | IF | 33.74 |
| IR | 16.91 | 52.24 | IFs | 26.49 |
| IAu | 0.28 | 0.01 | Ilam | 16.78 |
| II | 17.18 | 53.09 | TF | 11.3 |
| III | 0.02 | 7.46 | Kombewa | 0.27 |
| IV | 6.83 | 21.11 | Core edge | 1.47 |
| | | | Nat. backed | 6.82 |

**Table 6.66** Basic size measurements for complete flakes, scrapers, notched tools, other retouched tools, and cores from Level III-B

| | | Length | Width | Thickness | Weight |
|---|---|---|---|---|---|
| Flakes | *N* | 1290 | 1225 | 1228 | 1291 |
| | Mean | 38.89 | 26.82 | 7.39 | 11.05 |
| | S.D. | 12.64 | 8.7 | 3.55 | 12.03 |
| Scrapers | *N* | 166 | 166 | 167 | 170 |
| | Mean | 56.81 | 33.98 | 10.29 | 28.25 |
| | S.D. | 16.1 | 9.37 | 4 | 22.96 |
| Notches | *N* | 58 | 59 | 59 | 59 |
| | Mean | 44.26 | 29.28 | 8.82 | 16.25 |
| | S.D. | 15.1 | 9.79 | 4.76 | 14.75 |
| Other tools | *N* | 64 | 65 | 65 | 64 |
| | Mean | 44.33 | 29.93 | 10.27 | 18.27 |
| | S.D. | 11.22 | 6.72 | 4.03 | 13.18 |
| Cores | *N* | 159 | 155 | 155 | 158 |
| | Mean | 45.14 | 35.16 | 16.96 | 38.62 |
| | S.D. | 12.66 | 9.26 | 8.02 | 39.42 |

**Table 6.67** Type counts for Unit III-A

| Type | Name | Real | Percent | Essential | Ess. perc. |
|---|---|---|---|---|---|
| 1 | Levallois flake | 81 | 17.4 | 0 | 0 |
| 2 | Atypical Levallois flake | 45 | 9.7 | 0 | 0 |
| 3 | Levallois point | 4 | 0.9 | 0 | 0 |
| 4 | Retouched Levallois point | 1 | 0.2 | 1 | 0.5 |
| 5 | Pseudo-Levallois point | 11 | 2.4 | 0 | 0 |
| 6 | Mousterian point | 2 | 0.4 | 2 | 1.1 |
| 7 | Elongated mousterian point | 2 | 0.4 | 2 | 1.1 |
| 9 | Single straight scraper | 22 | 4.7 | 22 | 11.6 |
| 10 | Single convex scraper | 63 | 13.5 | 63 | 33.2 |
| 11 | Single concave scraper | 4 | 0.9 | 4 | 2.1 |
| 12 | Double straight scraper | 2 | 0.4 | 2 | 1.1 |
| 13 | Double straight-convex scraper | 4 | 0.9 | 4 | 2.1 |
| 15 | Double convex scraper | 2 | 0.4 | 2 | 1.1 |
| 17 | Double convex–concave scraper | 2 | 0.4 | 2 | 1.1 |
| 18 | Straight convergent scraper | 1 | 0.2 | 1 | 0.5 |
| 19 | Convex convergent scraper | 7 | 1.5 | 7 | 3.7 |
| 20 | Concave convergent scraper | 1 | 0.2 | 1 | 0.5 |
| 21 | Dejete scraper | 6 | 1.3 | 6 | 3.2 |
| 22 | Straight transverse scraper | 3 | 0.6 | 3 | 1.6 |
| 23 | Convex transverse scraper | 5 | 1.1 | 5 | 2.6 |
| 24 | Concave transverse scraper | 1 | 0.2 | 1 | 0.5 |
| 25 | Scraper on interior surface | 2 | 0.4 | 2 | 1.1 |
| 28 | Scraper with bifacial retouch | 1 | 0.2 | 1 | 0.5 |
| 30 | Typical endscraper | 2 | 0.4 | 2 | 1.1 |
| 32 | Typical burin | 2 | 0.4 | 2 | 1.1 |
| 33 | Atypical burin | 2 | 0.4 | 2 | 1.1 |
| 34 | Typical percoir | 2 | 0.4 | 2 | 1.1 |
| 35 | Atypical percoir | 1 | 0.2 | 1 | 0.5 |

(continued)

**Table 6.67** (continued)

| Type | Name | Real | Percent | Essential | Ess. perc. |
|---|---|---|---|---|---|
| 36 | Typical backed knife | 1 | 0.2 | 1 | 0.5 |
| 37 | Atypical backed knife | 1 | 0.2 | 1 | 0.5 |
| 38 | Naturally backed knife | 64 | 13.8 | 0 | 0 |
| 39 | Raclette | 2 | 0.4 | 2 | 1.1 |
| 42 | Notch | 19 | 4.1 | 19 | 10 |
| 43 | Denticulate | 7 | 1.5 | 7 | 3.7 |
| 45 | Flake with irregular retouch on | 4 | 0.9 | 0 | 0 |
| 48 | Flake with abrupt and alter | 66 | 14.2 | 0 | 0 |
| 50 | Bifacially retouched flake | 4 | 0.9 | 4 | 2.1 |
| 54 | End-notched flake | 3 | 0.6 | 3 | 1.6 |
| 58 | Stemmed tool | 1 | 0.2 | 1 | 0.5 |
| 64 | Truncated-facetted piece | 11 | 2.4 | 11 | 5.8 |
| 65 | Scraper on the platform | 1 | 0.2 | 1 | 0.5 |
| | Totals | 465 | 99.7 | 190 | 100.4 |

**Table 6.68** Artifact counts for Unit III-A

| | *N* |
|---|---|
| Complete and proximal flakes | 1087 |
| Other flake fragments | 417 |
| Complete and proximal tools | 136 |
| Other tool fragments | 55 |
| Handaxes | 2 |
| Handaxe fragments | 0 |
| Shatter | 118 |
| Complete cores | 46 |
| Core fragments | 29 |

**Table 6.69** Typological and technological indices for Unit III-A

| Typological indices | | | Technological indices | |
|---|---|---|---|---|
| | Real | Essential | IL | 12.02 |
| ILty | 28.17 | | IF | 34.53 |
| IR | 27.1 | 66.32 | IFs | 24.38 |
| IAu | 0.43 | 0.01 | Ilam | 21.3 |
| II | 27.96 | 68.42 | TF | 5.79 |
| III | 0.02 | 5.79 | Kombewa | 0.08 |
| IV | 5.59 | 13.68 | Core edge | 2.04 |
| | | | Nat. backed | 11.37 |

**Table 6.70** Basic size measurements for complete flakes, scrapers, notched tools, other retouched tools, and cores from Level III-A

| | | Length | Width | Thickness | Weight |
|---|---|---|---|---|---|
| Flakes | *N* | 757 | 758 | 758 | 752 |
| | Mean | 37.7 | 24.48 | 6.51 | 8.68 |
| | S.D. | 11.92 | 7.83 | 3.42 | 15.52 |
| Scrapers | *N* | 79 | 79 | 79 | 79 |
| | Mean | 59.43 | 35 | 9.2 | 26.54 |
| | S.D. | 13.02 | 8.84 | 3.28 | 17.44 |
| Notches | *N* | 14 | 14 | 14 | 13 |
| | Mean | 45.08 | 31.46 | 9.42 | 17.15 |
| | S.D. | 10.72 | 4.97 | 3.46 | 10.53 |
| Other tools | *N* | 17 | 17 | 17 | 17 |
| | Mean | 48.25 | 32.6 | 10.39 | 21.76 |
| | S.D. | 11.11 | 8.67 | 3.53 | 12.6 |
| Cores | *N* | 46 | 46 | 46 | 46 |
| | Mean | 50.67 | 38.74 | 18.49 | 47.46 |
| | S.D. | 13.31 | 9.91 | 7.36 | 37.97 |

**Table 6.71** Type counts for Unit II-C

| Type | Name | Real | Percent | Essential | Ess. perc. |
|---|---|---|---|---|---|
| 1 | Levallois flake | 264 | 12.1 | 0 | 0 |
| 2 | Atypical Levallois flake | 398 | 18.3 | 0 | 0 |
| 3 | Levallois point | 7 | 0.3 | 0 | 0 |
| 4 | Retouched levallois point | 7 | 0.3 | 7 | 0.7 |
| 5 | Pseudo-Levallois point | 31 | 1.4 | 0 | 0 |
| 6 | Mousterian point | 25 | 1.1 | 25 | 2.4 |
| 7 | Elongated mousterian point | 6 | 0.3 | 6 | 0.6 |
| 8 | Limace | 3 | 0.1 | 3 | 0.3 |
| 9 | Single straight scraper | 112 | 5.2 | 112 | 10.7 |
| 10 | Single convex scraper | 307 | 14.1 | 307 | 29.3 |
| 11 | Single concave scraper | 30 | 1.4 | 30 | 2.9 |
| 12 | Double straight scraper | 13 | 0.6 | 13 | 1.2 |
| 13 | Double straight-convex scraper | 21 | 1 | 21 | 2 |
| 14 | Double straight-concave scraper | 5 | 0.2 | 5 | 0.5 |
| 15 | Double convex scraper | 38 | 1.7 | 38 | 3.6 |
| 16 | Double concave scraper | 1 | 0 | 1 | 0.1 |
| 17 | Double convex–concave scraper | 12 | 0.6 | 12 | 1.1 |
| 18 | Straight convergent scraper | 2 | 0.1 | 2 | 0.2 |
| 19 | Convex convergent scraper | 86 | 4 | 86 | 8.2 |
| 20 | Concave convergent scraper | 4 | 0.2 | 4 | 0.4 |
| 21 | Dejete scraper | 39 | 1.8 | 39 | 3.7 |
| 22 | Straight transverse scraper | 4 | 0.2 | 4 | 0.4 |
| 23 | Convex transverse scraper | 35 | 1.6 | 35 | 3.3 |
| 24 | Concave transverse scraper | 1 | 0 | 1 | 0.1 |
| 25 | Scraper on interior surface | 5 | 0.2 | 5 | 0.5 |
| 26 | Abrupt scraper | 2 | 0.1 | 2 | 0.2 |
| 27 | Scraper with thinned back | 2 | 0.1 | 2 | 0.2 |

(continued)

**Table 6.71** (continued)

| Type | Name | Real | Percent | Essential | Ess. perc. |
|---|---|---|---|---|---|
| 28 | Scraper with bifacial retouch | 2 | 0.1 | 2 | 0.2 |
| 29 | Alternate scraper | 5 | 0.2 | 5 | 0.5 |
| 30 | Typical endscraper | 12 | 0.6 | 12 | 1.1 |
| 31 | Atypical endscraper | 5 | 0.2 | 5 | 0.5 |
| 32 | Typical burin | 11 | 0.5 | 11 | 1.1 |
| 33 | Atypical burin | 11 | 0.5 | 11 | 1.1 |
| 34 | Typical percoir | 4 | 0.2 | 4 | 0.4 |
| 35 | Atypical percoir | 1 | 0 | 1 | 0.1 |
| 36 | Typical backed knife | 2 | 0.1 | 2 | 0.2 |
| 37 | Atypical backed knife | 3 | 0.1 | 3 | 0.3 |
| 38 | Naturally backed knife | 248 | 11.4 | 0 | 0 |
| 39 | Raclette | 6 | 0.3 | 6 | 0.6 |
| 40 | Truncation | 8 | 0.4 | 8 | 0.8 |
| 42 | Notch | 76 | 3.5 | 76 | 7.3 |
| 43 | Denticulate | 62 | 2.9 | 62 | 5.9 |
| 44 | Alternate retouched bec | 4 | 0.2 | 4 | 0.4 |
| 45 | Flake with irregular retouch on | 10 | 0.5 | 0 | 0 |
| 48 | Flake with abrupt and alter | 170 | 7.8 | 0 | 0 |
| 50 | Bifacially retouched flake | 7 | 0.3 | 7 | 0.7 |
| 51 | Tayac point | 1 | 0 | 1 | 0.1 |
| 54 | End-notched flake | 10 | 0.5 | 10 | 1 |
| 61 | Chopping-tool | 2 | 0.1 | 2 | 0.2 |
| 62 | Miscellaneous | 17 | 0.8 | 17 | 1.6 |
| 64 | Truncated-facetted piece | 37 | 1.7 | 37 | 3.5 |
| | Totals | 2174 | 99.9 | 1046 | 100.2 |

**Table 6.72** Artifact counts for Unit II-C

| | *N* |
|---|---|
| Complete and proximal flakes | 5046 |
| Other flake fragments | 1764 |
| Complete and proximal tools | 767 |
| Other tool fragments | 287 |
| Handaxes | 1 |
| Handaxe fragments | 0 |
| Shatter | 640 |
| Complete cores | 166 |
| Core fragments | 64 |

**Table 6.73** Typological and technological indices for Unit II-C

| Typological indices | | | Technological indices | |
|---|---|---|---|---|
| | Real | Essential | IL | 12.11 |
| ILty | 31.09 | | IF | 36.11 |
| IR | 33.39 | 69.41 | IFs | 27.92 |
| IAu | 0.23 | 0 | Ilam | 24.13 |
| II | 34.96 | 72.66 | TF | 3.54 |
| III | 0.02 | 4.68 | Kombewa | 0.09 |
| IV | 6.35 | 13.19 | Core edge | 1.74 |
| | | | Nat. backed | 7.53 |

**Table 6.74** Basic size measurements for complete flakes, scrapers, notched tools, other retouched tools, and cores from Level II-C

| | | Length | Width | Thickness | Weight |
|---|---|---|---|---|---|
| Flakes | *N* | 3346 | 3235 | 3242 | 3330 |
| | Mean | 39.85 | 25.45 | 6.54 | 9.52 |
| | S.D. | 12.6 | 7.96 | 3.3 | 13.94 |
| Scrapers | *N* | 457 | 458 | 458 | 453 |
| | Mean | 61.56 | 35.62 | 9.16 | 28.55 |
| | S.D. | 15.3 | 10.11 | 4.08 | 23.22 |
| Notches | *N* | 72 | 72 | 72 | 73 |
| | Mean | 48.58 | 30.9 | 9.52 | 20.47 |
| | S.D. | 14.81 | 9.3 | 4.74 | 16.9 |
| Other tools | *N* | 84 | 85 | 85 | 84 |
| | Mean | 49.87 | 32.02 | 9.91 | 22.85 |
| | S.D. | 14.39 | 9.24 | 4.74 | 18.68 |
| Cores | *N* | 164 | 163 | 163 | 164 |
| | Mean | 52.3 | 39.91 | 19.39 | 55.4 |
| | S.D. | 12.55 | 9.19 | 8.11 | 59.49 |

**Table 6.75** Type counts for Unit II-B

| Type | Name | Real | Percent | Essential | Ess. perc. |
|---|---|---|---|---|---|
| 1 | Levallois flake | 17 | 7.4 | 0 | 0 |
| 2 | Atypical Levallois flake | 22 | 9.6 | 0 | 0 |
| 5 | Pseudo-Levallois point | 2 | 0.9 | 0 | 0 |
| 8 | Limace | 1 | 0.4 | 1 | 0.7 |
| 9 | Single straight scraper | 5 | 2.2 | 5 | 3.5 |
| 10 | Single convex scraper | 47 | 20.5 | 47 | 33.1 |
| 11 | Single concave scraper | 5 | 2.2 | 5 | 3.5 |
| 13 | Double straight-convex scraper | 4 | 1.7 | 4 | 2.8 |
| 14 | Double straight-concave scraper | 1 | 0.4 | 1 | 0.7 |
| 15 | Double convex scraper | 4 | 1.7 | 4 | 2.8 |
| 17 | Double convex–concave scraper | 3 | 1.3 | 3 | 2.1 |
| 19 | Convex convergent scraper | 11 | 4.8 | 11 | 7.7 |
| 21 | Dejete scraper | 7 | 3.1 | 7 | 4.9 |
| 22 | Straight transverse scraper | 1 | 0.4 | 1 | 0.7 |

(continued)

**Table 6.75** (continued)

| Type | Name | Real | Percent | Essential | Ess. perc. |
|---|---|---|---|---|---|
| 23 | Convex transverse scraper | 12 | 5.2 | 12 | 8.5 |
| 24 | Concave transverse scraper | 1 | 0.4 | 1 | 0.7 |
| 27 | Scraper with thinned back | 1 | 0.4 | 1 | 0.7 |
| 29 | Alternate scraper | 1 | 0.4 | 1 | 0.7 |
| 31 | Atypical endscraper | 3 | 1.3 | 3 | 2.1 |
| 32 | Typical burin | 1 | 0.4 | 1 | 0.7 |
| 33 | Atypical burin | 2 | 0.9 | 2 | 1.4 |
| 37 | Atypical backed knife | 2 | 0.9 | 2 | 1.4 |
| 38 | Naturally backed knife | 11 | 4.8 | 0 | 0 |
| 39 | Raclette | 2 | 0.9 | 2 | 1.4 |
| 40 | Truncation | 2 | 0.9 | 2 | 1.4 |
| 42 | Notch | 13 | 5.7 | 13 | 9.2 |
| 43 | Denticulate | 3 | 1.3 | 3 | 2.1 |
| 45 | Flake with irregular retouch on | 2 | 0.9 | 0 | 0 |
| 48 | Flake with abrupt and alter | 33 | 14.4 | 0 | 0 |
| 54 | End-notched flake | 1 | 0.4 | 1 | 0.7 |
| 59 | Chopper | 1 | 0.4 | 1 | 0.7 |
| 62 | Miscellaneous | 3 | 1.3 | 3 | 2.1 |
| 64 | Truncated-facetted piece | 5 | 2.2 | 5 | 3.5 |
| | Totals | 229 | 99.7 | 142 | 99.8 |

**Table 6.76** Artifact counts for Unit II-B

| | *N* |
|---|---|
| Complete and proximal flakes | 301 |
| Other flake fragments | 132 |
| Complete and proximal tools | 103 |
| Other tool fragments | 40 |
| Handaxes | 0 |
| Handaxe fragments | 1 |
| Shatter | 110 |
| Complete cores | 16 |
| Core fragments | 11 |

**Table 6.77** Typological and technological indices for Unit II-B

| Typological indices | | | Technological indices | |
|---|---|---|---|---|
| | Real | Essential | IL | 8.42 |
| ILty | 17.03 | | IF | 26.14 |
| IR | 44.98 | 72.54 | IFs | 20.06 |
| IAu | 0.87 | 0.01 | Ilam | 17.37 |
| II | 45.41 | 73.24 | TF | 3.52 |
| III | 0.03 | 5.63 | Kombewa | 0 |
| IV | 6.99 | 11.27 | Core edge | 1.49 |
| | | | Nat. backed | 5.2 |

**Table 6.78** Basic size measurements for complete flakes, scrapers, notched tools, other retouched tools, and cores from Level II-B

| | | Length | Width | Thickness | Weight |
|---|---|---|---|---|---|
| Flakes | *N* | 192 | 190 | 190 | 192 |
| | Mean | 38.52 | 27.15 | 7.62 | 12.85 |
| | S.D. | 12.11 | 10.27 | 4.69 | 19.12 |
| Scrapers | *N* | 64 | 64 | 64 | 64 |
| | Mean | 55.06 | 36.4 | 11.18 | 30.72 |
| | S.D. | 13.36 | 9.23 | 4.52 | 21.54 |
| Notches | *N* | 7 | 7 | 7 | 7 |
| | Mean | 44.39 | 34.25 | 9.48 | 22.14 |
| | S.D. | 18.14 | 9.06 | 4.04 | 16.5 |
| Other tools | *N* | 13 | 13 | 13 | 13 |
| | Mean | 52.42 | 37.73 | 10.16 | 31.69 |
| | S.D. | 14.72 | 15.59 | 4.2 | 22.92 |
| Cores | *N* | 16 | 16 | 16 | 16 |
| | Mean | 57.35 | 44.04 | 24.67 | 114.12 |
| | S.D. | 21.16 | 17.08 | 16.28 | 210.31 |

**Table 6.79** Type counts for Unit II-A

| Type | Name | Real | Percent | Essential | Ess. perc. |
|---|---|---|---|---|---|
| 1 | Levallois flake | 21 | 7.1 | 0 | 0 |
| 2 | Atypical Levallois flake | 22 | 7.5 | 0 | 0 |
| 3 | Levallois point | 2 | 0.7 | 0 | 0 |
| 5 | Pseudo-Levallois point | 7 | 2.4 | 0 | 0 |
| 6 | Mousterian point | 3 | 1 | 3 | 1.5 |
| 7 | Elongated mousterian point | 1 | 0.3 | 1 | 0.5 |
| 9 | Single straight scraper | 21 | 7.1 | 21 | 10.5 |
| 10 | Single convex scraper | 58 | 19.7 | 58 | 29 |
| 11 | Single concave scraper | 3 | 1 | 3 | 1.5 |
| 13 | Double straight-convex scraper | 1 | 0.3 | 1 | 0.5 |
| 15 | Double convex scraper | 4 | 1.4 | 4 | 2 |
| 18 | Straight convergent scraper | 2 | 0.7 | 2 | 1 |
| 19 | Convex convergent scraper | 5 | 1.7 | 5 | 2.5 |
| 21 | Dejete scraper | 8 | 2.7 | 8 | 4 |
| 22 | Straight transverse scraper | 4 | 1.4 | 4 | 2 |
| 23 | Convex transverse scraper | 22 | 7.5 | 22 | 11 |
| 25 | Scraper on interior surface | 2 | 0.7 | 2 | 1 |
| 27 | Scraper with thinned back | 4 | 1.4 | 4 | 2 |
| 30 | Typical endscraper | 3 | 1 | 3 | 1.5 |
| 31 | Atypical endscraper | 1 | 0.3 | 1 | 0.5 |
| 33 | Atypical burin | 2 | 0.7 | 2 | 1 |
| 34 | Typical percoir | 1 | 0.3 | 1 | 0.5 |
| 38 | Naturally backed knife | 19 | 6.4 | 0 | 0 |
| 42 | Notch | 21 | 7.1 | 21 | 10.5 |
| 43 | Denticulate | 21 | 7.1 | 21 | 10.5 |
| 48 | Flake with abrupt and alter | 24 | 8.1 | 0 | 0 |
| 51 | Tayac point | 1 | 0.3 | 1 | 0.5 |

(continued)

**Table 6.79** (continued)

| Type | Name | Real | Percent | Essential | Ess. perc. |
|---|---|---|---|---|---|
| 54 | End-notched flake | 3 | 1 | 3 | 1.5 |
| 59 | Chopper | 2 | 0.7 | 2 | 1 |
| 62 | Miscellaneous | 3 | 1 | 3 | 1.5 |
| 64 | Truncated-facetted piece | 4 | 1.4 | 4 | 2 |
| | Totals | 295 | 100 | 200 | 100 |

**Table 6.80** Artifact counts for Unit II-A

| | *N* |
|---|---|
| Complete and proximal flakes | 592 |
| Other flake fragments | 212 |
| Complete and proximal tools | 139 |
| Other tool fragments | 60 |
| Handaxes | 1 |
| Handaxe fragments | 1 |
| Shatter | 160 |
| Complete cores | 18 |
| Core fragments | 5 |

**Table 6.81** Typological and technological indices for Unit II-A

| Typological indices | | | Technological indices | |
|---|---|---|---|---|
| | Real | Essential | IL | 4.79 |
| ILty | 15.25 | | IF | 21.34 |
| IR | 45.42 | 67 | IFs | 15.19 |
| IAu | 0 | 0 | Ilam | 16.46 |
| II | 46.78 | 69 | TF | 2 |
| III | 0.02 | 3.5 | Kombewa | 0.14 |
| IV | 14.24 | 21 | Core edge | 1.09 |
| | | | Nat. backed | 4.1 |

An application of this approach to the Pech IV collection has already been published (Lin et al. 2015) and is summarized here. In this study, the expected amount of cortex was modeled by estimating the number of nodules of a given size and shape that would be required to account for the total mass of the assemblage. A set of experimental data suggested that the longest flake in an assemblage is the best predictor of the original nodule width. A survey of nodules around Pech IV was then used to determine the relationship between nodule size and weight gave the shape characteristics, and the largest Pech IV flake was then put into the width component of this relationship to estimate nodule

**Table 6.82** Basic size measurements for complete flakes, scrapers, notched tools, other retouched tools, and cores from Level II-A

| | | Length | Width | Thickness | Weight |
|---|---|---|---|---|---|
| Flakes | *N* | 408 | 407 | 407 | 402 |
| | Mean | 35.59 | 24.86 | 6.66 | 9.27 |
| | S.D. | 11.9 | 8.91 | 3.67 | 12.11 |
| Scrapers | *N* | 82 | 77 | 81 | 82 |
| | Mean | 52.9 | 39.9 | 12.52 | 41.17 |
| | S.D. | 13.57 | 12.35 | 4.95 | 35.12 |
| Notches | *N* | 26 | 26 | 26 | 27 |
| | Mean | 42.87 | 29.13 | 12.87 | 24.26 |
| | S.D. | 13.95 | 8.06 | 7.18 | 26.29 |
| Other tools | *N* | 7 | 7 | 7 | 7 |
| | Mean | 58.63 | 38.23 | 19.11 | 76.71 |
| | S.D. | 17.51 | 13.33 | 15.87 | 112.02 |
| Cores | *N* | 18 | 18 | 18 | 18 |
| | Mean | 53.8 | 42.47 | 22.43 | 71.33 |
| | S.D. | 15.89 | 10.14 | 9.73 | 71.55 |

**Table 6.83** Type counts for Unit I-B

| Type | Name | Real | Percent | Essential | Ess. perc. |
|---|---|---|---|---|---|
| 1 | Levallois flake | 250 | 12.8 | 0 | 0 |
| 2 | Atypical Levallois flake | 231 | 11.9 | 0 | 0 |
| 3 | Levallois point | 7 | 0.4 | 0 | 0 |
| 5 | Pseudo-Levallois point | 88 | 4.5 | 0 | 0 |
| 6 | Mousterian point | 4 | 0.2 | 4 | 0.4 |
| 8 | Limace | 1 | 0.1 | 1 | 0.1 |
| 9 | Single straight scraper | 48 | 2.5 | 48 | 5.2 |
| 10 | Single convex scraper | 103 | 5.3 | 103 | 11.1 |
| 11 | Single concave scraper | 23 | 1.2 | 23 | 2.5 |
| 12 | Double straight scraper | 3 | 0.2 | 3 | 0.3 |
| 13 | Double straight-convex scraper | 7 | 0.4 | 7 | 0.8 |
| 14 | Double straight-concave scraper | 1 | 0.1 | 1 | 0.1 |
| 15 | Double convex scraper | 4 | 0.2 | 4 | 0.4 |
| 17 | Double convex–concave scraper | 3 | 0.2 | 3 | 0.3 |
| 18 | Straight convergent scraper | 1 | 0.1 | 1 | 0.1 |
| 19 | Convex convergent scraper | 7 | 0.4 | 7 | 0.8 |
| 20 | Concave convergent scraper | 2 | 0.1 | 2 | 0.2 |
| 21 | Dejete scraper | 4 | 0.2 | 4 | 0.4 |
| 22 | Straight transverse scraper | 3 | 0.2 | 3 | 0.3 |
| 23 | Convex transverse scraper | 26 | 1.3 | 26 | 2.8 |
| 24 | Concave transverse scraper | 3 | 0.2 | 3 | 0.3 |
| 25 | Scraper on interior surface | 18 | 0.9 | 18 | 1.9 |
| 26 | Abrupt scraper | 5 | 0.3 | 5 | 0.5 |
| 27 | Scraper with thinned back | 3 | 0.2 | 3 | 0.3 |
| 28 | Scraper with bifacial retouch | 1 | 0.1 | 1 | 0.1 |
| 29 | Alternate scraper | 2 | 0.1 | 2 | 0.2 |
| 30 | Typical endscraper | 13 | 0.7 | 13 | 1.4 |

(continued)

**Table 6.83** (continued)

| Type | Name | Real | Percent | Essential | Ess. perc. |
|---|---|---|---|---|---|
| 31 | Atypical endscraper | 5 | 0.3 | 5 | 0.5 |
| 32 | Typical burin | 15 | 0.8 | 15 | 1.6 |
| 33 | Atypical burin | 2 | 0.1 | 2 | 0.2 |
| 34 | Typical percoir | 10 | 0.5 | 10 | 1.1 |
| 35 | Atypical percoir | 8 | 0.4 | 8 | 0.9 |
| 36 | Typical backed knife | 7 | 0.4 | 7 | 0.8 |
| 37 | Atypical backed knife | 4 | 0.2 | 4 | 0.4 |
| 38 | Naturally backed knife | 148 | 7.6 | 0 | 0 |
| 39 | Raclette | 7 | 0.4 | 7 | 0.8 |
| 40 | Truncation | 17 | 0.9 | 17 | 1.8 |
| 42 | Notch | 230 | 11.8 | 230 | 24.8 |
| 43 | Denticulate | 249 | 12.8 | 249 | 26.9 |
| 44 | Alternate retouched bec | 3 | 0.2 | 3 | 0.3 |
| 45 | Flake with irregular retouch on | 11 | 0.6 | 0 | 0 |
| 48 | Flake with abrupt and alter | 286 | 14.7 | 0 | 0 |
| 50 | Bifacially retouched flake | 1 | 0.1 | 1 | 0.1 |
| 51 | Tayac point | 12 | 0.6 | 12 | 1.3 |
| 52 | Notched triangle | 2 | 0.1 | 2 | 0.2 |
| 54 | End-notched flake | 31 | 1.6 | 31 | 3.3 |
| 56 | Rabot | 1 | 0.1 | 1 | 0.1 |
| 59 | Chopper | 1 | 0.1 | 1 | 0.1 |
| 62 | Miscellaneous | 15 | 0.8 | 15 | 1.6 |
| 64 | Truncated-facetted piece | 19 | 1 | 19 | 2.1 |
| 65 | Scraper on the platform | 2 | 0.1 | 2 | 0.2 |
| | Totals | 1947 | 101 | 926 | 99.6 |

**Table 6.84** Artifact counts for Unit I-B

| | *N* |
|---|---|
| Complete and proximal flakes | 6024 |
| Other flake fragments | 1748 |
| Complete and proximal tools | 656 |
| Other tool fragments | 278 |
| Handaxes | 5 |
| Handaxe fragments | 5 |
| Shatter | 925 |
| Complete cores | 216 |
| Core fragments | 130 |

mass. The nodule mass was then used to estimate the number of nodules required to produce each assemblage, and this, in turn, was converted into the amount of surface cortex by using the surface area formula for a cylinder (which the Pech IV survey indicated was the best estimate of shape). Because nodules cannot be always expected to be 100%

**Table 6.85** Typological and technological indices for Unit I-B

| Typological indices | | | Technological indices | |
|---|---|---|---|---|
| | Real | Essential | IL | 6.59 |
| ILty | 25.06 | | IF | 21.42 |
| IR | 13.71 | 28.83 | IFs | 13.2 |
| IAu | 0.56 | 0.01 | Ilam | 13.75 |
| II | 13.97 | 29.37 | TF | 2.05 |
| III | 0.03 | 6.91 | Kombewa | 0.18 |
| IV | 24.6 | 51.73 | Core edge | 2.31 |
| | | | Nat. backed | 7.02 |

**Table 6.86** Basic size measurements for complete flakes, scrapers, notched tools, other retouched tools, and cores from Level I-B

| | | Length | Width | Thickness | Weight |
|---|---|---|---|---|---|
| Flakes | *N* | 4250 | 3969 | 3977 | 4235 |
| | Mean | 34.38 | 24.75 | 7.17 | 8.28 |
| | S.D. | 10.24 | 7.76 | 3.37 | 8.84 |
| Scrapers | *N* | 143 | 143 | 141 | 143 |
| | Mean | 47.83 | 33.41 | 10.61 | 26.25 |
| | S.D. | 16.7 | 11.55 | 4.6 | 26.81 |
| Notches | *N* | 296 | 294 | 296 | 299 |
| | Mean | 39.6 | 28.31 | 10.11 | 14.53 |
| | S.D. | 10.92 | 7.32 | 5.39 | 10.67 |
| Other tools | *N* | 77 | 77 | 78 | 77 |
| | Mean | 40.03 | 28.65 | 10.13 | 17.88 |
| | S.D. | 12.06 | 9.42 | 4.14 | 16.16 |
| Cores | *N* | 208 | 206 | 207 | 208 |
| | Mean | 48.49 | 37.76 | 20.4 | 51.78 |
| | S.D. | 13.3 | 11.36 | 9.17 | 69.58 |

**Table 6.87** Type counts for Unit I-A

| Type | Name | Real | Percent | Essential | Ess. perc. |
|---|---|---|---|---|---|
| 1 | Levallois flake | 58 | 5.5 | 0 | 0 |
| 2 | Atypical Levallois flake | 106 | 10 | 0 | 0 |
| 3 | Levallois point | 4 | 0.4 | 0 | 0 |
| 4 | Retouched Levallois point | 2 | 0.2 | 2 | 0.4 |
| 5 | Pseudo-Levallois point | 108 | 10.2 | 0 | 0 |
| 6 | Mousterian point | 1 | 0.1 | 1 | 0.2 |
| 9 | Single straight scraper | 18 | 1.7 | 18 | 3.8 |
| 10 | Single convex scraper | 20 | 1.9 | 20 | 4.2 |
| 11 | Single concave scraper | 11 | 1 | 11 | 2.3 |
| 12 | Double straight scraper | 1 | 0.1 | 1 | 0.2 |
| 15 | Double convex scraper | 1 | 0.1 | 1 | 0.2 |
| 17 | Double convex–concave scraper | 1 | 0.1 | 1 | 0.2 |
| 19 | Convex convergent scraper | 2 | 0.2 | 2 | 0.4 |
| 20 | Concave convergent scraper | 1 | 0.1 | 1 | 0.2 |
| 22 | Straight transverse scraper | 1 | 0.1 | 1 | 0.2 |

(continued)

**Table 6.87** (continued)

| Type | Name | Real | Percent | Essential | Ess. perc. |
|---|---|---|---|---|---|
| 23 | Convex transverse scraper | 2 | 0.2 | 2 | 0.4 |
| 25 | Scraper on interior surface | 6 | 0.6 | 6 | 1.3 |
| 26 | Abrupt scraper | 4 | 0.4 | 4 | 0.8 |
| 30 | Typical endscraper | 7 | 0.7 | 7 | 1.5 |
| 31 | Atypical endscraper | 3 | 0.3 | 3 | 0.6 |
| 32 | Typical burin | 6 | 0.6 | 6 | 1.3 |
| 33 | Atypical burin | 3 | 0.3 | 3 | 0.6 |
| 34 | Typical percoir | 3 | 0.3 | 3 | 0.6 |
| 35 | Atypical percoir | 2 | 0.2 | 2 | 0.4 |
| 36 | Typical backed knife | 25 | 2.4 | 25 | 5.3 |
| 37 | Atypical backed knife | 7 | 0.7 | 7 | 1.5 |
| 38 | Naturally backed knife | 170 | 16 | 0 | 0 |
| 40 | Truncation | 8 | 0.8 | 8 | 1.7 |
| 42 | Notch | 127 | 12 | 127 | 26.7 |
| 43 | Denticulate | 168 | 15.8 | 168 | 35.3 |
| 44 | Alternate retouched bec | 3 | 0.3 | 3 | 0.6 |
| 45 | Flake with irregular retouch on | 11 | 1 | 0 | 0 |
| 48 | Flake with abrupt and alter | 127 | 12 | 0 | 0 |
| 51 | Tayac point | 2 | 0.2 | 2 | 0.4 |
| 52 | Notched triangle | 1 | 0.1 | 1 | 0.2 |
| 54 | End-notched flake | 6 | 0.6 | 6 | 1.3 |
| 61 | Chopping-tool | 1 | 0.1 | 1 | 0.2 |
| 62 | Miscellaneous | 8 | 0.8 | 8 | 1.7 |
| 64 | Truncated-facetted piece | 24 | 2.3 | 24 | 5 |
| 65 | Scraper on the platform | 1 | 0.1 | 1 | 0.2 |
| | Totals | 1060 | 100.5 | 476 | 99.9 |

**Table 6.88** Artifact counts for Unit I-A

| | *N* |
|---|---|
| Complete and proximal flakes | 3995 |
| Other flake fragments | 944 |
| Complete and proximal tools | 355 |
| Other tool fragments | 130 |
| Handaxes | 3 |
| Handaxe fragments | 5 |
| Shatter | 695 |
| Complete cores | 182 |
| Core fragments | 91 |

cortical when collected, and additional correction (again based on the survey data) of 9.2% was applied to the amount of expected cortex.

Following this procedure, the results are reproduced here in Table 6.49. The lower layers of the sequence (ignoring Layer 7 for its post-depositional issues) show cortex ratios

**Table 6.89** Typological and technological indices for Unit I-A

| Typological indices | | | Technological indices | |
|---|---|---|---|---|
| | Real | Essential | IL | 3.15 |
| ILty | 16.04 | | IF | 19.44 |
| IR | 6.42 | 14.29 | IFs | 9.96 |
| IAu | 3.02 | 0.07 | Ilam | 14.87 |
| II | 6.51 | 14.5 | TF | 5.04 |
| III | 0.05 | 11.76 | Kombewa | 0.14 |
| IV | 27.83 | 61.97 | Core edge | 4.09 |
| | | | Nat. backed | 8.92 |

**Table 6.90** Basic size measurements for complete flakes, scrapers, notched tools, other retouched tools, and cores from Level I-A

| | | Length | Width | Thickness | Weight |
|---|---|---|---|---|---|
| Flakes | *N* | 3063 | 2899 | 2895 | 3053 |
| | Mean | 33.46 | 23.44 | 7.69 | 8.47 |
| | S.D. | 9.53 | 7.32 | 3.36 | 9.48 |
| Scrapers | *N* | 30 | 28 | 29 | 30 |
| | Mean | 45.76 | 29.67 | 10.38 | 18.5 |
| | S.D. | 13.33 | 9.34 | 3.65 | 18.06 |
| Notches | *N* | 156 | 154 | 155 | 156 |
| | Mean | 42 | 28.69 | 11.23 | 19.4 |
| | S.D. | 13 | 8.96 | 3.98 | 20.13 |
| Other tools | *N* | 83 | 83 | 83 | 82 |
| | Mean | 45.14 | 30.1 | 11.65 | 20.02 |
| | S.D. | 13.28 | 8.45 | 5.04 | 17.3 |
| Cores | *N* | 181 | 178 | 178 | 180 |
| | Mean | 45.98 | 35.45 | 22.47 | 49.24 |
| | S.D. | 12.29 | 9.63 | 8.56 | 47.14 |

close to 1 meaning that there is no evidence for artifact transport based on this measure. Layer 5A shows a more elevated cortex ratio indicating an over representation of cortical pieces such as would occur if non-cortical artifacts were transported off site after manufacture. Layers 4C and 4B are again close to 1, but Layer 4A has an under representation of cortex. Then both of the Layer 3 assemblages show over representation.

One difficulty with the cortex ratio, however, is judging whether the computed value is sufficiently distinct from 1 to be significant. This issue was addressed in Lin et al. (2015) by resampling experimental assemblages of two different nodule sizes to assess how frequently an assemblage with this cortex ratio can be expected given the assemblage size. Using the less than 2 kg model, which seems to be the most likely for Pech IV based on the survey work and nodule size estimates, Layers 5A, 3B, and 3A have cortex ratios significantly above 1. The lower average in Layer 4A, given its small sample size, could occur as a result of sampling bias.

**Table 6.91** Core types (completes and broken) by layer

| Unit | Biface | Chopper | Chopping-tool | Disc-core | Globular | Inform | Kombewa | Levallois | Moust-disk | Other | Prismatic | Pyramidal | Sing-rem | Sing-surf | Tested | Twosurf |
|---|---|---|---|---|---|---|---|---|---|---|---|---|---|---|---|---|
| I-A | 0 | 0 | 17 | 4 | 18 | 118 | 5 | 4 | 0 | 2 | 1 | 2 | 0 | 69 | 5 | 1 |
| I-B | 2 | 2 | 10 | 7 | 21 | 43 | 7 | 7 | 3 | 61 | 1 | 1 | 0 | 149 | 4 | 0 |
| II-A | 1 | 0 | 0 | 1 | 1 | 7 | 0 | 0 | 0 | 0 | 0 | 0 | 0 | 9 | 2 | 0 |
| II-B | 1 | 0 | 0 | 1 | 1 | 3 | 0 | 2 | 0 | 7 | 0 | 0 | 0 | 9 | 1 | 0 |
| II-C | 1 | 1 | 1 | 4 | 10 | 36 | 8 | 8 | 9 | 17 | 1 | 0 | 3 | 116 | 1 | 1 |
| III-A | 0 | 0 | 0 | 1 | 1 | 9 | 5 | 5 | 2 | 2 | 0 | 0 | 0 | 34 | 0 | 0 |
| III-B | 0 | 1 | 1 | 2 | 2 | 40 | 10 | 21 | 3 | 17 | 0 | 0 | 0 | 112 | 1 | 3 |
| IV-A | 2 | 0 | 1 | 4 | 9 | 151 | 64 | 118 | 28 | 42 | 2 | 2 | 7 | 534 | 5 | 1 |
| IV-B | 2 | 0 | 1 | 3 | 3 | 63 | 23 | 33 | 4 | 28 | 1 | 0 | 7 | 223 | 1 | 0 |
| V | 0 | 0 | 1 | 5 | 2 | 44 | 4 | 9 | 3 | 4 | 0 | 0 | 3 | 85 | 1 | 4 |

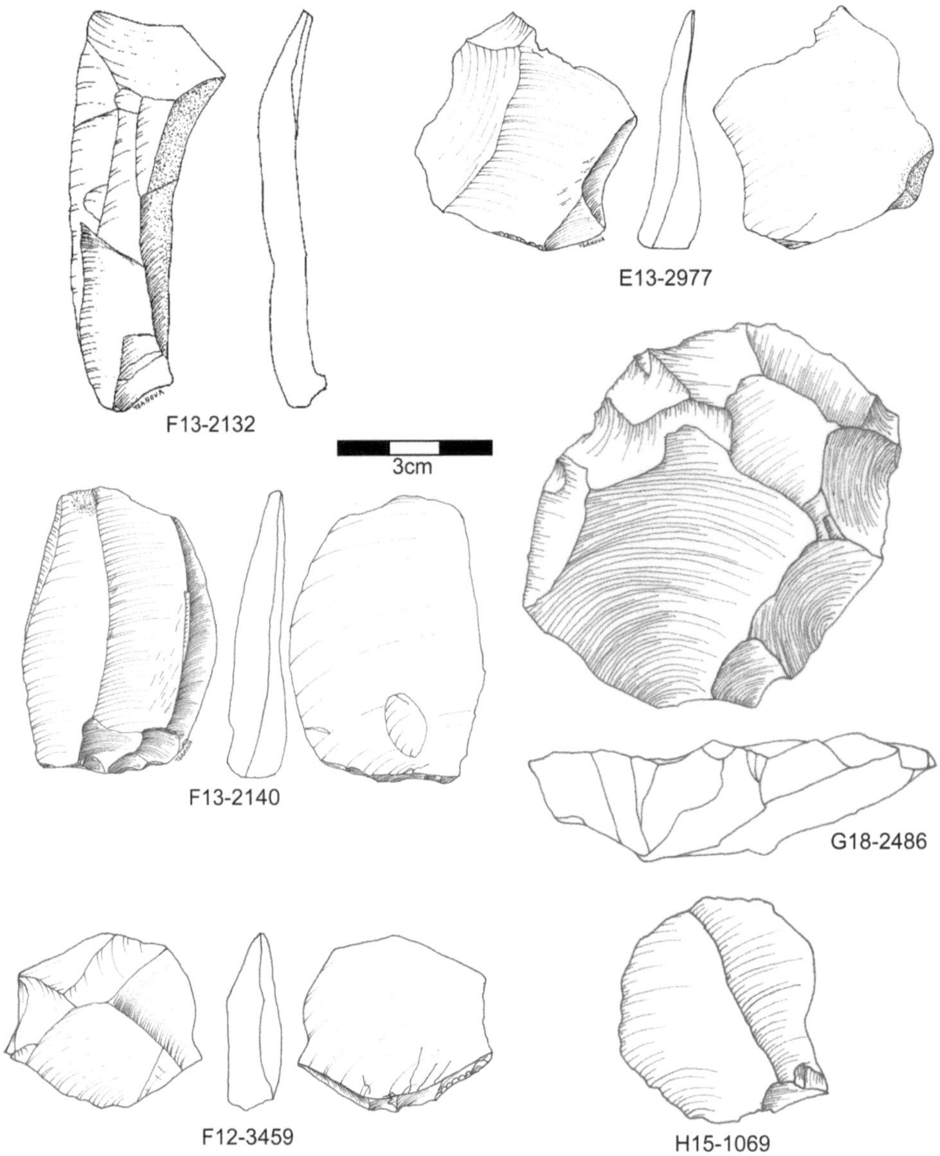


**Fig. 6.3** Lithics from Combined Unit V. F13-2132: Complete flake; E13-2977: Levallois flake; E13-2140: Levallois flake; G18-2486: Levallois core; F12-3459: Levallois flake; H15-1069: flake

## Combining Bordes Collection and the New Collection

This summary of the assemblages coming from our excavations is not unlike what we previously published (Turq et al. 2011) for our assemblages, what Bordes (1975) published for the first four years of his excavations, and what we published from the complete Bordes collection (McPherron and Dibble 2000). This, plus observations on the stratigraphy (see Chap. 2) suggest that the two sets of assemblages can be combined into one set (Table 6.50). In this combined system, the larger units from our excavations (e.g., Layer 3, 4, 5, etc.) corresponds to at least one of Bordes' larger units (e.g., Layer F, G, etc.). Bordes divided Layer J into 1–3 and

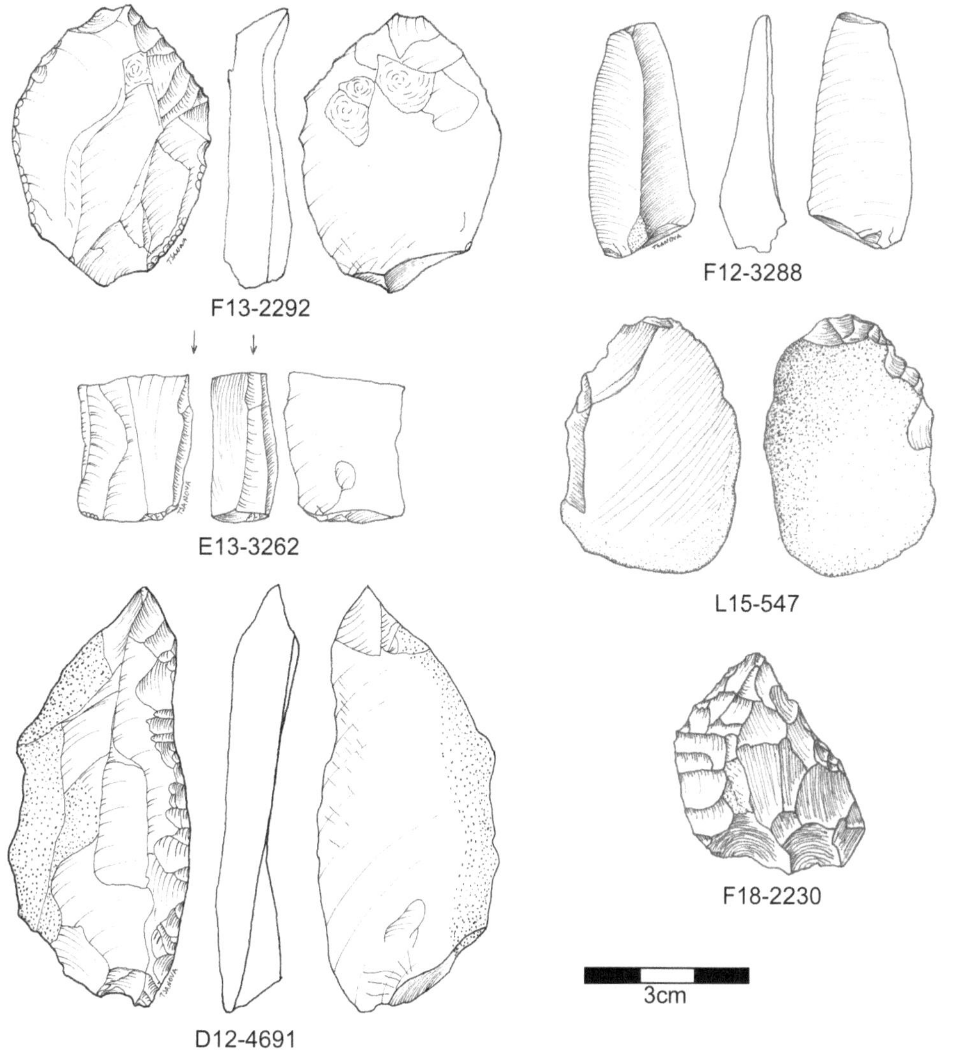


**Fig. 6.4** Lithics from Combined Unit V. F13-2292: Convergent scraper on Levallois flake; F12-3288: blade; E13-3262: Burin; L15-547: Truncated-faceted piece; D12-4691: Single scraper; F18-2230: Convergent scraper

then divided Layer J3 into A, B, and C. Here, the latter corresponds to our Layer 6 which becomes Unit IV in the unified stratigraphy, while the rest of Layer J corresponds to our Layer 5 and becomes Unit III in the unified stratigraphy. No attempt to bring Layer 7 into the new system is made here. Layer 7 mostly corresponds to Bordes' Layer X but circumstances of excavation and the fact that the assemblage is so heavily altered post-depositionally make it unsuitable for further analysis.

Of course the corresponding layers in the two sets of lithic assemblages are not exactly the same in terms of various measures that can be made (e.g., proportions of various tool types, tool, and flakes sizes, core sizes, platforms, cortex, etc.), but it is difficult to know whether these

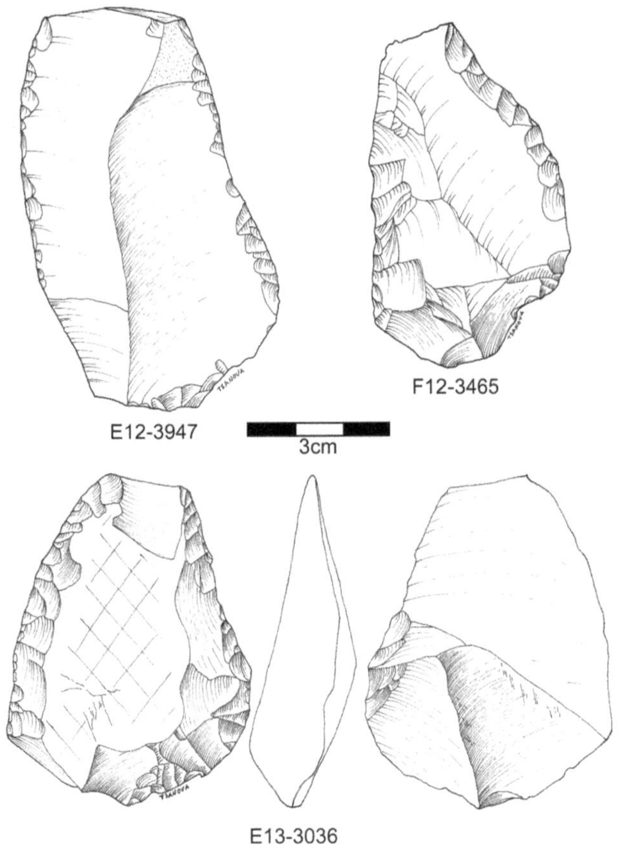


**Fig. 6.5** Lithics from Combined Unit V. E12-3947: Double scraper; F12-3465: Single scraper; E13-3036: Double scraper

differences reflect random variation, biases resulting from the way the two assemblages were excavated and curated, systematic horizontal variations perhaps from spatially patterned behavior, etc., and whether these differences have any behavioral meaning at all. That said, a quick look at differences in typology is provided in Fig. 6.2. In these figures, lines of a similar color are group together in the unified stratigraphy. In most cases, the differences are slight. However, Unit III does show considerable variability in Unit III-B (Bordes' Layers J2 and J3 combined with our Layer 5B), especially in terms of the notched tools (Layer 5B having a higher relative percentage than Bordes' Layers J1 or J2). For this, it is worth noting that Layer 5B where we excavated was characterized by a pocket of reworked material (probably solifluction) that was not separated during excavation (see Chap. 2). This may account for the elevated percentages of notched tools. With the two stratigraphies combined, we present here results for the entire site.

## Lithic Descriptions

What follows in Tables 6.51, 6.52, 6.53, 6.54, 6.55, 6.56, 6.57, 6.58, 6.59, 6.60, 6.61, 6.62, 6.63, 6.64, 6.65, 6.66, 6.67, 6.68, 6.69, 6.70, 6.71, 6.72, 6.73, 6.74, 6.75, 6.76, 6.77, 6.78, 6.79, 6.80, 6.81, 6.82, 6.83, 6.84, 6.85, 6.86, 6.87, 6.88, 6.89, 6.90, 6.91 and Figs. 6.3, 6.4, 6.5, 6.6, 6.7, 6.8, 6.9, 6.10, 6.11, 6.12, 6.13, 6.14, 6.15, 6.16, 6.17, 6.18, 6.19, 6.20, 6.21, 6.22, 6.23, 6.24, 6.25, 6.26, 6.27, 6.28, 6.29, 6.30, 6.31, 6.32, 6.33, 6.34, 6.35, 6.36, 6.37, 6.38, 6.39, 6.40, 6.41, 6.42, 6.43, 6.44, 6.45, 6.46, 6.47, 6.48, 6.49 and 6.50 are the final counts based on the previously published study of Bordes' collection (McPherron and Dibble 2000) combined with the results presented above.

## Blank Production

The assemblages of the combined layers show the same general trends as seen before with only the newly excavation assemblage (Table 6.92). Again, several changes can be seen throughout the sequence. First, Units V, III, and II are characterized by high proportions of scrapers among the retouched tools (Fig. 6.51). Second, Unit IV has instead roughly equal proportions of notched tools and

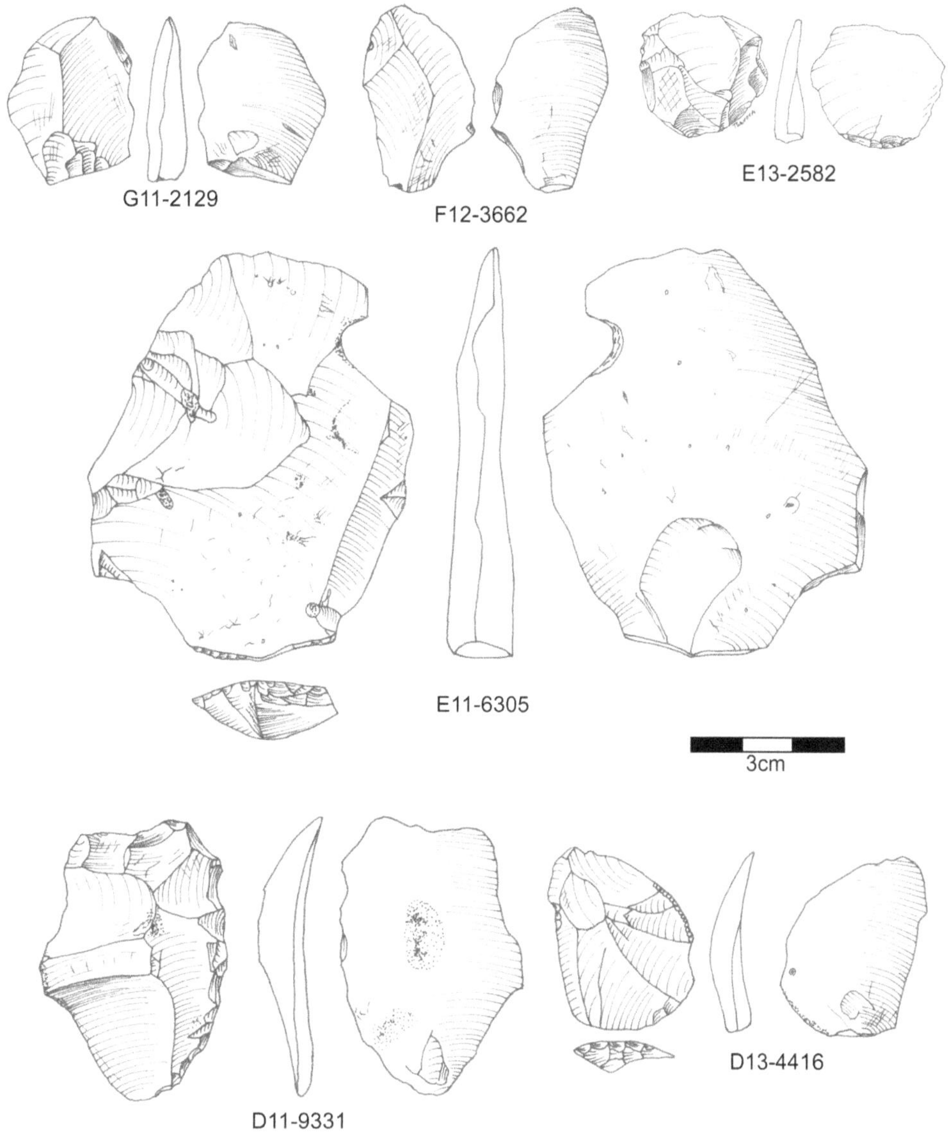


**Fig. 6.6** Lithics from Combined Unit IV-B. G11-2129: Levallois flake; F12-3662: Levallois flake; e13-2582: Levallois flake; E11-6305: Levallois flake; D11-9331: Levallois flake; D13-4416: Levallois flake

truncated-facetted artifacts, with scrapers being less frequent (Fig. 6.51). Third, Kombewa flakes are also more frequent in Unit IV (Fig. 6.52). They are also frequent in the layers immediately above and below Unit IV, and in these layers truncated-facetted pieces are also common (Fig. 6.52). Fourth, Levallois blank production is relatively common and constant in the lower half of the sequence but decreases starting in Unit II (Fig. 6.53). Facetted platforms follow the same pattern (Fig. 6.53). Fifth, Levallois is least frequent in the top of the sequence, Unit I-A, where pseudo-Levallois points, a type common is discoidal technologies, are frequent. Unit I-A is the only layer in the Pech IV sequence with this character. Sixth, Unit I-A also has an elevated percentage of Upper Paleolithic types, of which typical backed knives (Type 36) are the most common, and Unit I has some handaxes.

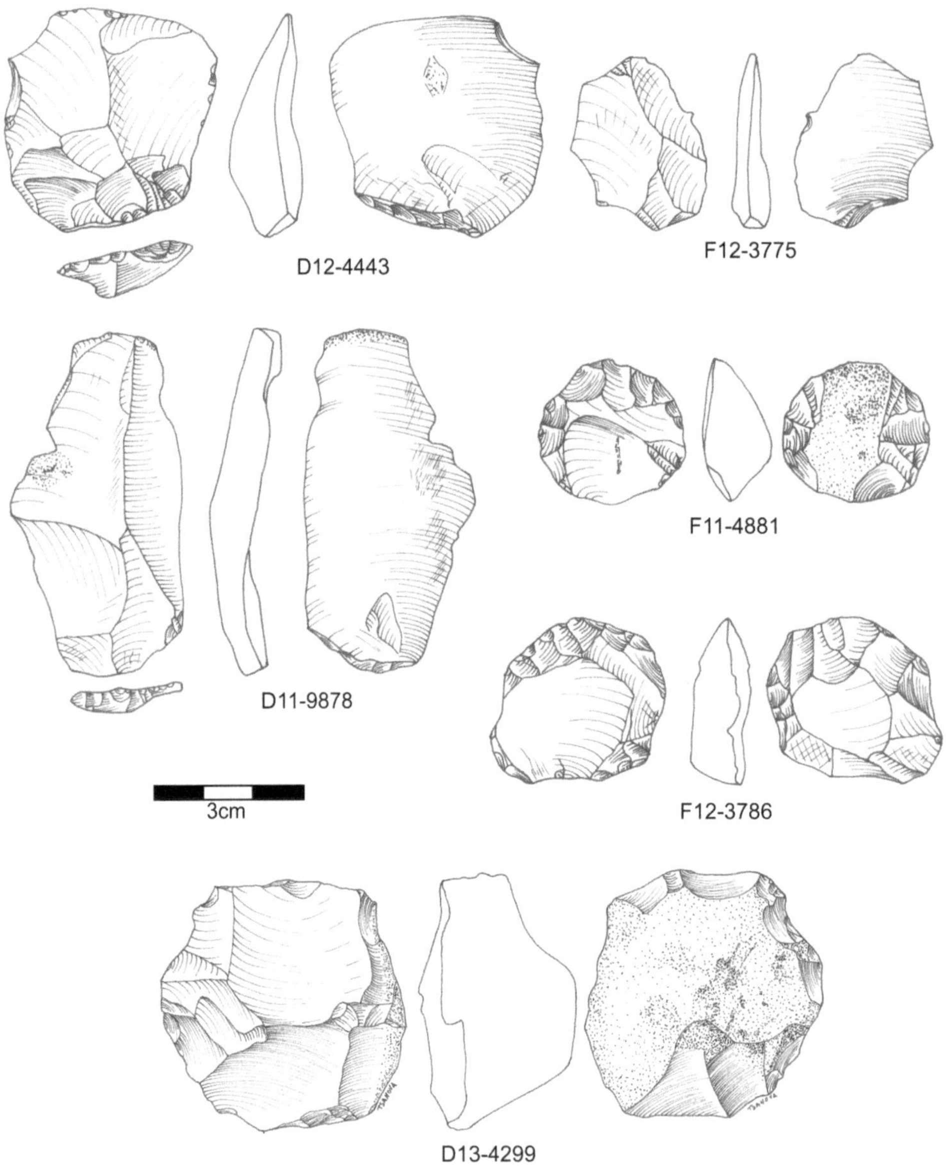


**Fig. 6.7** Lithics from Combined Unit IV-B. D12-4443: Flake; F12-3775: Levallois flake; D11-9878: Levallois flake; F11-4881: Levallois core; F12-3786: Levallois core; D13-4299: Levallois core

Some of these same differences between layers are seen when looking at measures of the intensity of raw material utilization as measured by the number of flakes produced on average from each core, the number of retouched tools made on average from a core, and the relative number of flake blanks turned into tools (Table 6.93; Fig. 6.54). In terms of blank production, Unit II-A has a very large number of blanks for the relatively few cores in this assemblage, and Unit IV has very few blanks produced from each core. Unit IV also has relatively few tools, whereas Unit II-A has the highest ratio of tools per cores. Unit I-A also again stands out as having low levels of tool production even though the number of blanks produced per core is average for the site. It is worth noting too that for reasons of consistency in reporting, the counts presented here do not combine truncated-facetted artifacts with the cores, even though we

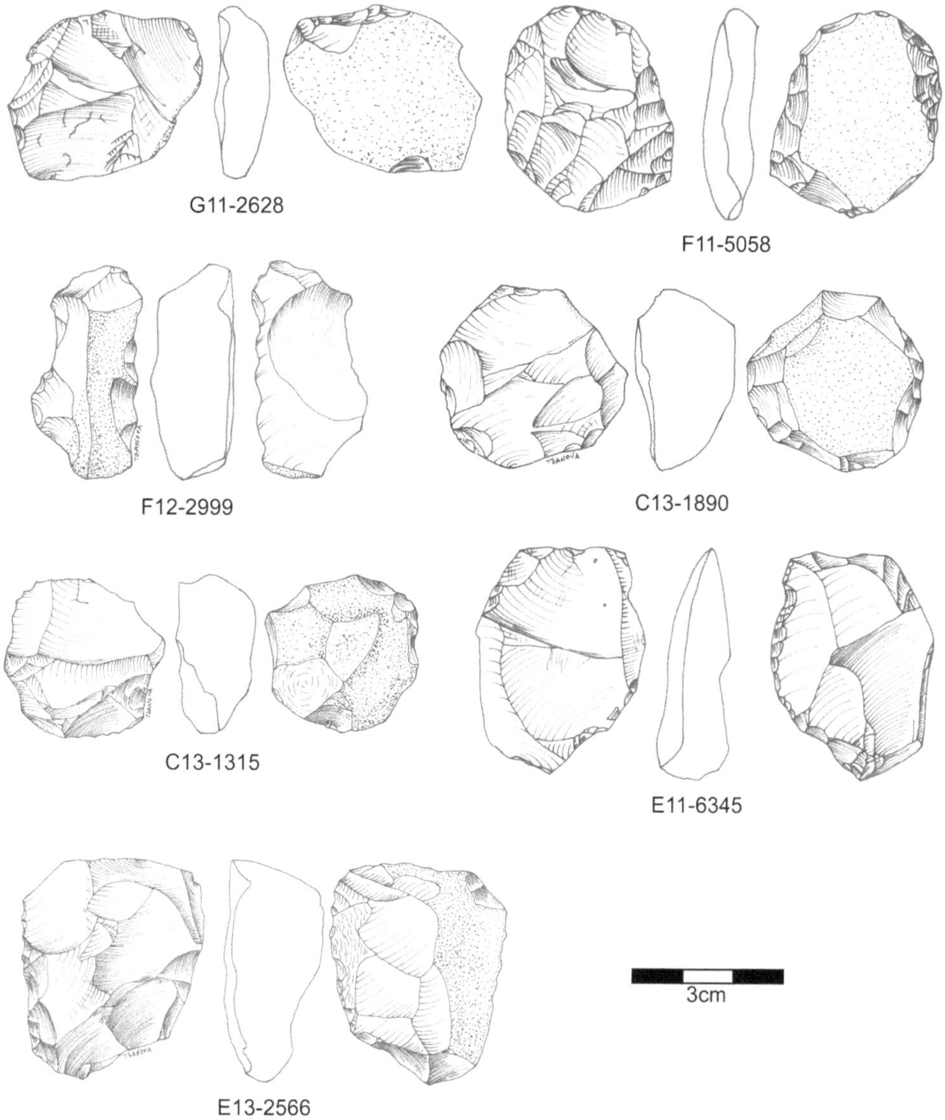


**Fig. 6.8** Lithics from Combined Unit IV-B. G11-2628: Core; F11-5058: Core; F12-2999: Core; C13-1890: Core; C13-1315: Core; E11-6345: Core; E13-2566: Core

have argued elsewhere that they are, in fact, cores (Dibble and McPherron 2006). Adding them to the totals would be a 38% increase in the number of cores and would bring the blank to core ratio down to under 5 for Unit IV. This number is not surprising in that truncated-facetted and Kombewa type cores do not yield many flakes per core. At the same time, this number is a minimum because many of the flakes that come from these cores fall below our size cut-off

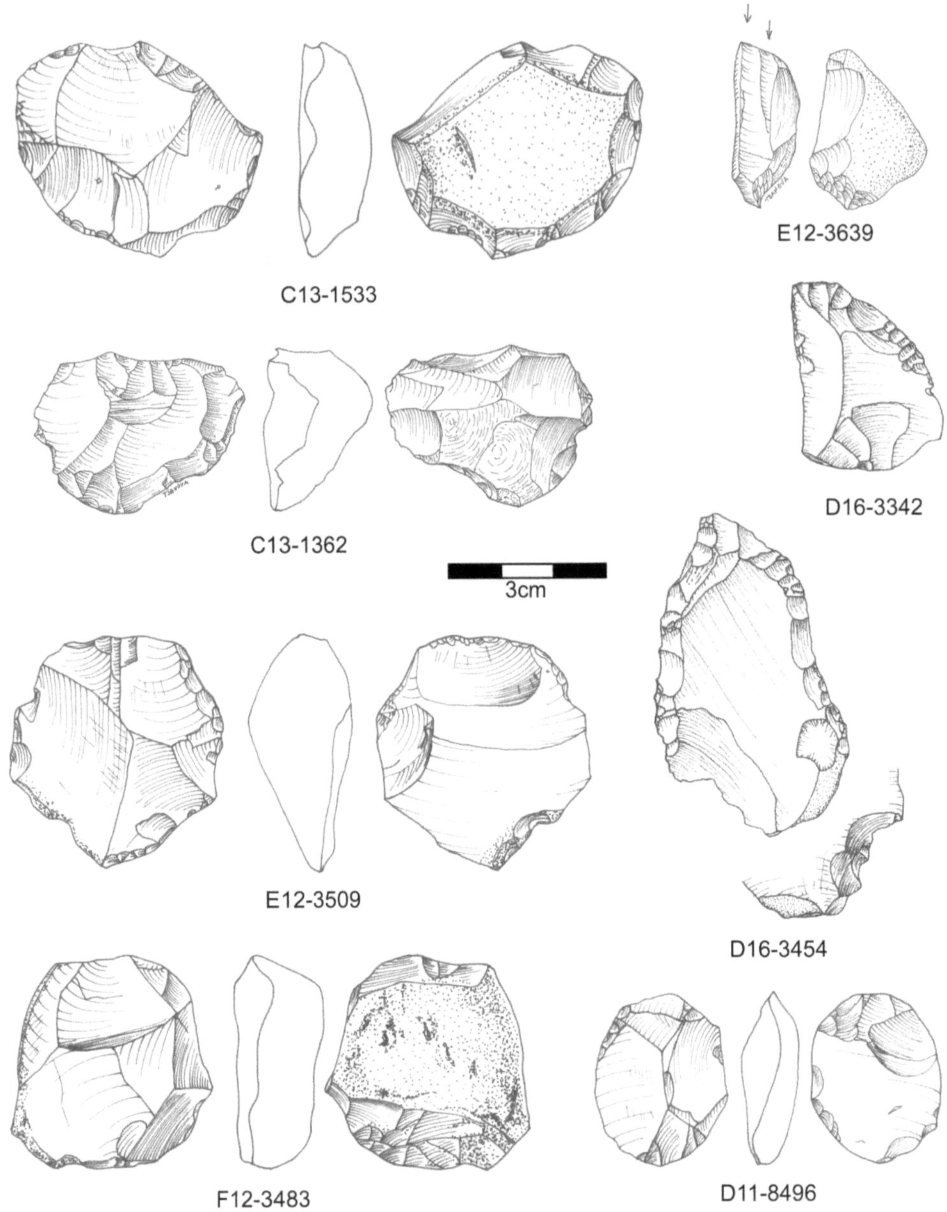


**Fig. 6.9** Lithics from Combined Unit IV-B. C13-1533: Core; E12-3639: Burin; C13-1362: Core; D16-3342: Déjeté scraper; E12-3509: Core; D16-3454: Convergent scraper; F12-3483: Truncated-faceted piece; D11-8496: Truncated-faceted piece

(25 mm) and are not counted here. Still, Unit IV has proportionately a lot of cores, and these cores have not been intensively exploited.

The sizes of Levallois and single-surface cores—as reflected by length, width, thickness, and weight—are provided in Table 6.94. Again, Unit IV stands apart with by far

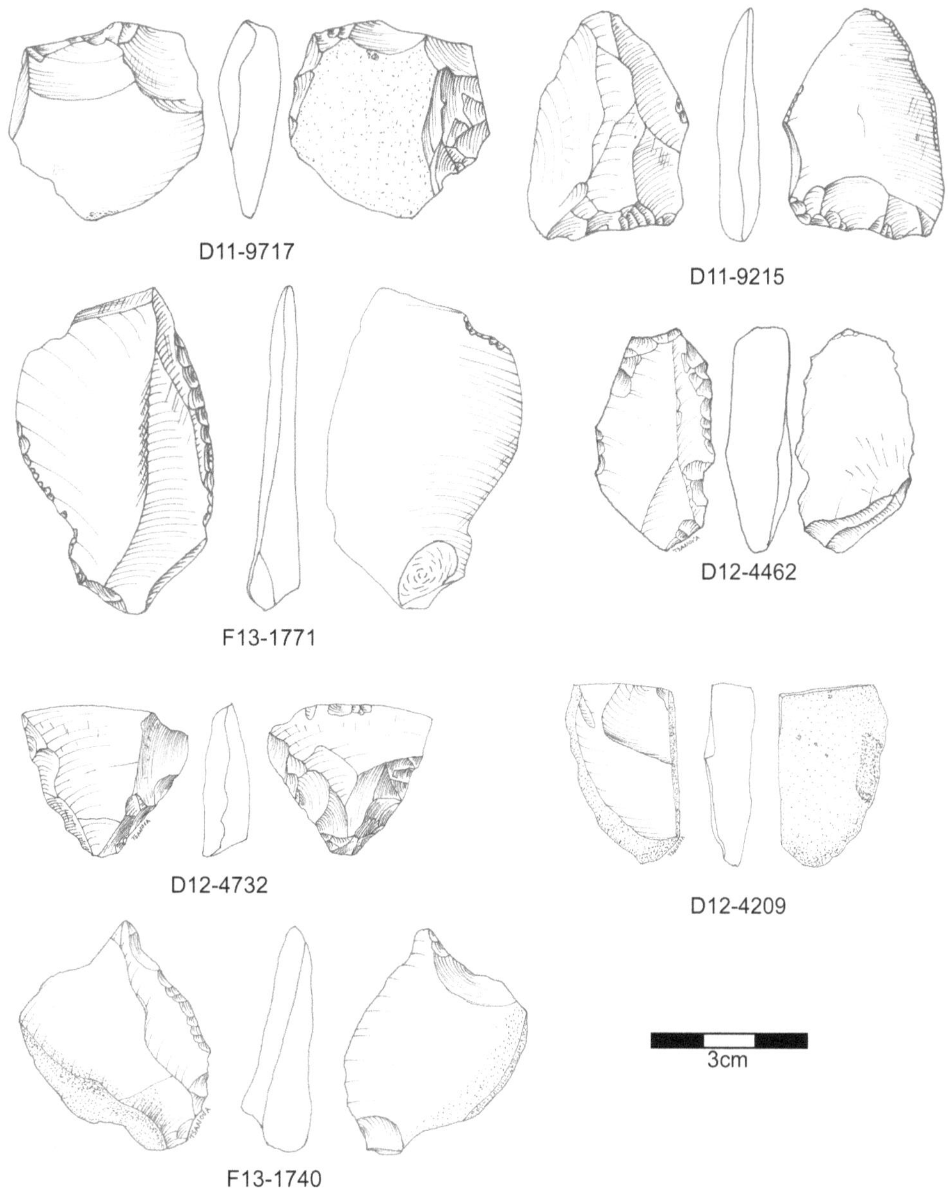


**Fig. 6.10** Lithics from Combined Unit IV-B. D11-9717: Truncated-faceted piece; D11-9215: Truncated-faceted piece; F13-1771: Single scraper; D12-4462: Endscraper; D12-4732: Tanged tool; D12-4209: Truncated-faceted piece; F13-1740: Alternate retouched beak

the smallest cores on all measures. Normally, in a model where blank to core ratios reflect the intensity of core reduction and with initial nodule size assumed to be constant across the sequence, the very low blank to core ratios of Unit IV should be associated with large cores. Instead, the largest cores come from Unit II, where the blank to core ratio

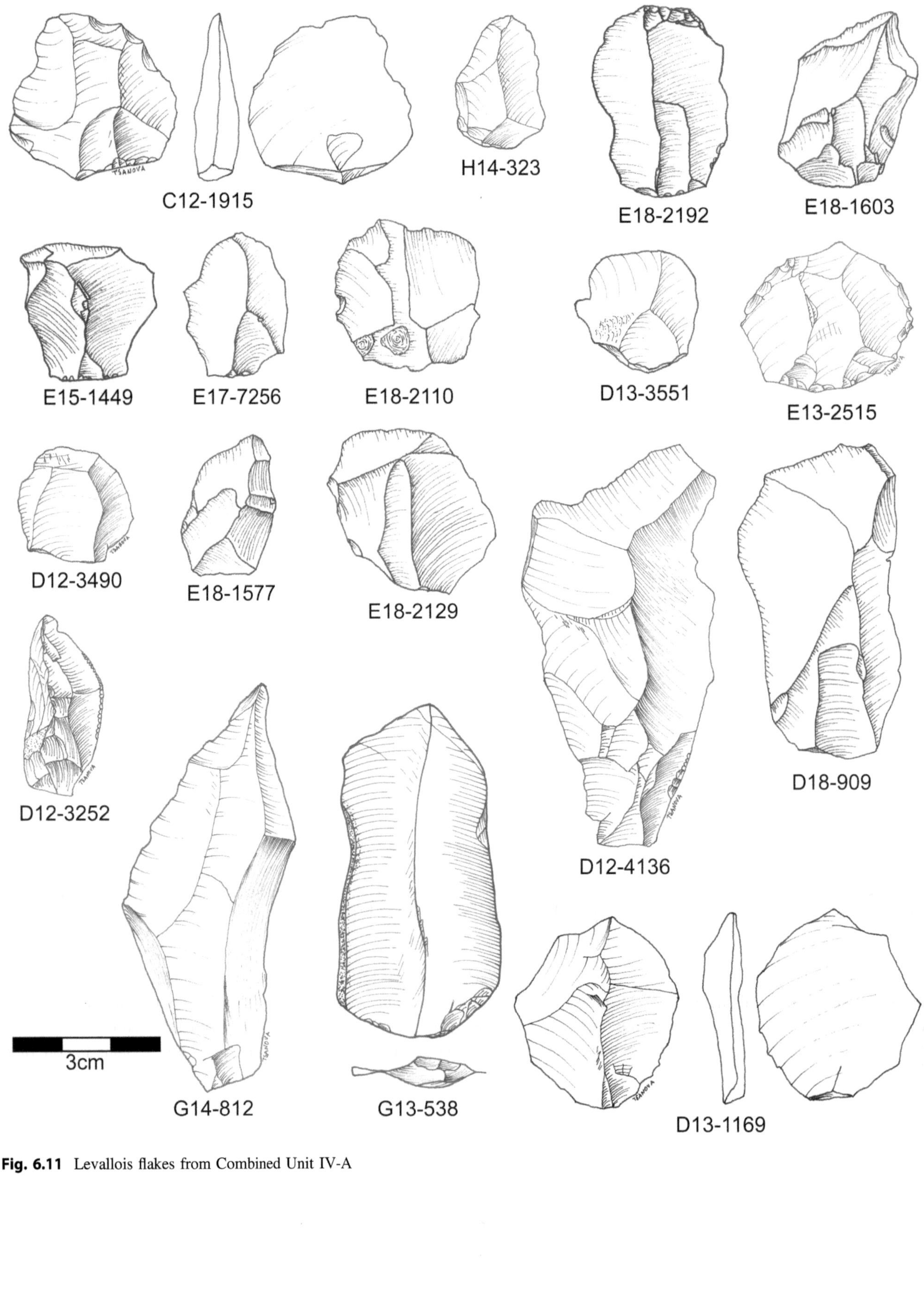


**Fig. 6.11** Levallois flakes from Combined Unit IV-A

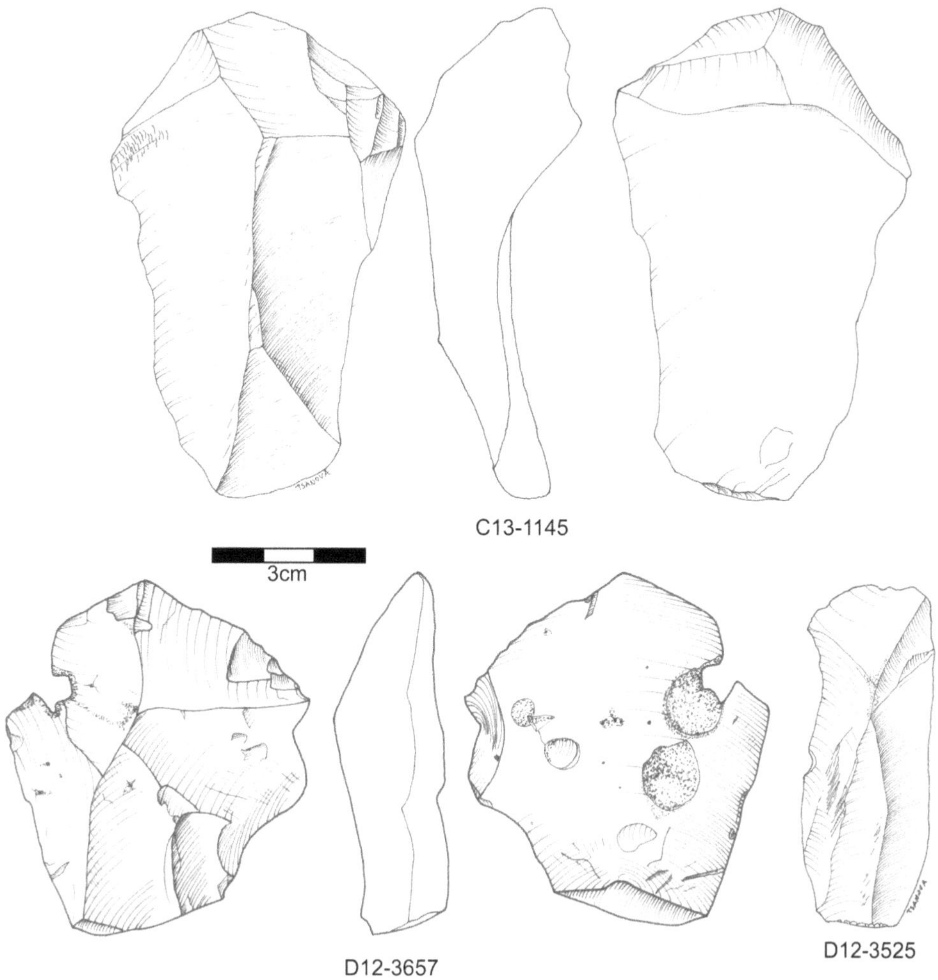


**Fig. 6.12** Lithics from Combined Unit IV-A. C13-1145: Overshot Levallois flake; D12-3657: Levallois flake; : D12-3525: Levallois flake

is the highest and yet the median core weight is three times those of Unit IV. In the case of Unit II, this suggests that flakes were imported, and for Unit IV these patterns strongly suggest the intentional production of small flakes from numerous, lightly reduced, small cores.

Another way of looking at these patterns is to consider the number of flakes produced per unit of mass of raw material (that is, per kg) as a measure of how intensively flakes were extracted from the available resources (Fig. 6.55). Here this value is plotted against the more

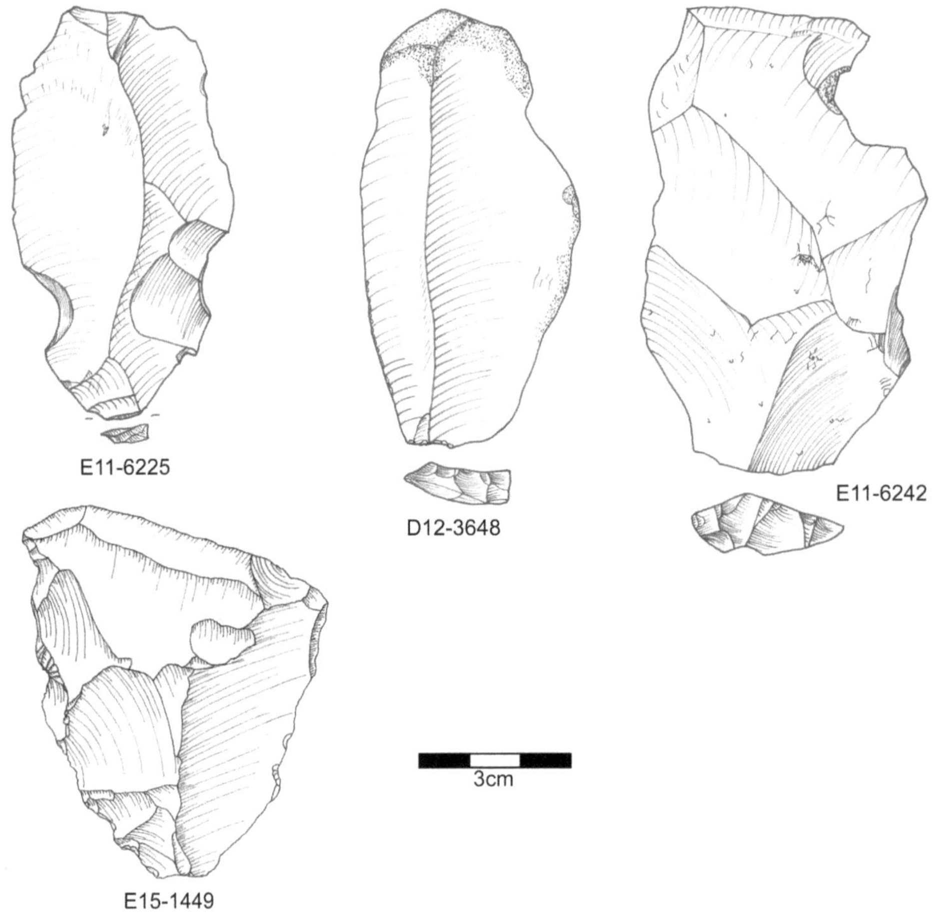


**Fig. 6.13** Levallois flakes from Combined Unit IV-A

traditional blank to core ratio. The expectation is that these two ratios will be linearly related: as the number of blanks extracted from a core increases, the number of blanks per unit material should increase as well. While there is overall a linear relationship in the Pech IV assemblages, there is also some variability. This variability, however, is mostly a result of Layers II-A and II-B. If we disregard these assemblages, the remaining assemblages show a clear pattern. At one end are the low blank to core and low blank per kilo assemblages of Unit IV plus Unit III-B which in many other measures (see above) also shows a continuation of the Unit IV patterns. Next, Layers I-A and I-B plot together with Unit III-A and not far away is Unit II-C. Unit V has the highest blank to core ratio and a blank per kg ratio similar to Unit I-B. Unit

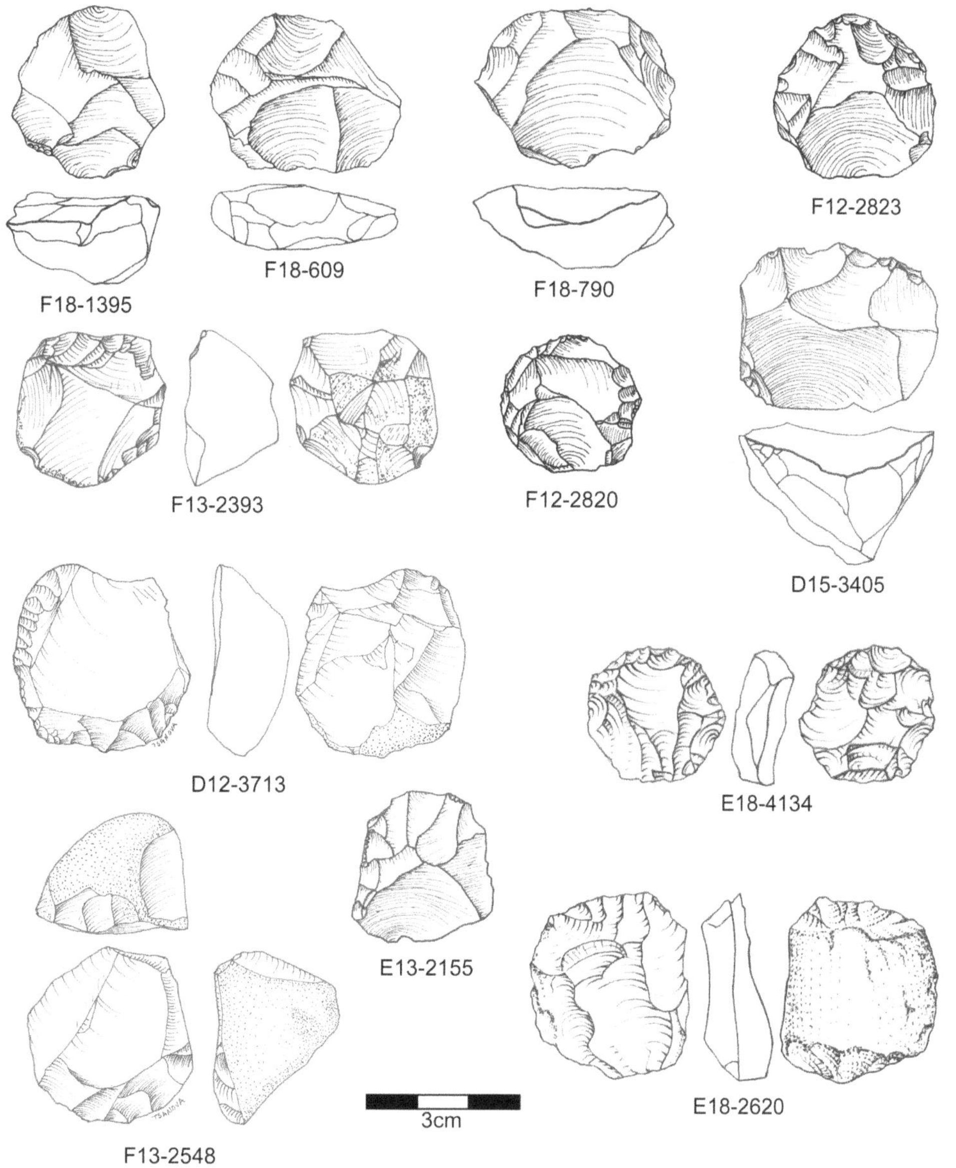


**Fig. 6.14** Small cores from Combined Unit IV-A

II-A and II-B, on the other hand, show low intensities of utilization as measured by blanks extracted per unit of raw material but, in the case of Unit II-A, high intensity of utilization in terms of the number of blanks per core. One explanation for these assemblages is again that flakes have been imported or cores have been exported. In both cases, doubling the number of cores would bring the blank to core ratio more in line with the blank to unit mass ratio.

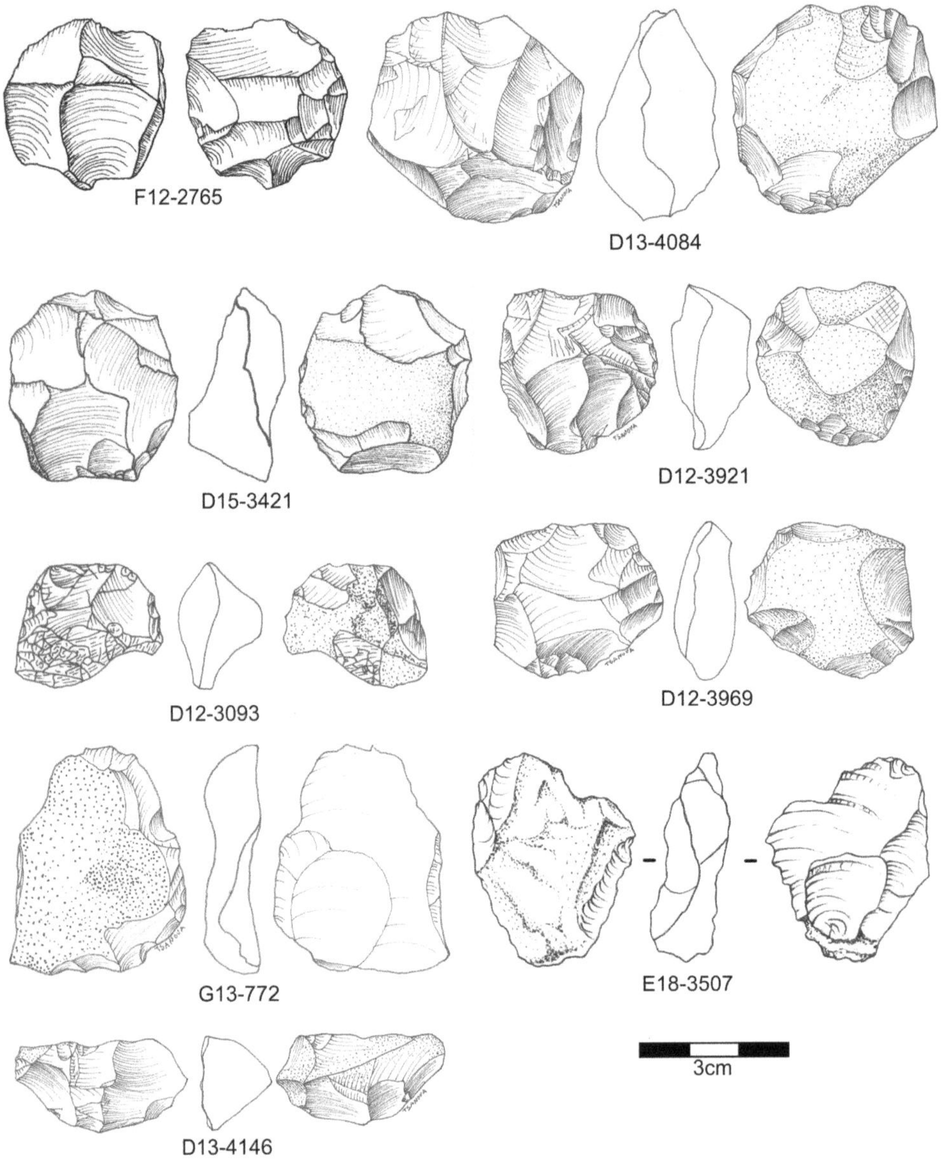


**Fig. 6.15** Small cores from Combined Unit IV-A

The types of blanks produced throughout the sequence, the types of platforms used to produce these blanks and the forms of the blanks are presented in Tables 6.95, 6.96 and 6.97. In terms of blank production technology, these results match what was expected based on the indices and core forms already presented. Levallois is consistently present in the lower half of the sequence but decreases during Unit II and is at its lowest at the top of the

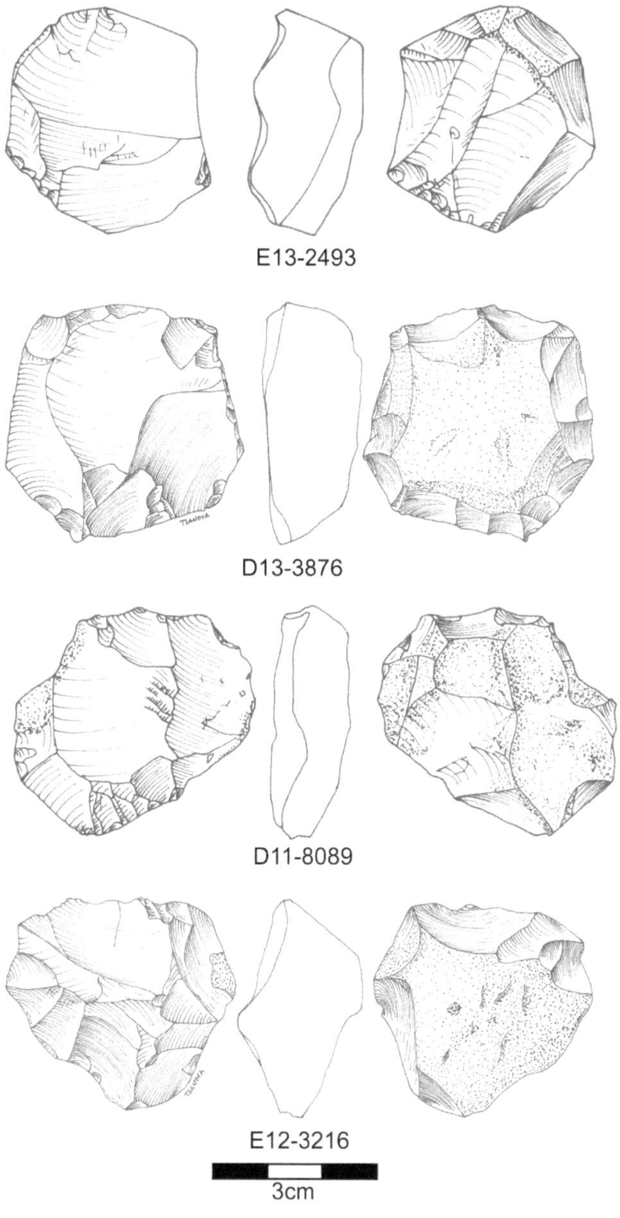


**Fig. 6.16** Cores from Combined Unit IV-A

sequence in Unit I-A. The presence of relatively frequent Clactonian and discoidal flakes in Unit I and Unit II-A and II-B also speak to a shift in technology in these layers. Kombewa flakes are, relatively speaking, never common but they are most frequent in Unit IV-B and the layers above and below Unit IV-B. Biface retouch flakes

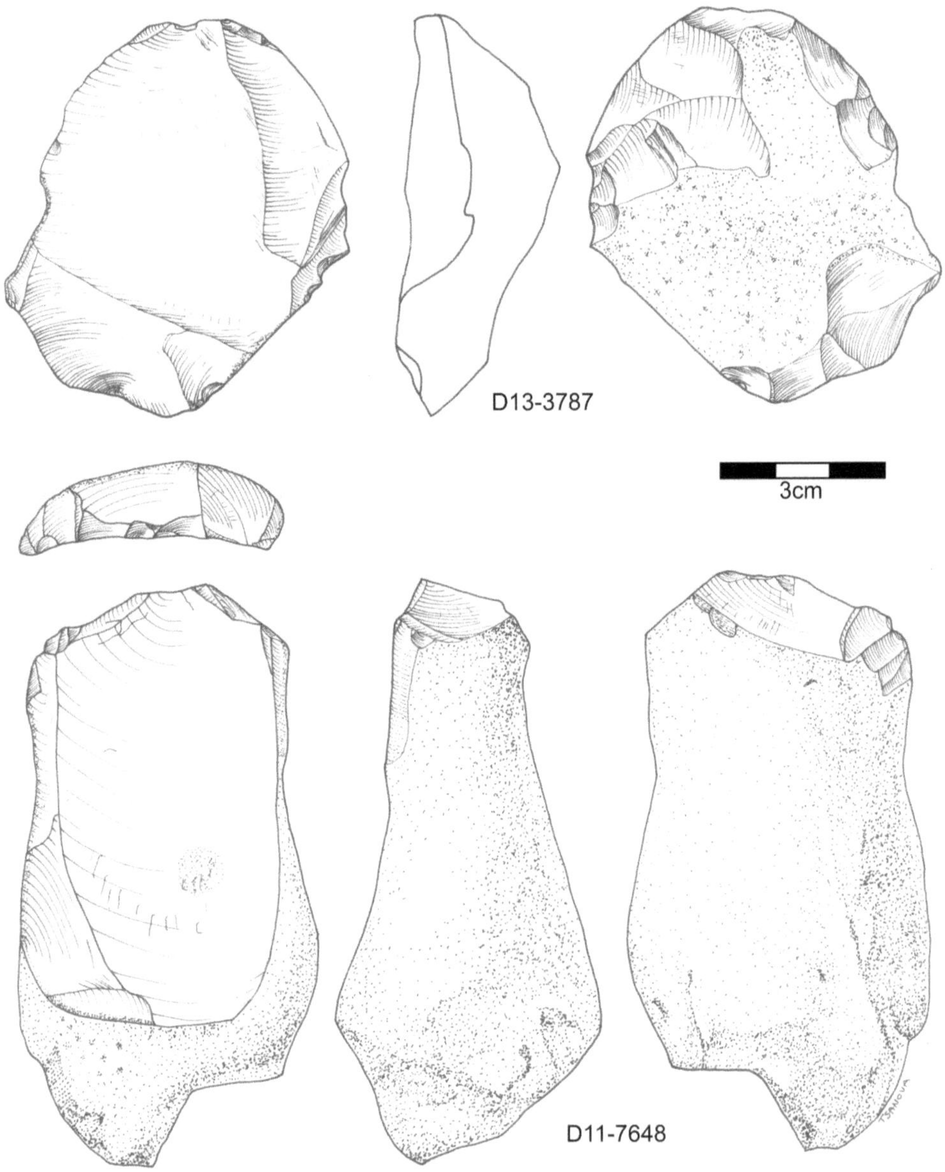


**Fig. 6.17** Cores from Combined Unit IV-A

are quite common in Unit II-A and relatively frequent in Unit I-B. Both of these layers have handaxes, though they are more common in Unit I. Unit II-A also has, again relatively speaking, more retouch flakes. The patterning in the platform data mirrors the technology in that facetted and dihedral platforms are more common in the Levallois

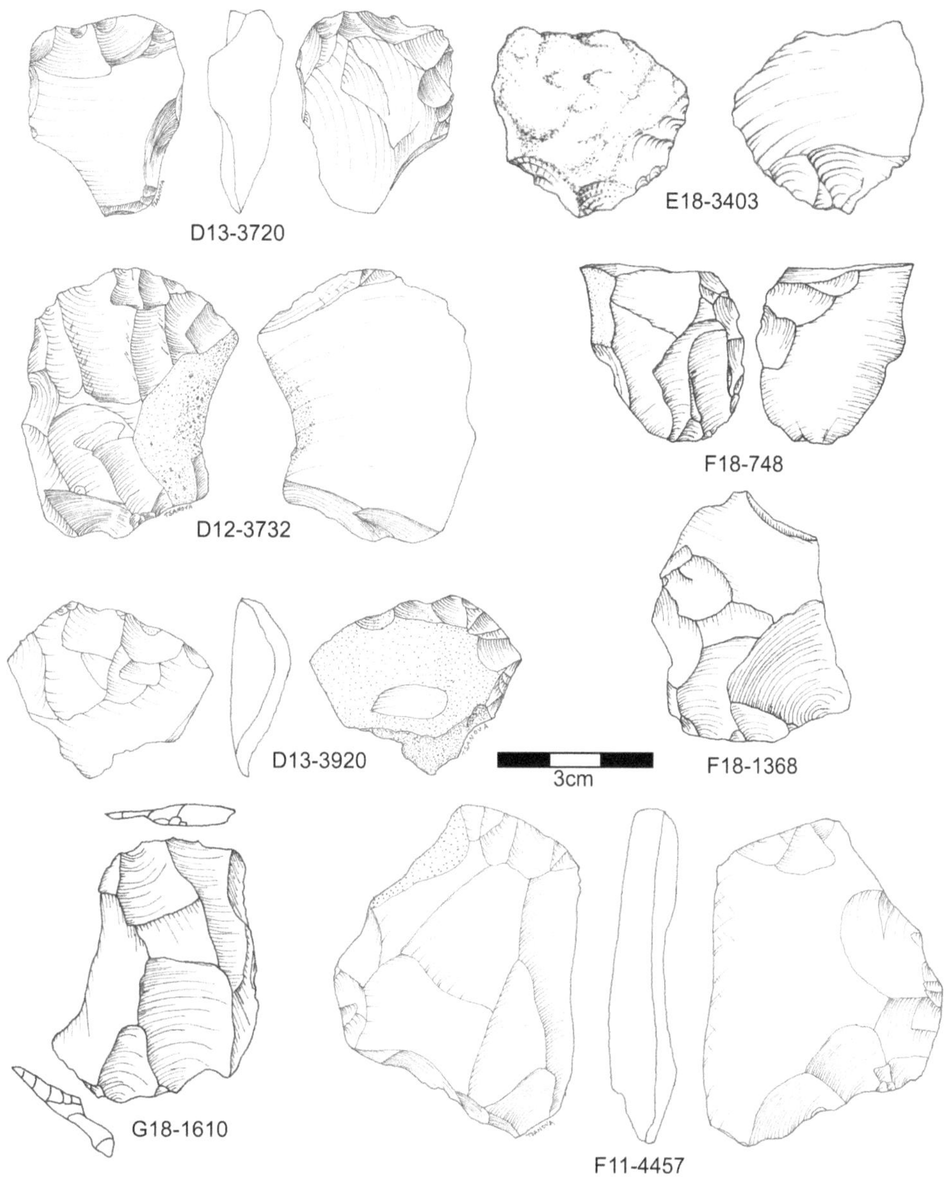


**Fig. 6.18** Truncated-faceted pieces from Combined Unit IV-A

layers. Patterning in cortical platforms is generally related to the blank to unit mass ratio with proportionally fewer cortical platforms in layers with higher blank to unit mass ratios (e.g., Units V and I-A). Unit I-B is an exception with a high percentage of cortical platforms and a high ratio of blanks to unit mass.

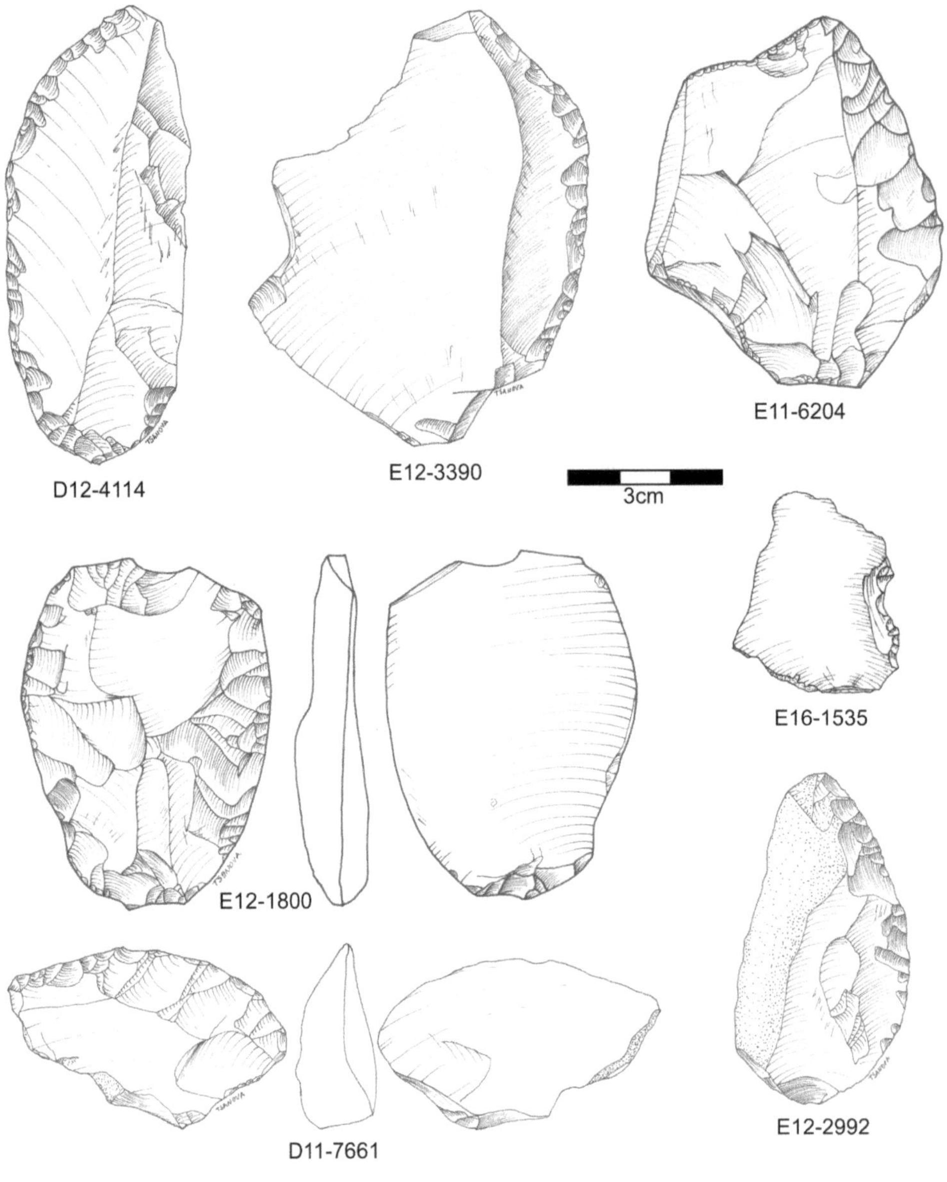


**Fig. 6.19** Retouched lithics from Combined Unit IV-A. D12-4114: Single scraper on Levallois flake; E12-3390: Single scraper on Levallois flake; E11-6204: Single scraper on Levallois flake; E16-1535: Notch; D11-7661: Transverse scraper; E12-2992: Single scraper

Blank form shows a few patterns of note, with some of the same layers showing highs and lows in particular categories. First, naturally backed artifacts are relatively rare in II-A and quite common in Unit I-A. They are also common in Unit III. Expanding flakes are common in Unit II-A (roughly double the overall site average). Core edge flakes are quite common in Unit I-A (again roughly double the site average). Broad flakes are common in Unit II-B and angular

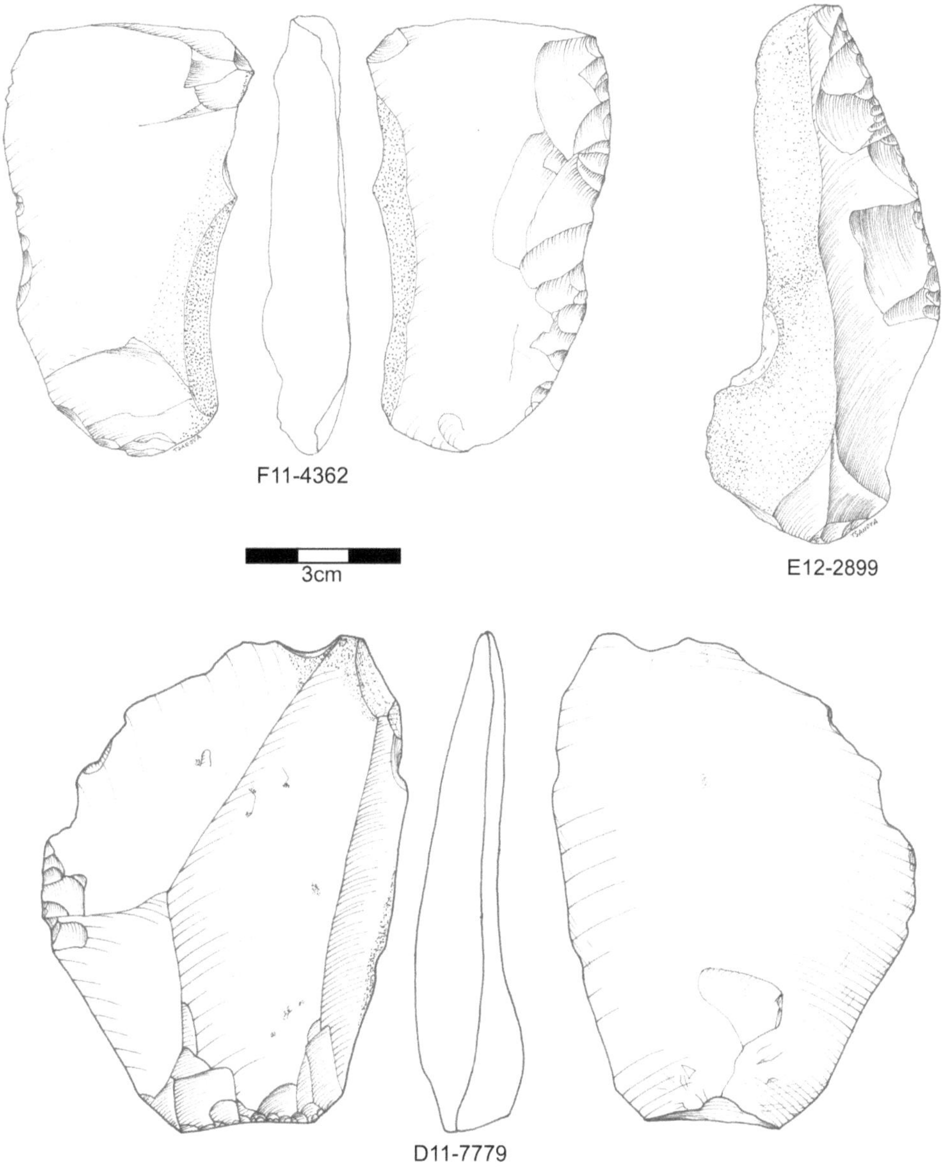


**Fig. 6.20** Retouched lithics from Combined Unit IV-A. F11-4362: Scraper on the interior; E12-2899: Single scraper; D11-7779: Single scraper

flakes are much more frequent in Unit I-A. Blade form shows no increase towards the top of the sequence, and in fact peaks at two other times in the lower part of the sequence.

The lack of a trend toward elongated blanks towards the top of the sequence is perhaps more effectively demonstrated using caliper measurements on complete flakes (Table 6.98). This does not fit with the pattern

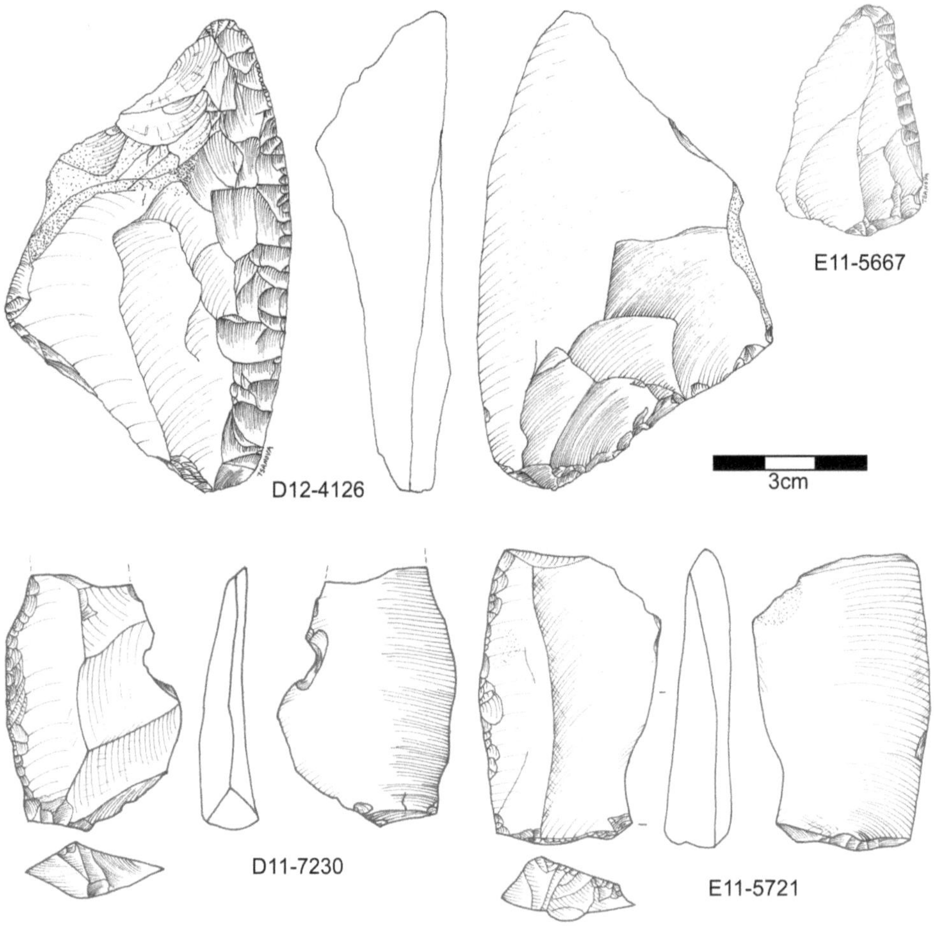


**Fig. 6.21** Retouched lithics from Combined Unit IV-A. D12-4126: Scraper with thinned back; E11-5667: Single scraper; D11-7230: Single scraper with interior notch; E11-5721: Single scraper on Levallois flake

described by Soressi (2002, 2004, 2005) for the nearby Pech I and other assemblages typed as Mousterian of Acheulian Tradition (MTA), but it does fit with the fact that the technology of this layer is less Levallois and more discoidal. Soressi (2005) has also argued that there is an increase in backed elements in the MTA, something that shows continuity with the subsequent Chatelperronian. The Pech IV data show this as well (see also McPherron et al. 2005) (Table 6.99). The relative proportion of backed blanks, as a result of various techniques, varies

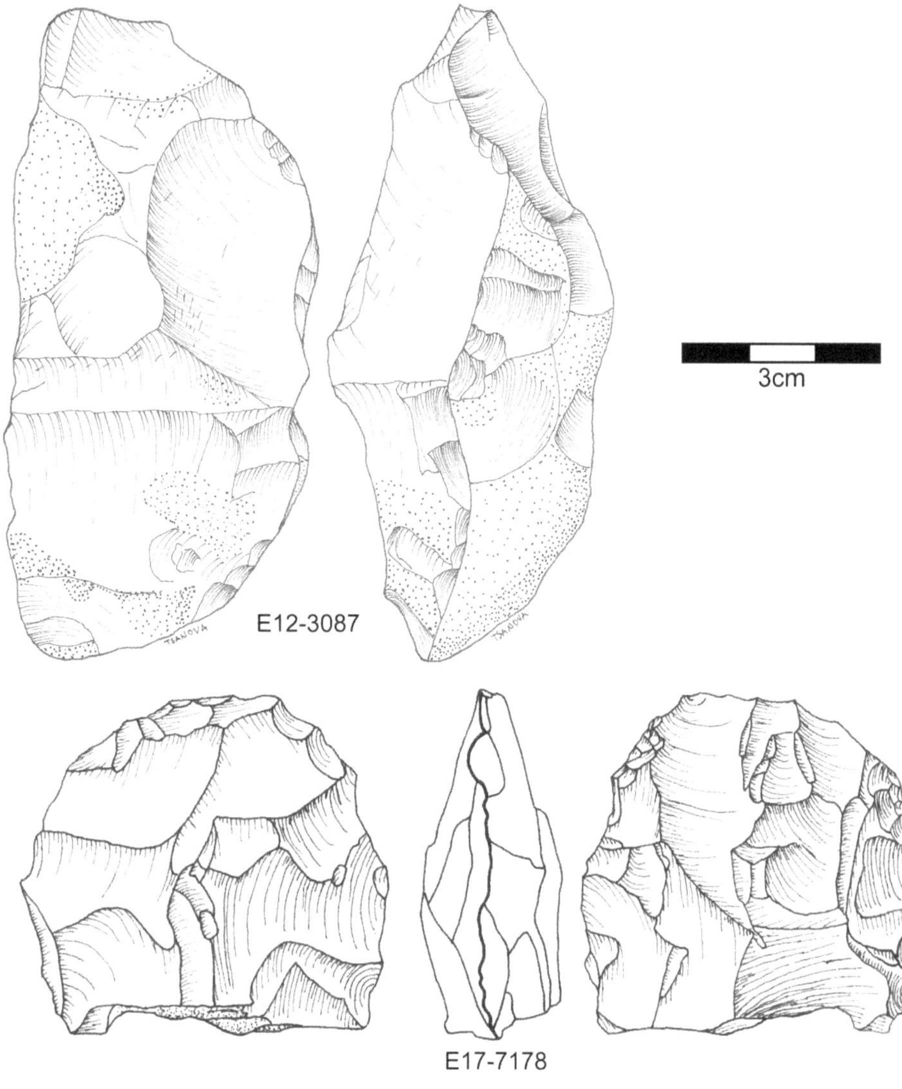


**Fig. 6.22** Partial bifaces from Combined Unit IV-A

throughout the sequence and then peaks in Unit I-A. This increase comes from the numerous pseudo-Levallois points (Type 5), backed knives (Type 36) and core edge flakes in the assemblage.

## Blank Selection

As is typical for Middle Paleolithic assemblages (Dibble 1988), scrapers and notched tools together are the most

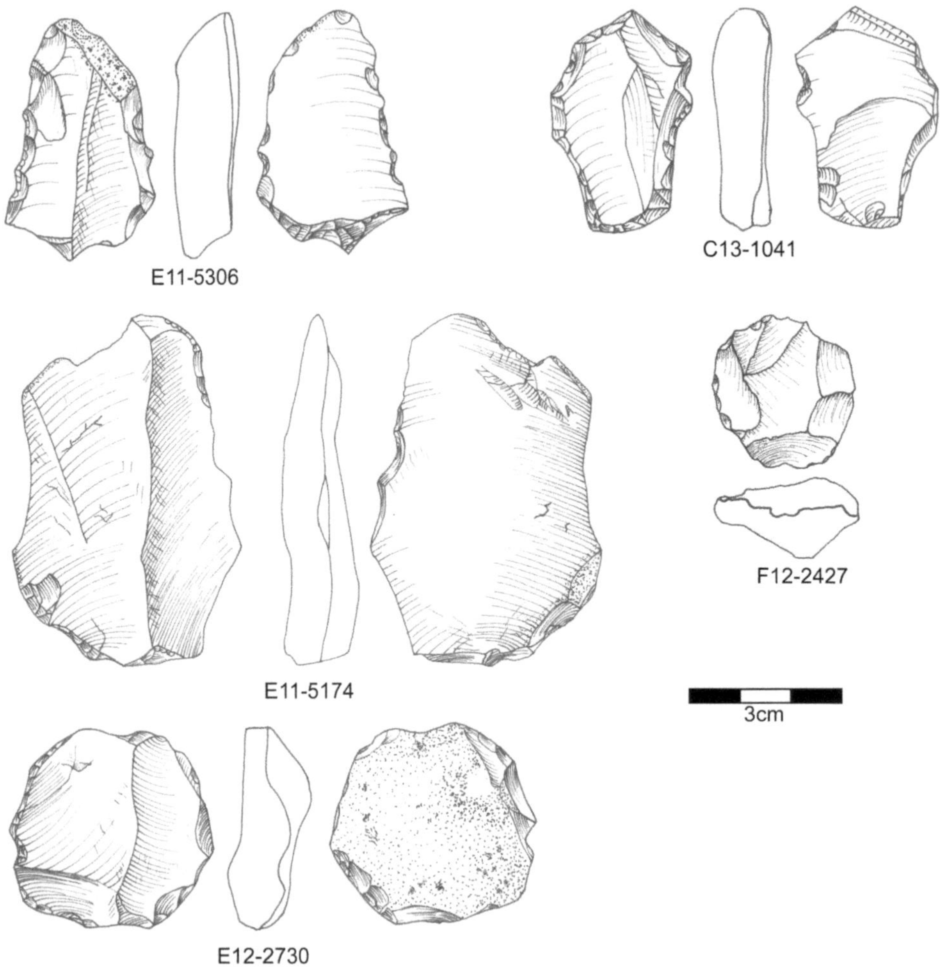


**Fig. 6.23** Lithics from Combined Unit III-B. E11-5306: Flake with alternating and abrupt retouch; C13-1041: Flake with alternating and abrupt retouch; E11-5174: Levallois flake; F12-2427: Core; E12-2730: Core

frequent tools in the Pech IV assemblage with some variability between the two through the sequence (Fig. 6.51). Scrapers are more frequent except in the Units IV and I where they are considerably less frequent. Of the scraper types, single scrapers are by far the most frequent (Fig. 6.56) in all layers followed by double, convergent, and transverse

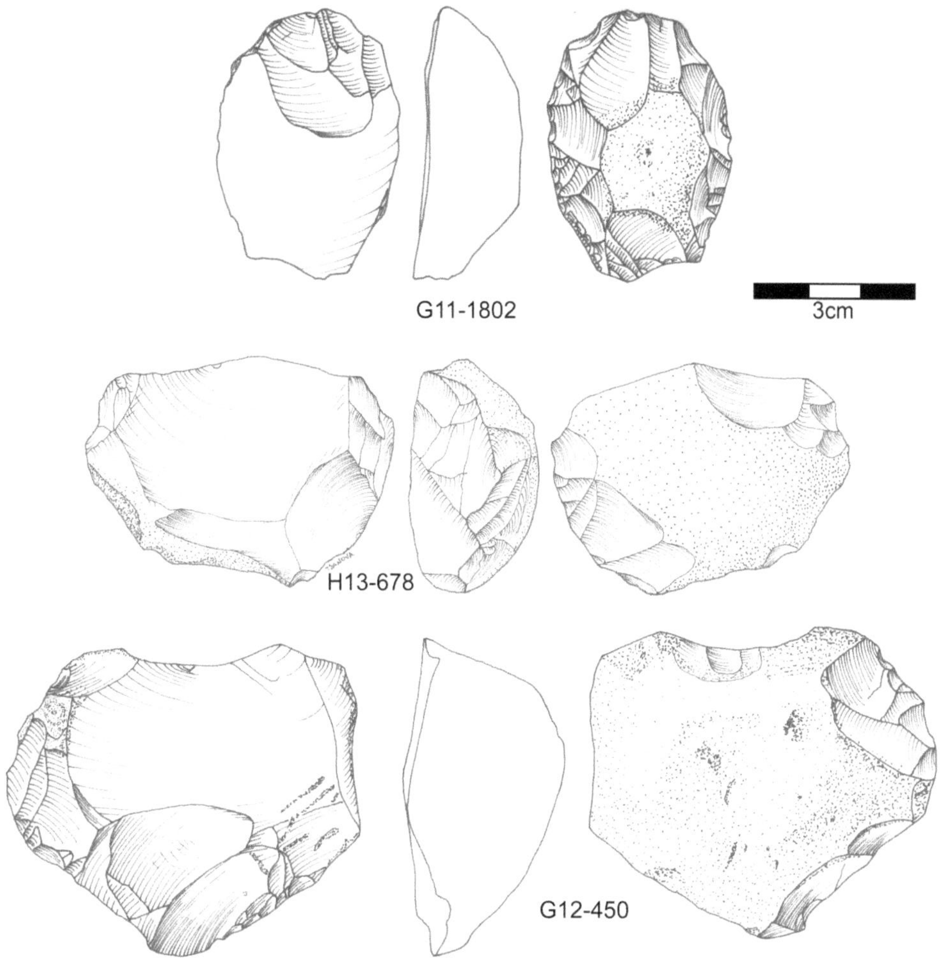


**Fig. 6.24** Cores from Combined Unit III-B. G11-1802: Kombewa core; H13-678: Levallois core; G12-450: Levallois core

scrapers in most layers. The primary exception is Unit II, but especially Units II-B and II-A, where transverse scrapers become the second most frequent type after single scrapers.

There are two complicating factors in this overall picture. First, some layers have a very high percentage of truncated-facetted pieces and because we count them as tools the relative percentages of scrapers and notched tools in the assemblage is somewhat distorted, especially in Unit IV-A. Second, Upper Paleolithic type tools are more frequent in Unit I and especially in Unit I-A. In both these case, the

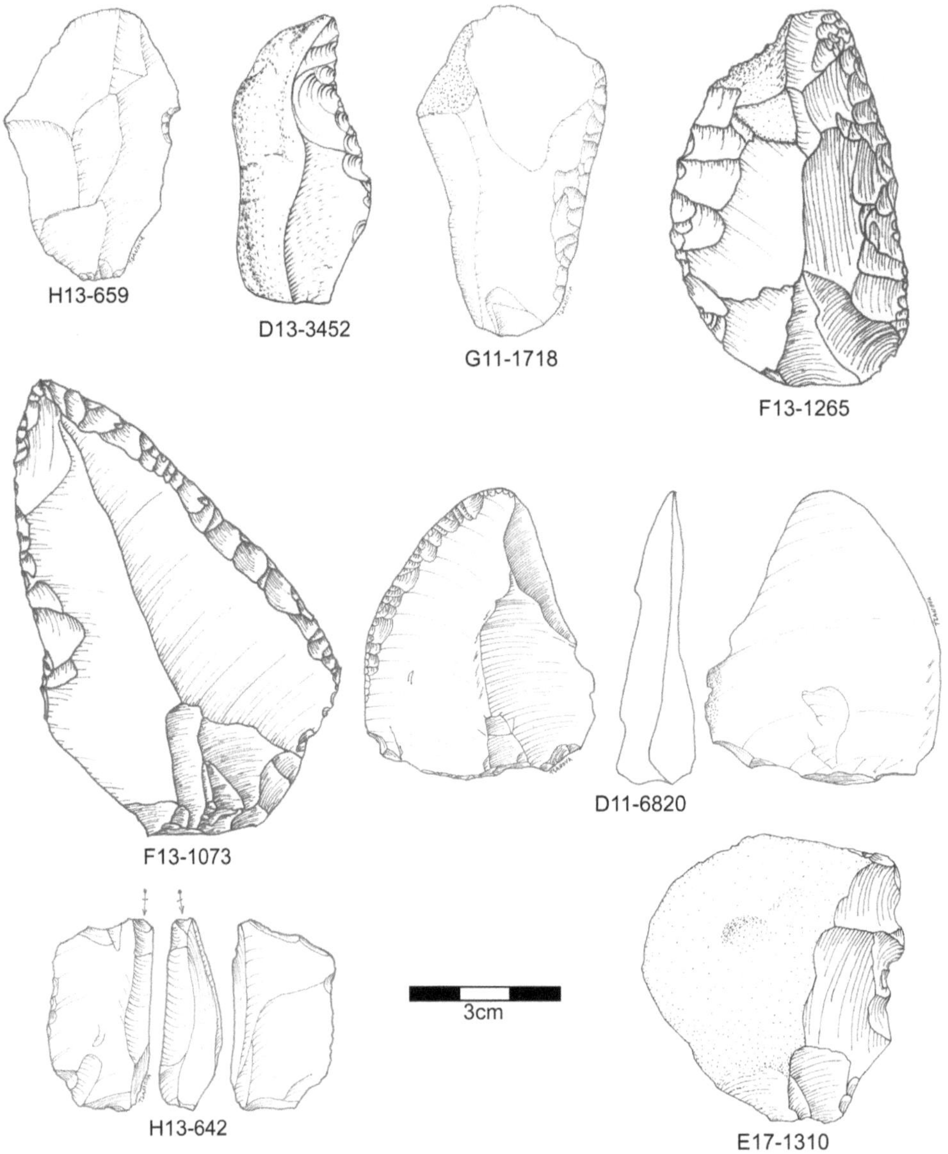


**Fig. 6.25** Retouched pieces from Combined Unit III-B. H13-659: Notch; D13-3452: Single scraper; G11-1708: Single scraper; F13-1265: Double scraper; F13-1073: Mousterian Point (convergent scraper); D11-6820: Single scraper; H13-642: Burin; E17-13010: Denticulate

relative proportions of scrapers to notched pieces remains the same, but the percentages of both are reduced in comparison to the other layers in proportion to the frequency of Upper Paleolithic and truncated-facetted pieces.

To examine patterns of blank selection for tool production, particularly scrapers and notched tools, we modeled whether an artifact is a scraper as binomial response to changes in blank size. For blank size, we initially used

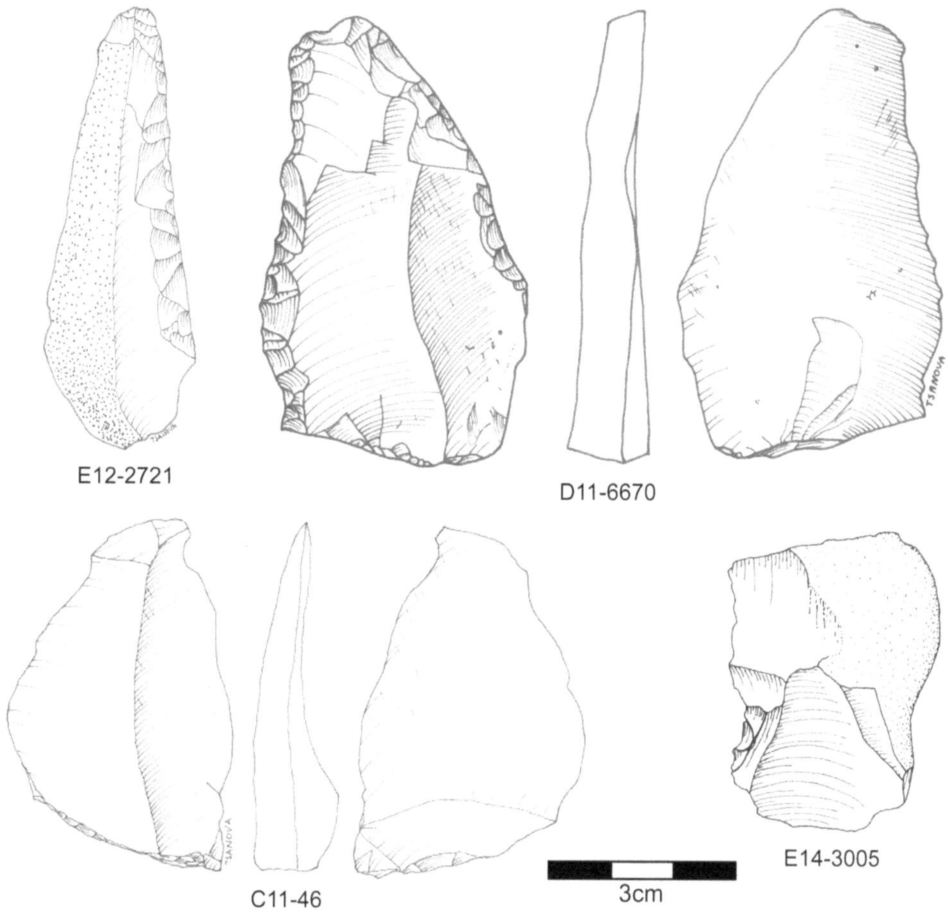


**Fig. 6.26** Retouched pieces from Combined Unit III-B. E12-2721: Single scraper; D11-6670: Double scraper; C11-46: Proximal truncation; E14-3005: Notch

platform width and platform thickness rather than length, width, and thickness because these platform variables, along with exterior platform angle, are known to correlate with blank size (Dibble and Rezek 2009) and because blank size measures are typically impacted by retouch whereas platform size measures are not. Platform measurements were log-transformed to remove skew and normalized to make their effect size more interpretable in the model. In this model, with a sample of 13,818 blanks and scrapers, platform width ($N = 965/13{,}819$, Estimate = 0.478, Std.

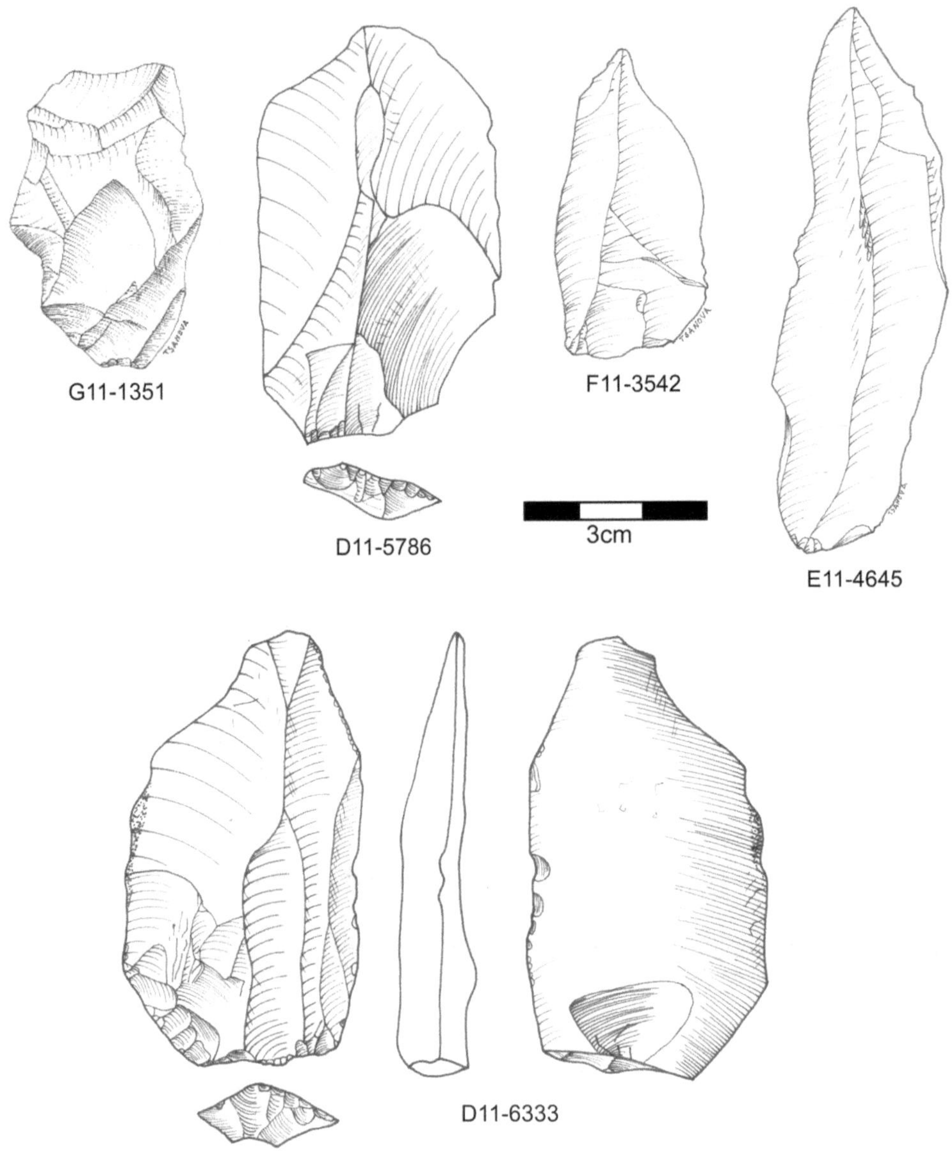


**Fig. 6.27** Levallois flakes from Combined Unit III-A

Error = 0.047, $Z = 10.133$, $p = 0.000$) and thickness ($N = 965/13{,}819$, Estimate = 0.442, Std. Error = 0.046, $Z = 9.657$, $p = 0.000$) increased the probability of a blank being converted into a scraper by a factor of about 0.45. Despite the fact that retouch reduced basic size measurements, we next modeled the effect of length, width and thickness on whether a blank enters the archaeological record as a scraper to have a better idea the relative

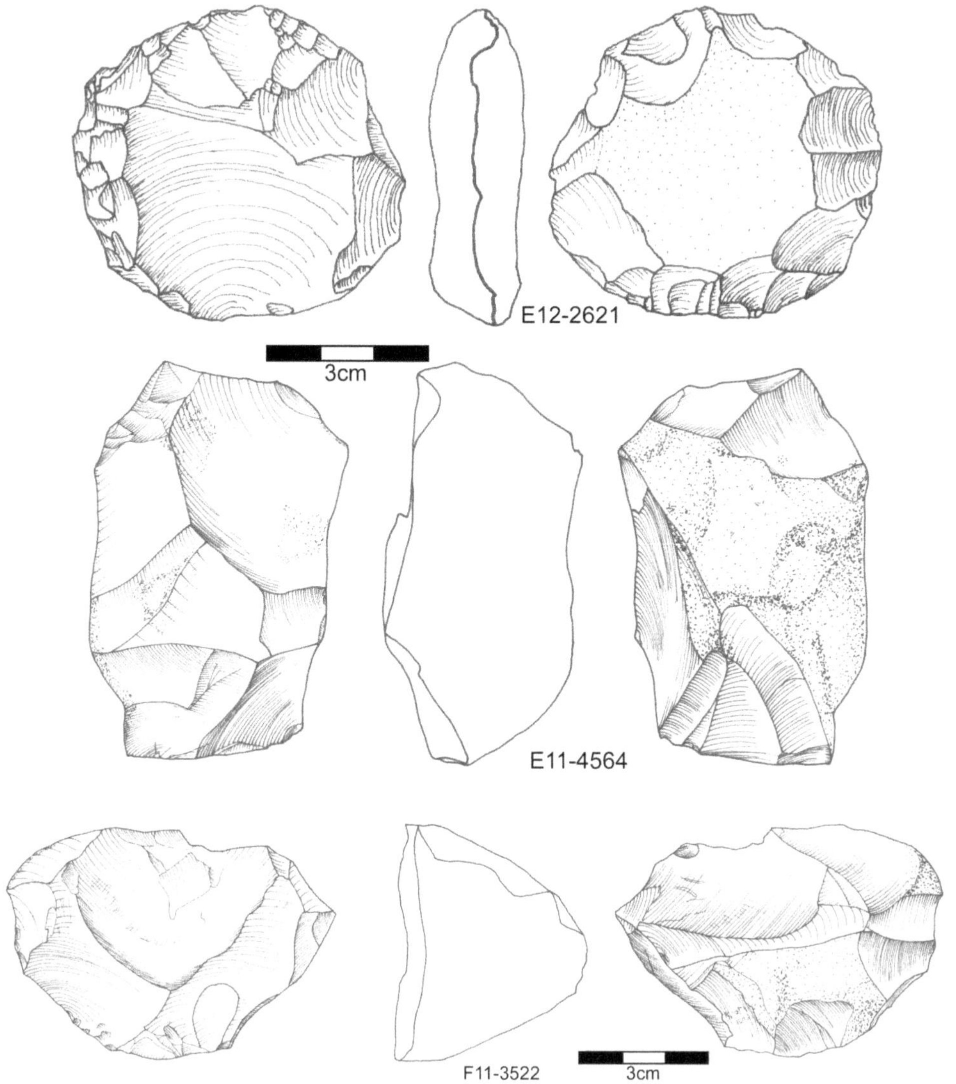


**Fig. 6.28** Cores from Combined Unit III-A

importance of each. Again, the measures were log-transformed and normalized. Length has the greatest effect ($N$ = 1293/20,475, Estimate = 1.322, Std. Error = 0.145, $Z$ = 9.137, $p$ = 0.000), more than doubling the likelihood that a blank has been converted into a scraper for an increase in size of one standard deviation. Width comes

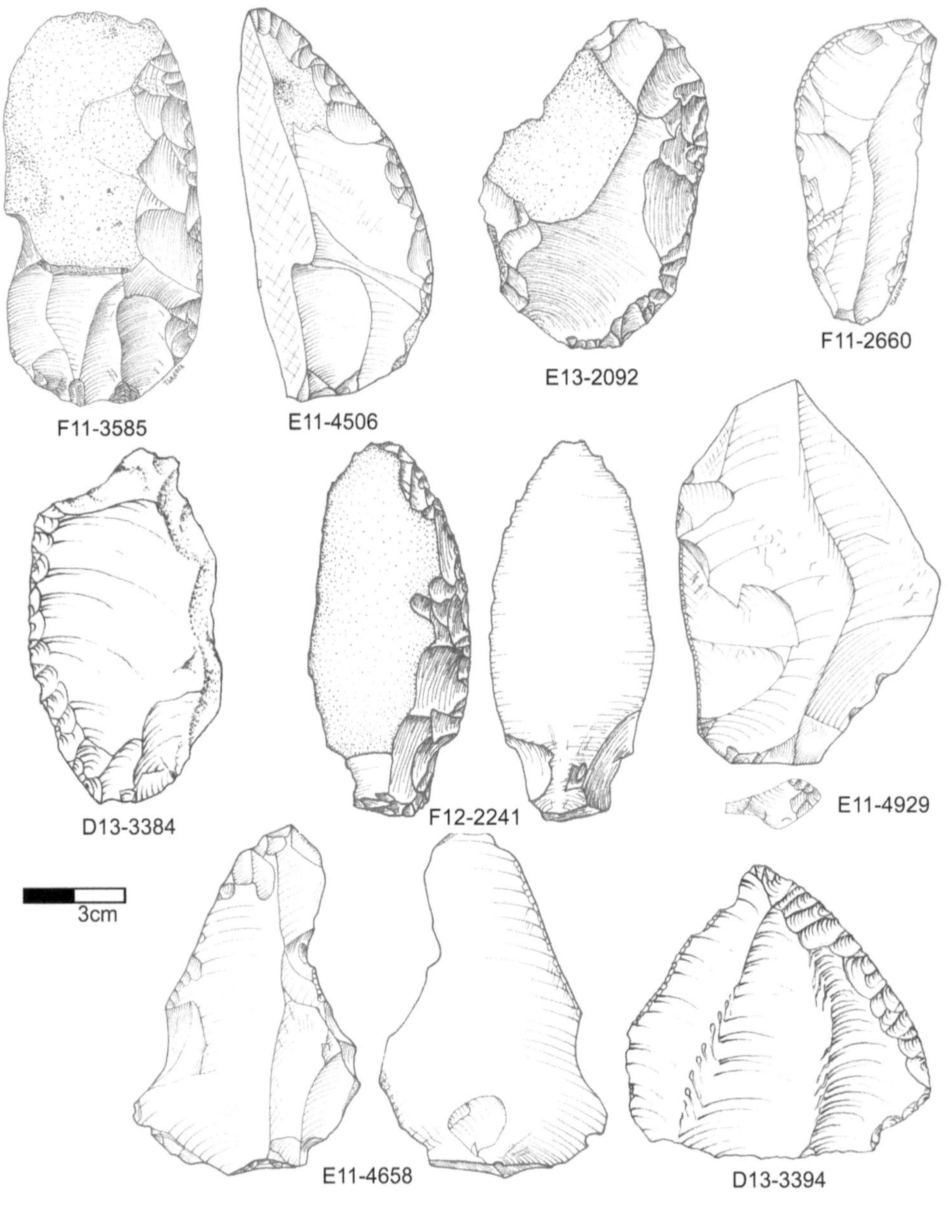


**Fig. 6.29** Retouched pieces from Combined Unit III-A. F11-3585: Single scraper; E11-4506: Single scraper; E13-2092: Single scraper; F11-2660: Backed knife D13-3384: Single scraper; F12-2241: Tanged tool; E11-4929: Single scraper; E11-4658: Scraper on interior; D13-3394: Retouched Levallois point

next with an effect size of 0.46 ($N = 1293/20{,}475$, Estimate = 0.463, Std. Error = 0.188, $Z = 2.460$, $p = 0.014$). The effect of thickness is weakly negative, meaning thinner flakes are more likely to be selected, but this is not significant at the 0.05 level ($N = 1293/20{,}475$, Estimate = −0.242, Std. Error = 0.147, $Z = -1.652$, $p = 0.099$). When the same is done for notched tools, platform thickness has an effect size ($N = 623/13{,}477$, Estimate = 0.499, Std.

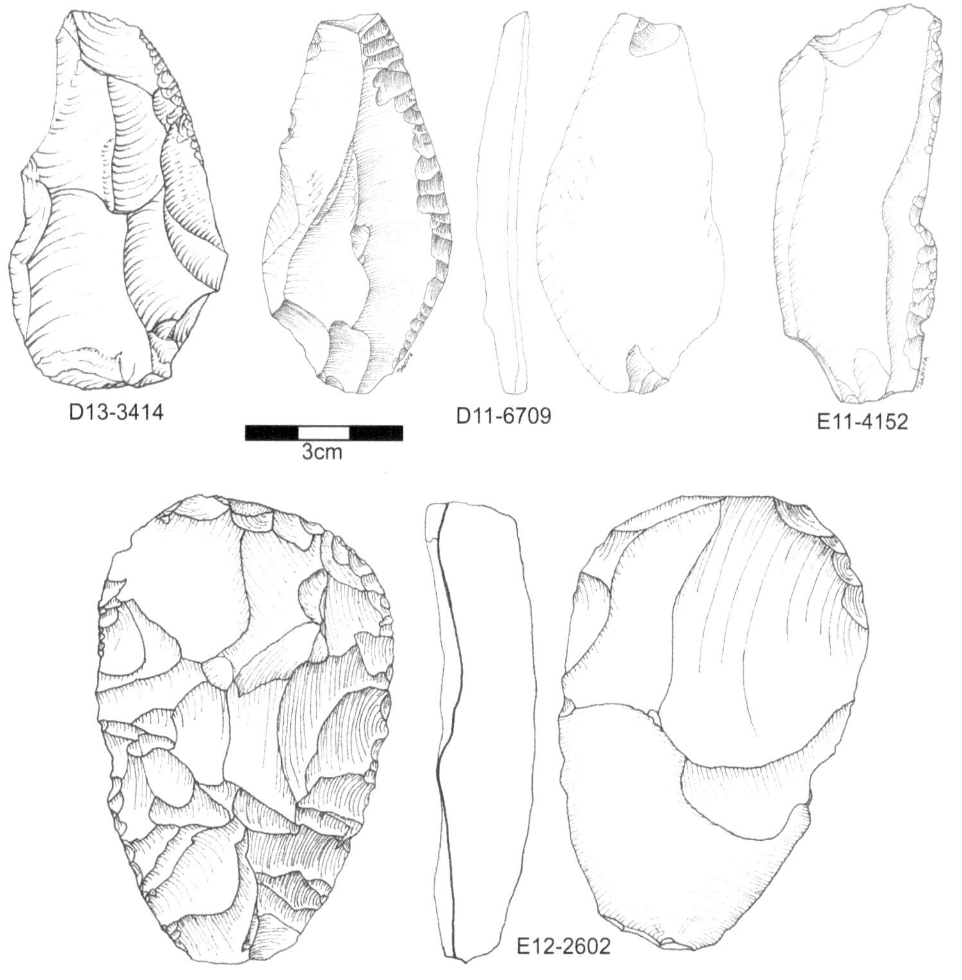


**Fig. 6.30** Retouched pieces from Combined Unit III-A. D13-3414: Single scraper; D11-5709: Single scraper; E11-4152: Single scraper; E12-2602: Bifacially-flaked flake

Error = 0.055, $Z = 9.134$, $p = 0.000$) roughly three times greater than platform width ($N = 623/13{,}477$, Estimate = 0.157, Std. Error = 0.055, $Z = 2.830$, $p = 0.005$), and for overall size thickness has a large effect ($N = 868/20{,}050$, Estimate = 0.674, Std. Error = 0.137, $Z = 4.926$, $p = 0.000$), whereas width has a small and not significant effect ($N = 868/20{,}050$, Estimate = 0.270, Std. Error = 0.167, $Z = 1.615$, $p = 0.106$) and length has no

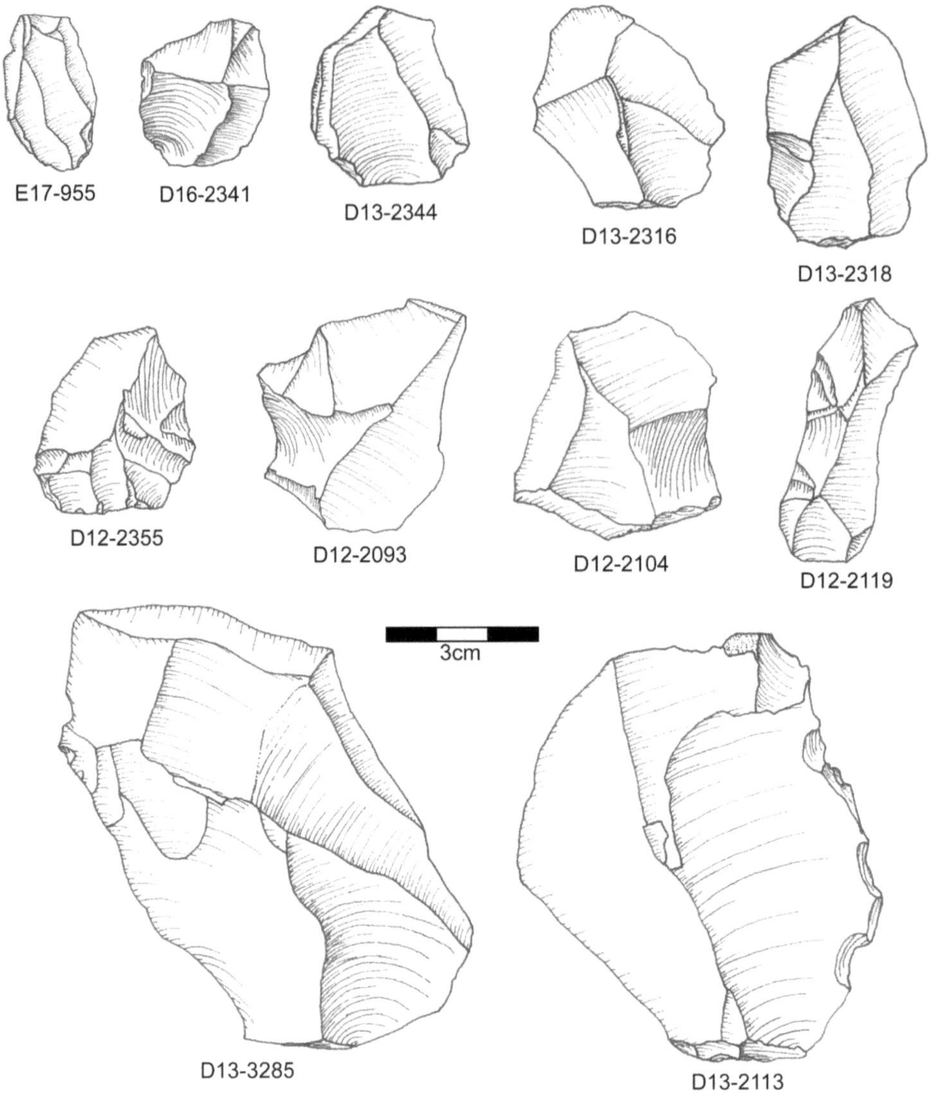


**Fig. 6.31** Levallois flakes from Combined Unit II-C

effect ($N$ = 868/20,050, Estimate = 0.010, Std. Error = 0.118, $Z$ = 0.084, $p$ = 0.933).

These models confirm the effect of size on blank selection for scraper and notch production; however, they also highlight a difference in selection criteria with length being the primary factor in scraper blank selection and thickness the primary factor in notched tool blank selection. To examine the role that blank production technology might

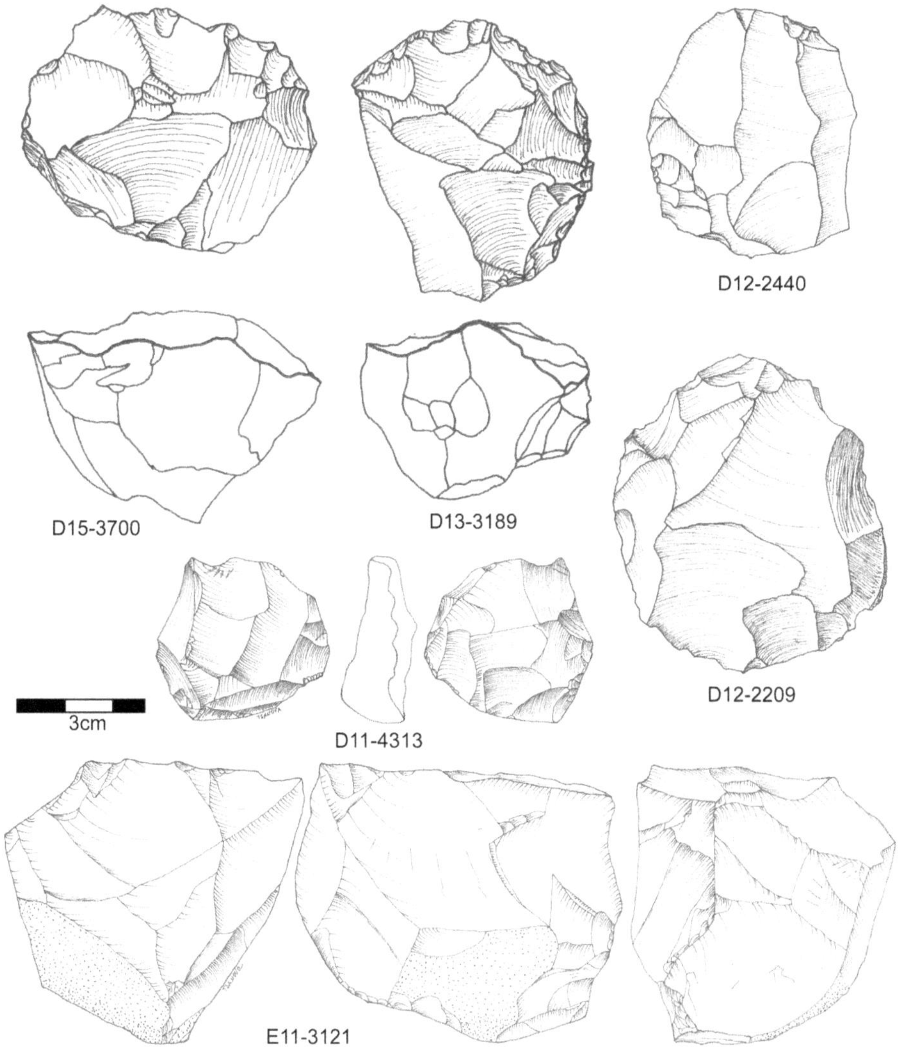


**Fig. 6.32** Cores from Combined Unit II-C

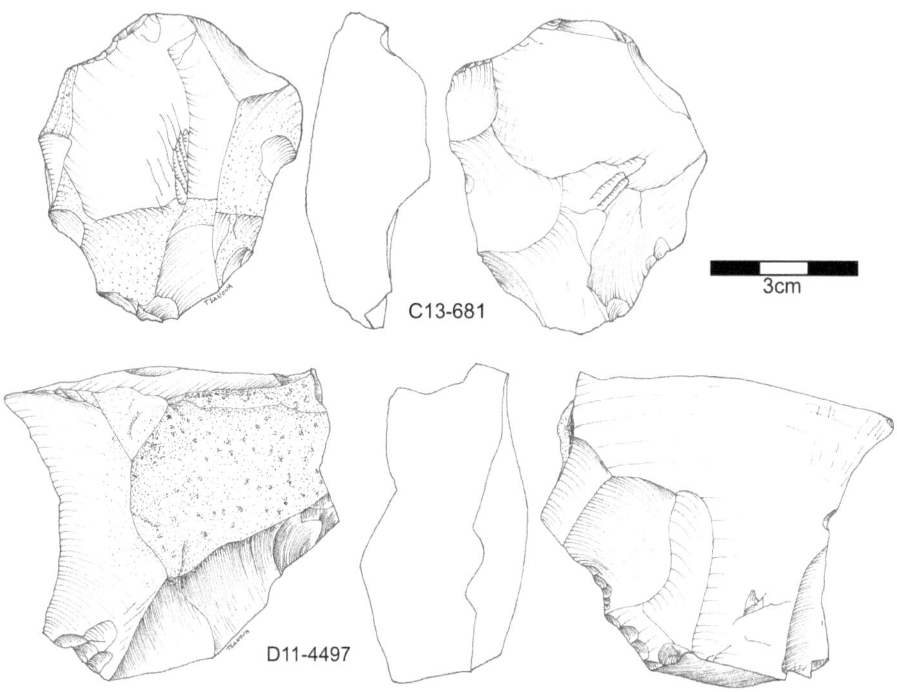


**Fig. 6.33** Cores from Combined Unit II-C

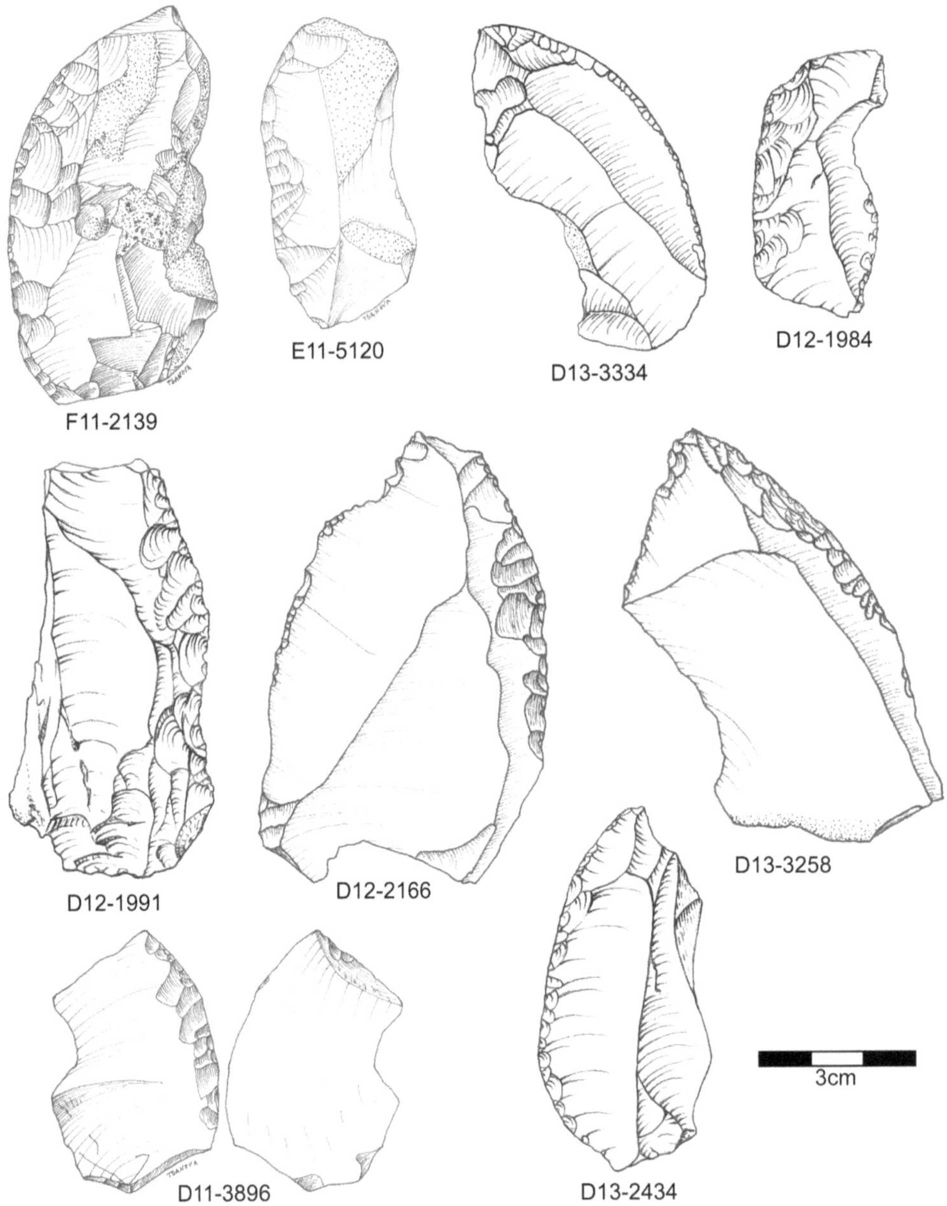


**Fig. 6.34** Single scrapers from Combined Unit II-C

play, Units V, IV-A, and III-A are examined because they have high frequencies of Levallois but varying proportions of scrapers and notches and because Unit IV-A stands in contrast to the other two in many measures but especially for its emphasis on small flake production (Fig. 6.57). Thus, two binary predictors (Levallois or not and facetted or not) were

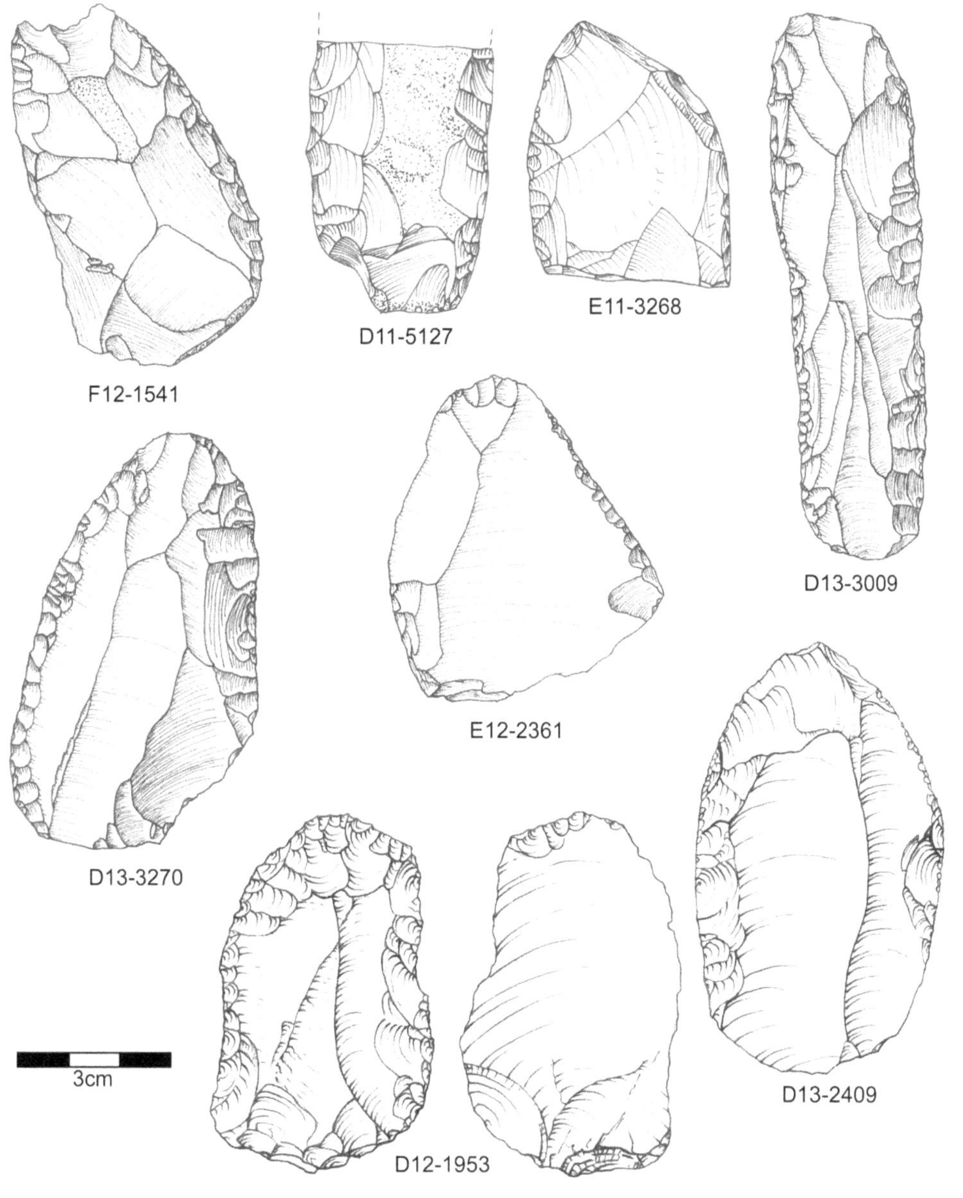


**Fig. 6.35** Double scrapers from Combined Unit II-C

added to the models described above. In Unit V, platform thickness ($N$ = 107/1039, Estimate = 0.405, Std. Error = 0.147, $Z$ = 2.763, $p$ = 0.006) and faceting ($N$ = 107/1039, Estimate = 0.556, Std. Error = 0.227, $Z$ = 2.448, $p$ = 0.014) have a significant effect on whether a tool is a scraper, and for notches only platform thickness has a significant effect

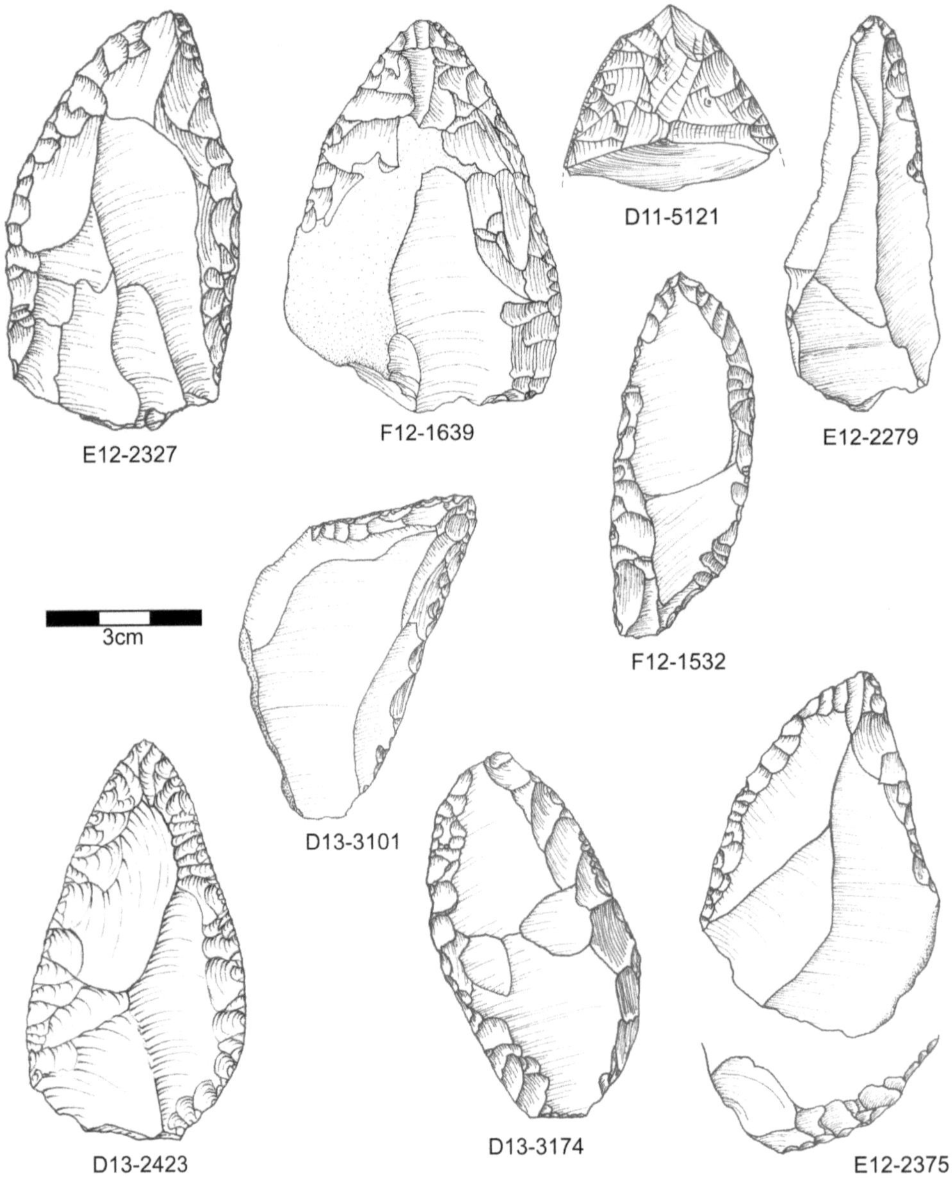


**Fig. 6.36** Convergent scrapers from Combined Unit II-C

($N$ = 28/960, Estimate = 0.507, Std. Error = 0.256, $Z$ = 1.983, $p$ = 0.047). In Unit IV-A, only platform thickness has an effect on scrapers ($N$ = 39/2315, Estimate = 0.603, Std. Error = 0.227, $Z$ = 2.653, $p$ = 0.008) and notched tools ($N$ = 77/2353, Estimate = 0.458, Std. Error = 0.159, $Z$ = 2.890, $p$ = 0.004). And in Unit III-A,

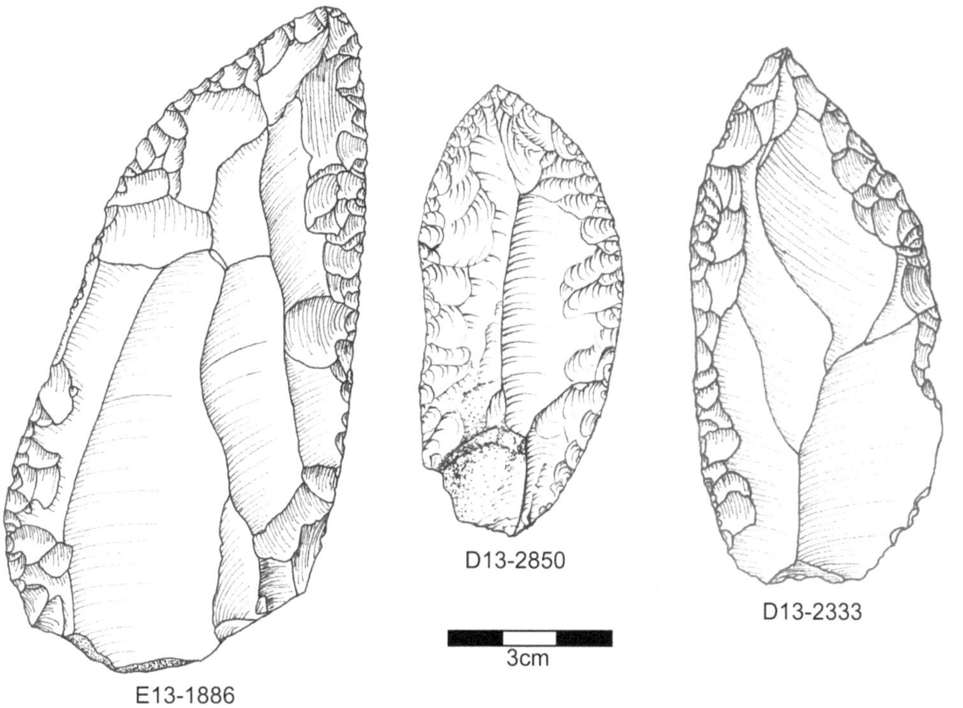


**Fig. 6.37** Convergent scrapers from Combined Unit II-C

platform thickness is again significant in scrapers ($N = 50/431$, Estimate = 0.811, Std. Error = 0.221, $Z = 3.663$, $p = 0.000$), but for the notches platform width ($N = 9/390$, Estimate = 1.118, Std. Error = 0.545, $Z = 2.052$, $p = 0.040$) reaches significance (though the sample size is too small to be reliable). So for these three layers, all with relatively high percentages of Levallois flakes and facetted platforms, blank size predictors, particularly platform thickness, have the greatest effect on whether a blank is converted into a scraper or a notch. These data mean that Levallois blanks are not being produced specifically for tool production but are instead driven by other technological decisions. Why some layers have more Levallois than others remains to be explained.

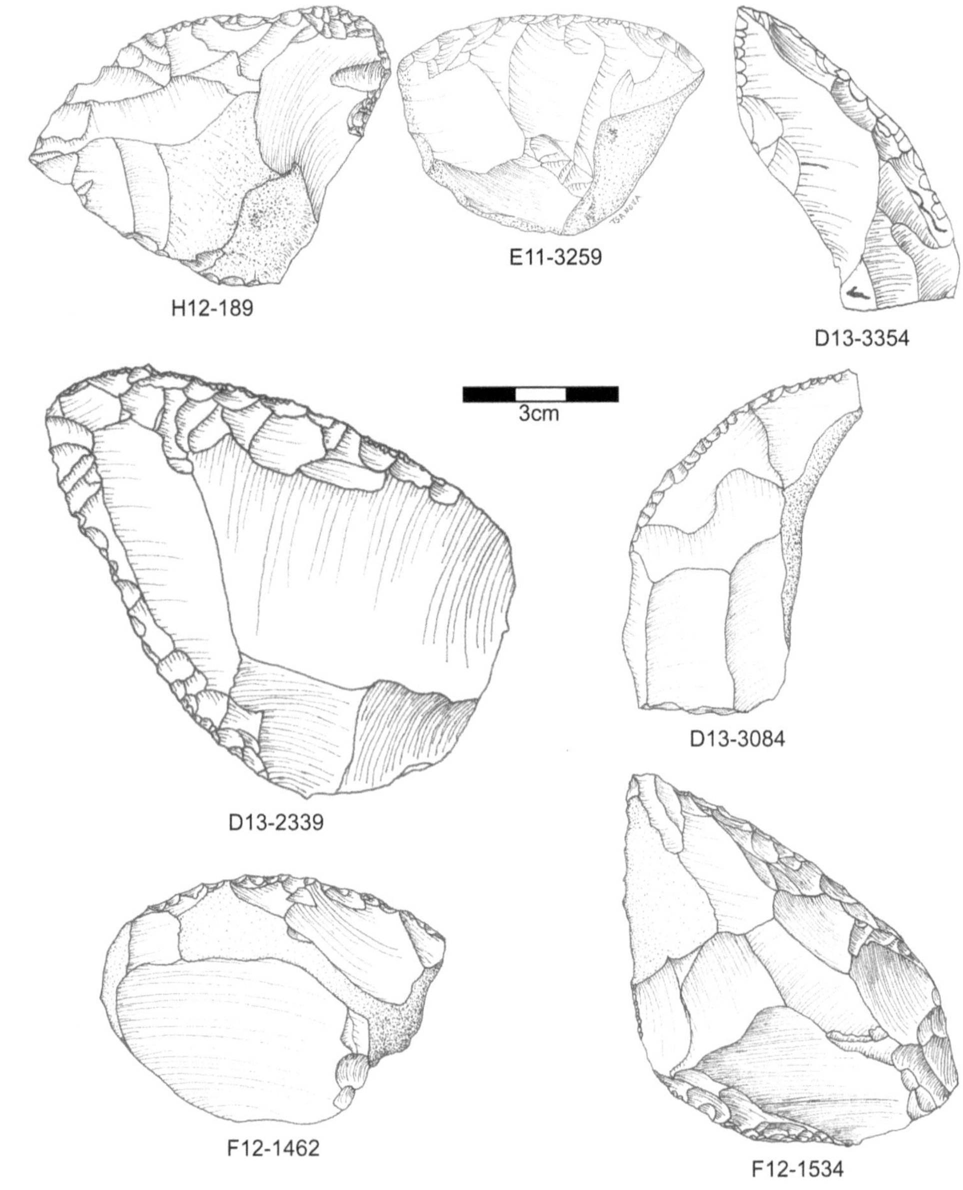


**Fig. 6.38** Scrapers from Combined Unit II-C. H12-189: Transverse scraper; E11-3259: Transverse scraper; D13-3354: Déjeté scraper; D13-2339: Déjeté scraper; D13-3084: Transverse scraper; F12-1462: Transverse scraper; F12-1534: Transverse scraper

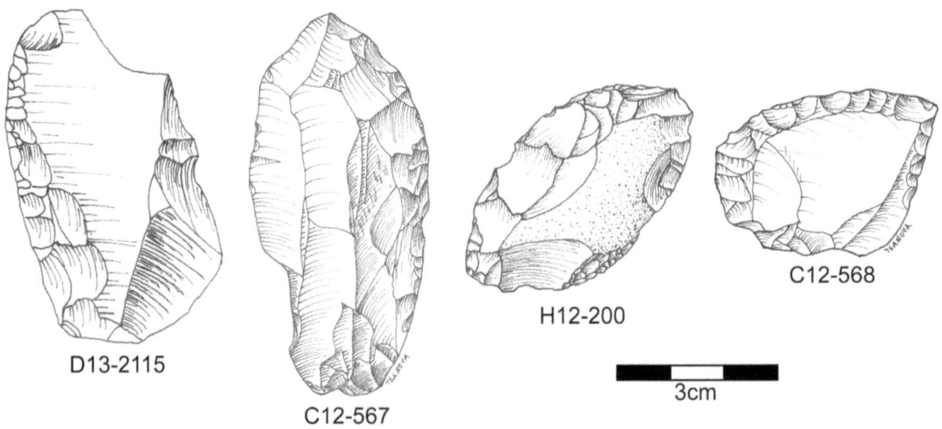


**Fig. 6.39** Retouched pieces from Combined Unit II-C. D13-2115: Single scraper; C12-567: Single scraper; H12-200: Oblique limace; C12-568: Déjeté scraper

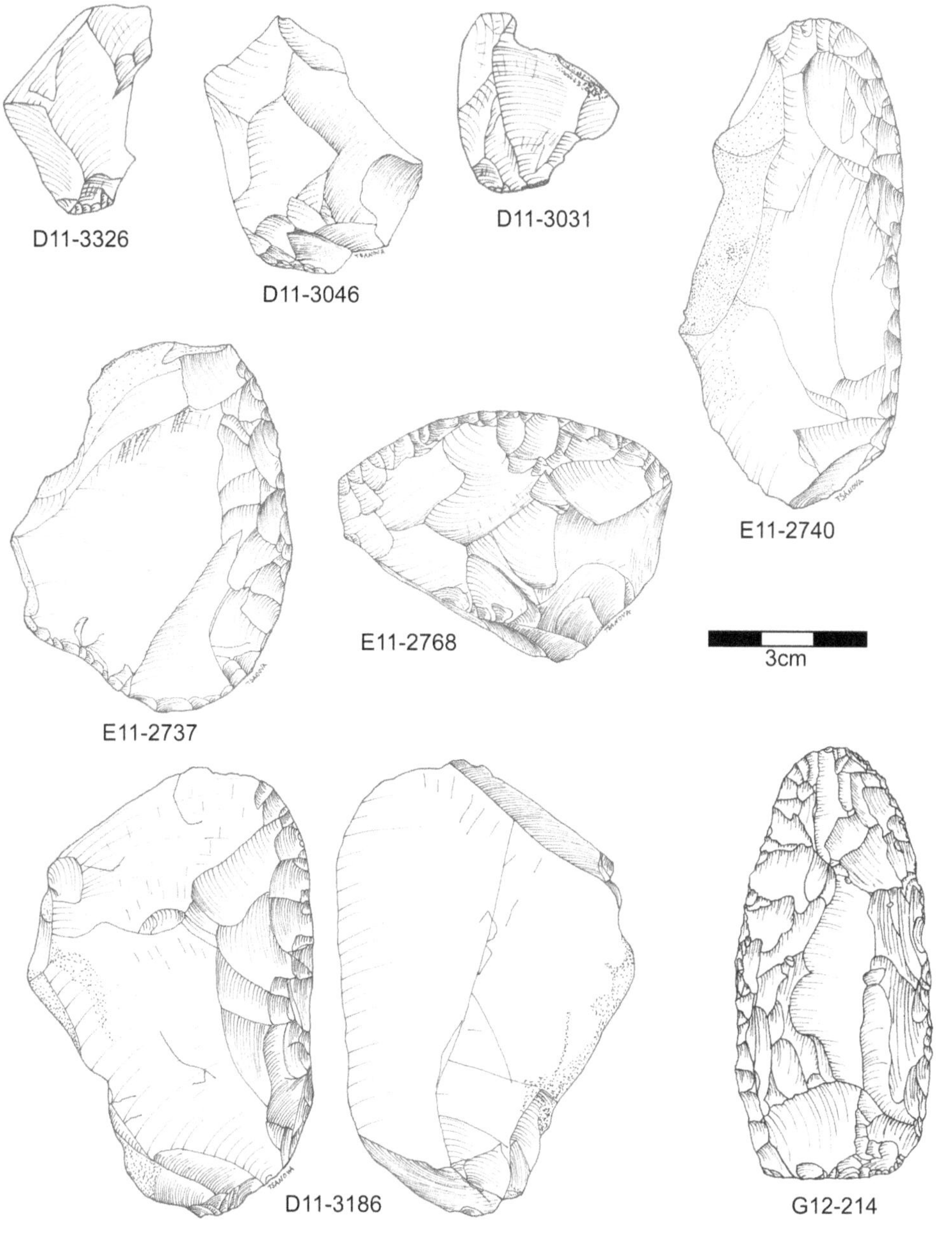


**Fig. 6.40** Lithics from Combined Unit II-C. D11-3326: Levallois flake; D11-3046: Levallois flake; D11-3031: Levallois flake; E11-2740: Endscraper; E11-2737: Single scraper; E11-2768: Transverse scraper; D11-3186: Single scraper; G12-214: Convergent scraper

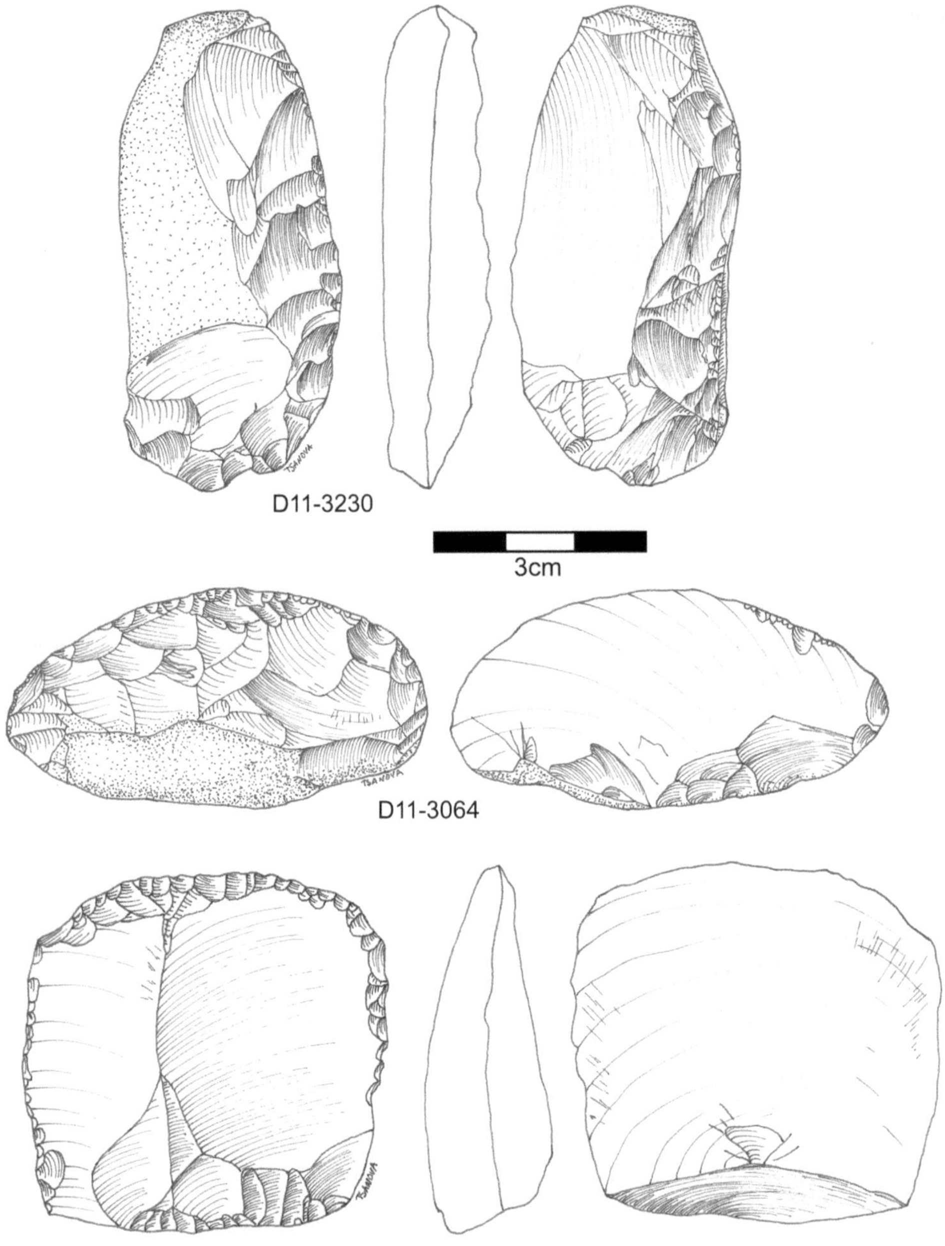


**Fig. 6.41** Retouched pieces from Combined Unit II-C. D11-3230: Scraper with thinned back; D11-3064: Transverse scraper; D11-3347: Transverse scraper

## Discussion

The Pech IV excavations confirmed the overall picture of assemblage variability, which is substantial, outlined by Bordes (1975) based on a partial analysis of his collections and later by ourselves based on an analysis of his complete collection (McPherron and Dibble 2000). However, the new excavations did bring some additional insights on the Pech IV lithic collection. For one, it was possible to get a better understanding of the relatively less dense layers just

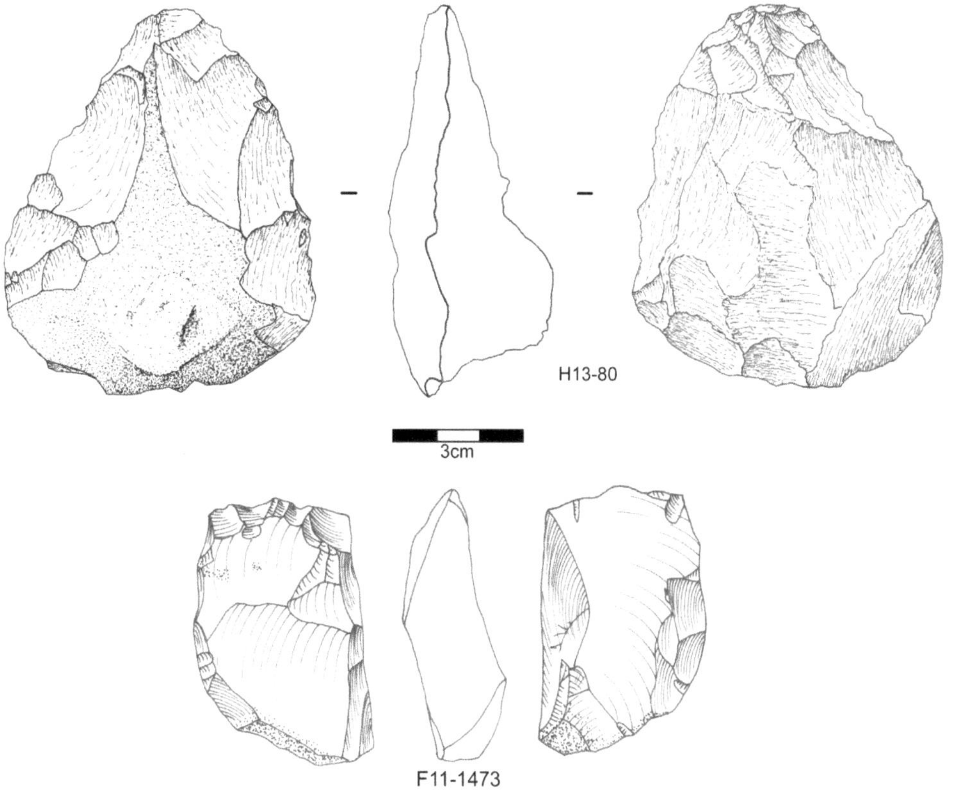


**Fig. 6.42** Retouched pieces from Combined Unit II-C. H13-80: Biface; F11-1473: Truncated-faceted piece

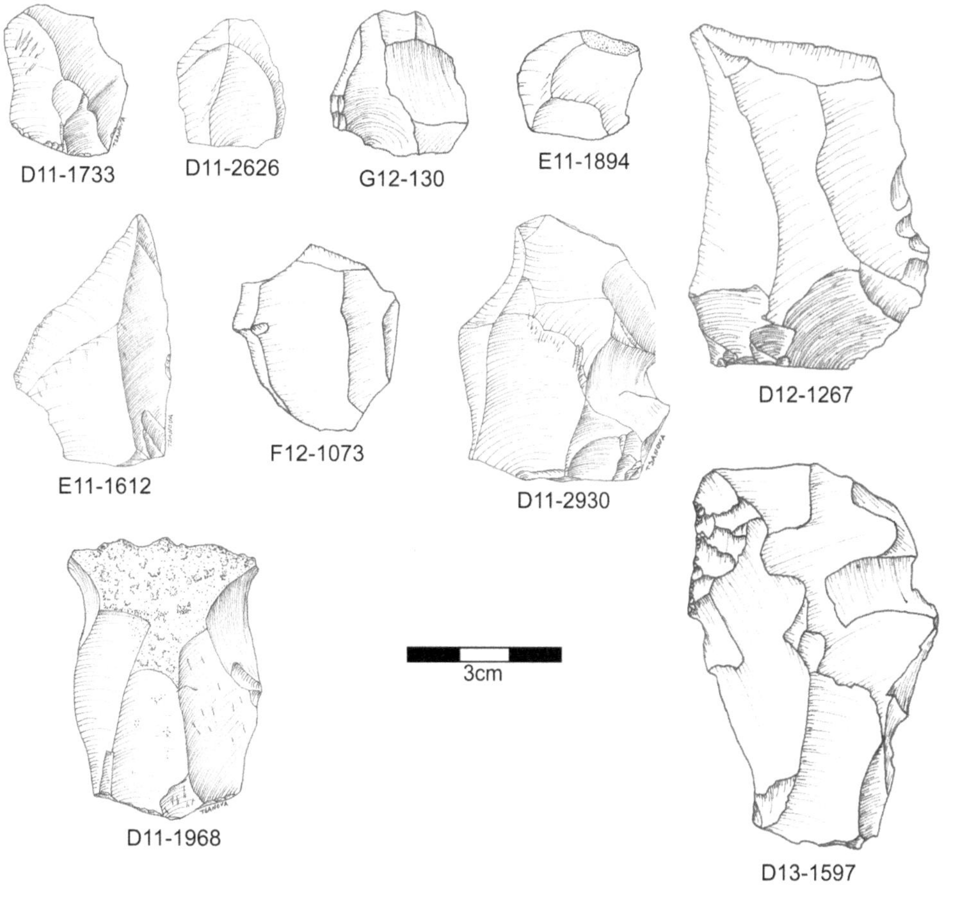


**Fig. 6.43** Levallois flakes from Combined Unit I-B

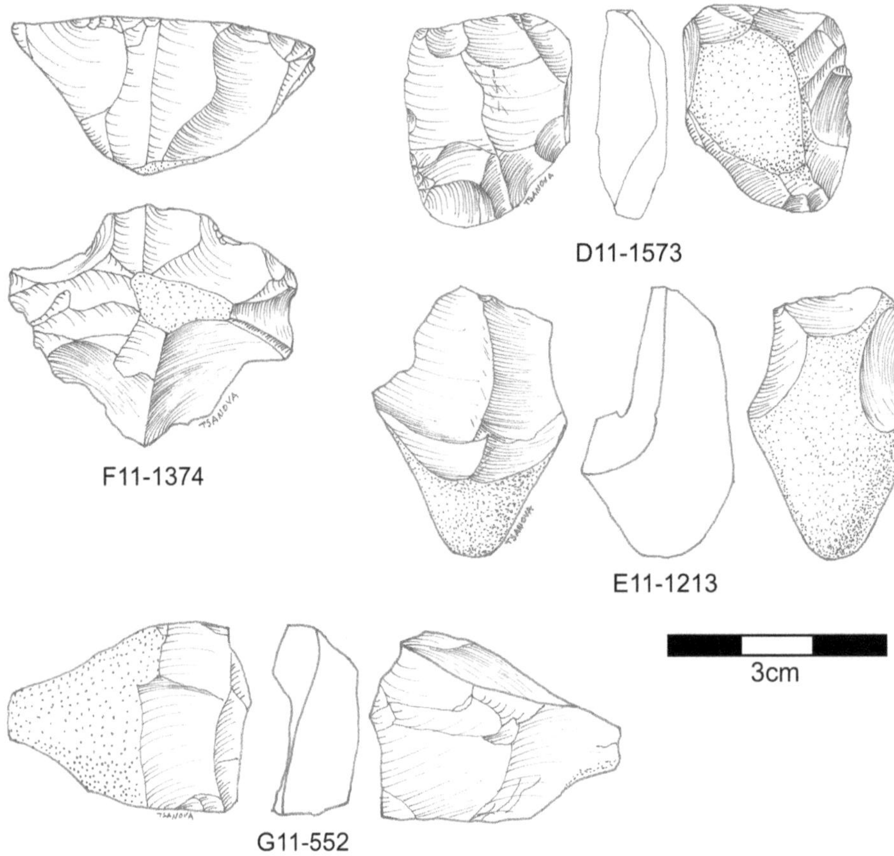


**Fig. 6.44** Cores from Combined Unit I-B

under the top of the sequence (Bordes' Layers G and H, our Layer 4A, and combined Unit II-A). Bordes had been unsure how to categorize the variability in these layers; however, it is now clear that Layer 4A shows Quina-type blank production techniques that are not seen elsewhere in the sequence. The character of these layers too is quite different from the rest of the sequence. What Bordes took as low density is in fact a series of quite dense but thin layers where fauna are far more frequent than lithics. In this way, this portion of the sequence is quite similar to what is seen in parts of the lower portion of La Quina and Jonzac where also Quina techniques for blank production predominate and the fauna is dominated by reindeer (Chase et al. 1994; Jaubert et al. 2008; Jelinek 2013).

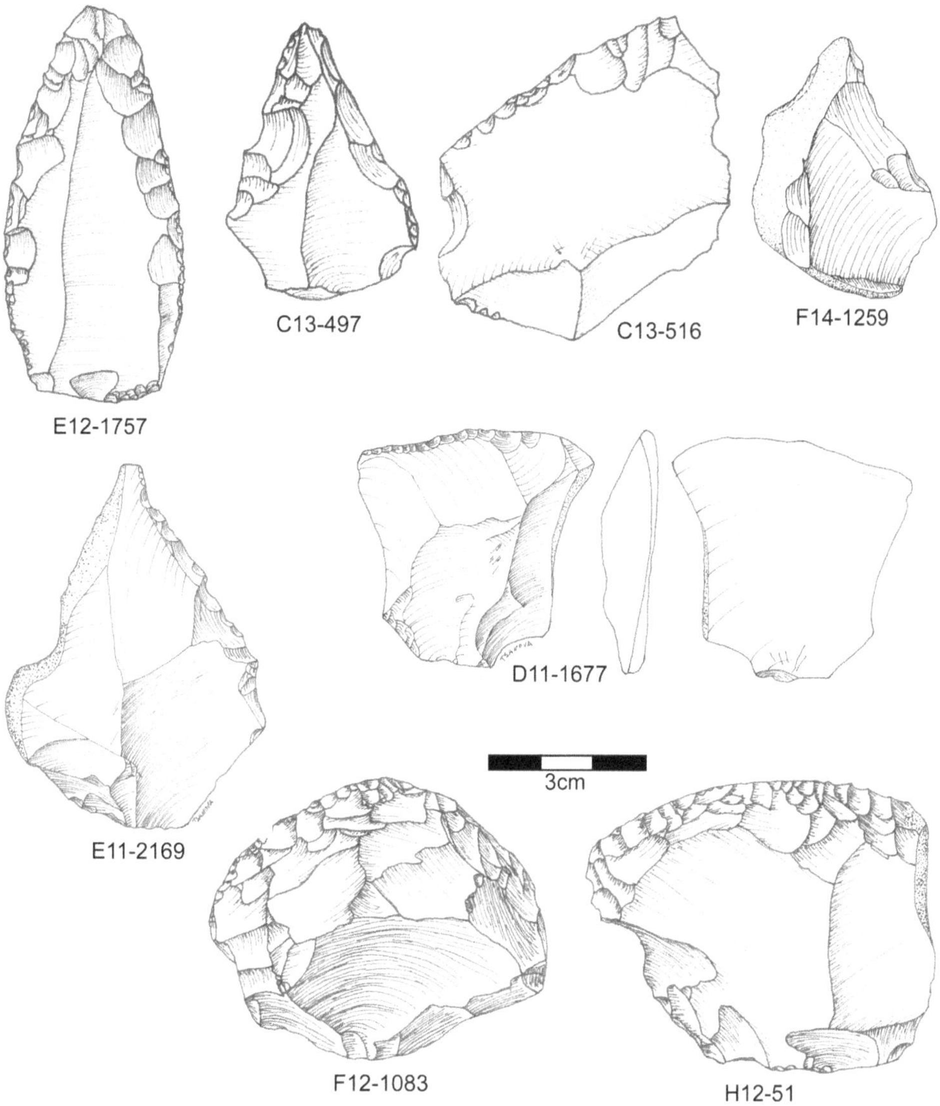


**Fig. 6.45** Retouched pieces from Combined Unit I-B. E12: 1757: Convergent scraper; C13-497: Convergent scraper; C13-516: Transverse scraper; F14-1259: Notch; E11-2169: Side scraper; D11-1677: Transverse scraper; F12-1083: Transverse scraper; H12-51: Transverse scraper

Second, while nearly all of the raw material present in the site comes from the immediate vicinity (including meters from the site), there are some exotic raw materials including Bergerac flint and chalcedony from the areas around Domme. The highest percentage of exotic flint (7.1% and double the second highest percentage) comes from Layer 4A. This feature too makes this layer comparable to the site of La Quina, where Quina blank production is associated with a higher diversity of raw materials (Park 2007).

Third, during the course of the Pech IV artifact analysis, we developed a new method to use cortex to assess the movement of material into and out of the site (Dibble et al. 2005). An application of this method to the Pech IV assemblage and to the subsequently excavated Roc de Marsal (France) assemblages (Lin et al. 2015) showed

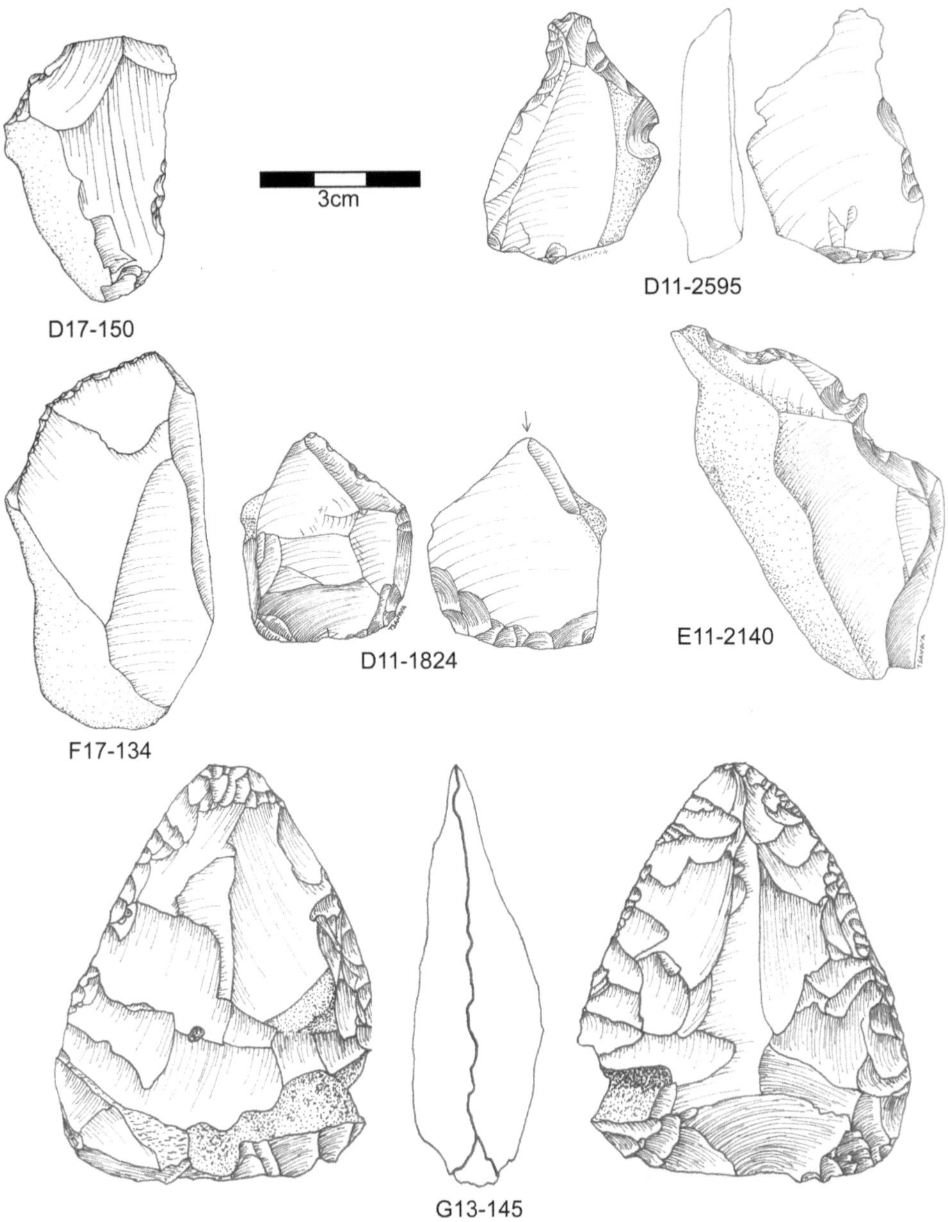


**Fig. 6.46** Retouched pieces from Combined Unit I-B. D17-150: Notch; D11-2595: D11-2595: Backed knife; F17-134: Backed knife; D11-1824: Burin; E11-2140: Percoir; G13-145: Biface

differences among the layers and similarities between the two sites. Many of the Pech IV layers (Layers 8, 7, 6B, 6A, 5B, and 4B + C) show cortex ratios close to 1, meaning that the assemblages are consistent with the importation and on site reduction of fully cortical nodules with little subsequent transport of material (though this cannot be excluded). The remaining layers (5A, 4A, 3B, and 3A) show cortex ratios that differ statistically from 1. Layer 4A has a ratio below 1 (cortex is on average missing from the assemblage), and Layers 3A, 3B, and 5A have cortex ratios above 1 (cortex is

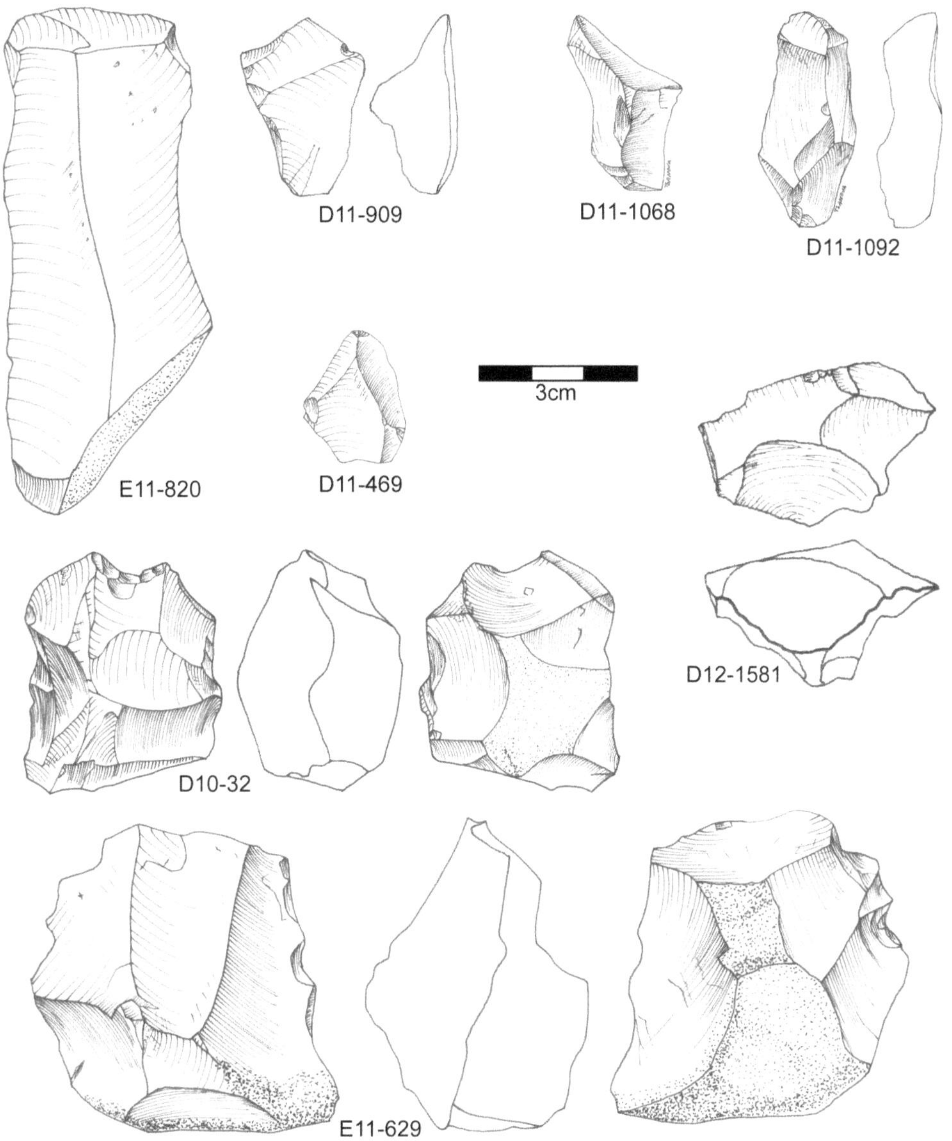


**Fig. 6.47** Lithics from Combined Unit I-A. E11-820: Flake; D11-909: Pseudo-Levallois point; D11-1068: Pseudo-Levallois point; D11-1092: Flake; D11-469: Levallois point; D10-32: Core; D12-1581: Core; E11-629: Levallois core

over-represented). In both cases transport is suggested. For Layer 4A this is again consistent with the raw material findings. For Layer 3, where bifacial technologies are also present, it is consistent with suggestions that this type of technology is particularly well suited to highly mobile groups (e.g., Turq et al. 2013). Additionally, all of the layers with cortex ratios differing from 1, at Pech IV and at Roc de Marsal, have reindeer dominated assemblages. However, the

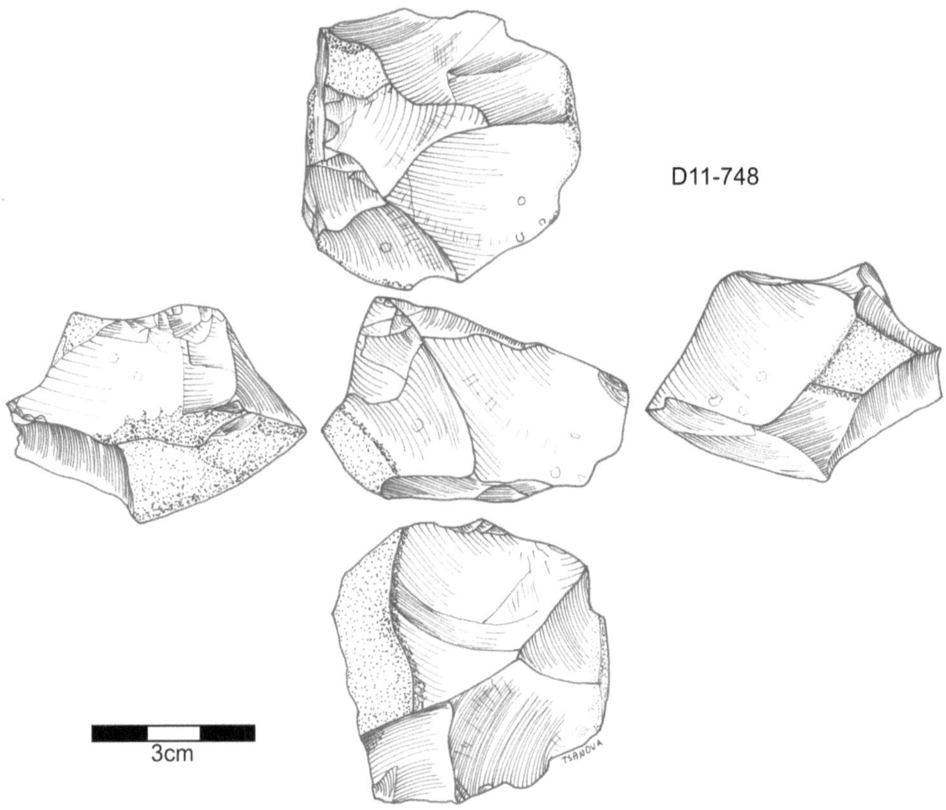


**Fig. 6.48** Core from Combined Unit I-A

results from the study of cortex and the identification of raw material diversity do not match in Layer 8.

In addition, we were able to confirm a number of trends we had already noticed in the Bordes collection. First, the particular nature of small flake production in Unit IV could be confirmed. In addition to small flake production via Kombewa cores and small Levallois cores, we had previously noted also the frequency of truncated-facetted artifacts in this layer (McPherron and Dibble 2000; Dibble and McPherron 2006, 2007). While we remain convinced that small flake production is part of most Mousterian assemblages, and other Middle Paleolithic assemblages from elsewhere (e.g., Dibble and McPherron 2006, 2007), it is also true that it is quite intensively expressed in this particular layer and for reasons that are difficult to explain. All three measures of small flake production, the presence of Kombewa cores, the presence of truncated-facetted pieces, and the diminutive size of the cores, are at their peak in this layer. Normally one explanation that could be evoked is raw material stress. However, this is difficult to demonstrate in

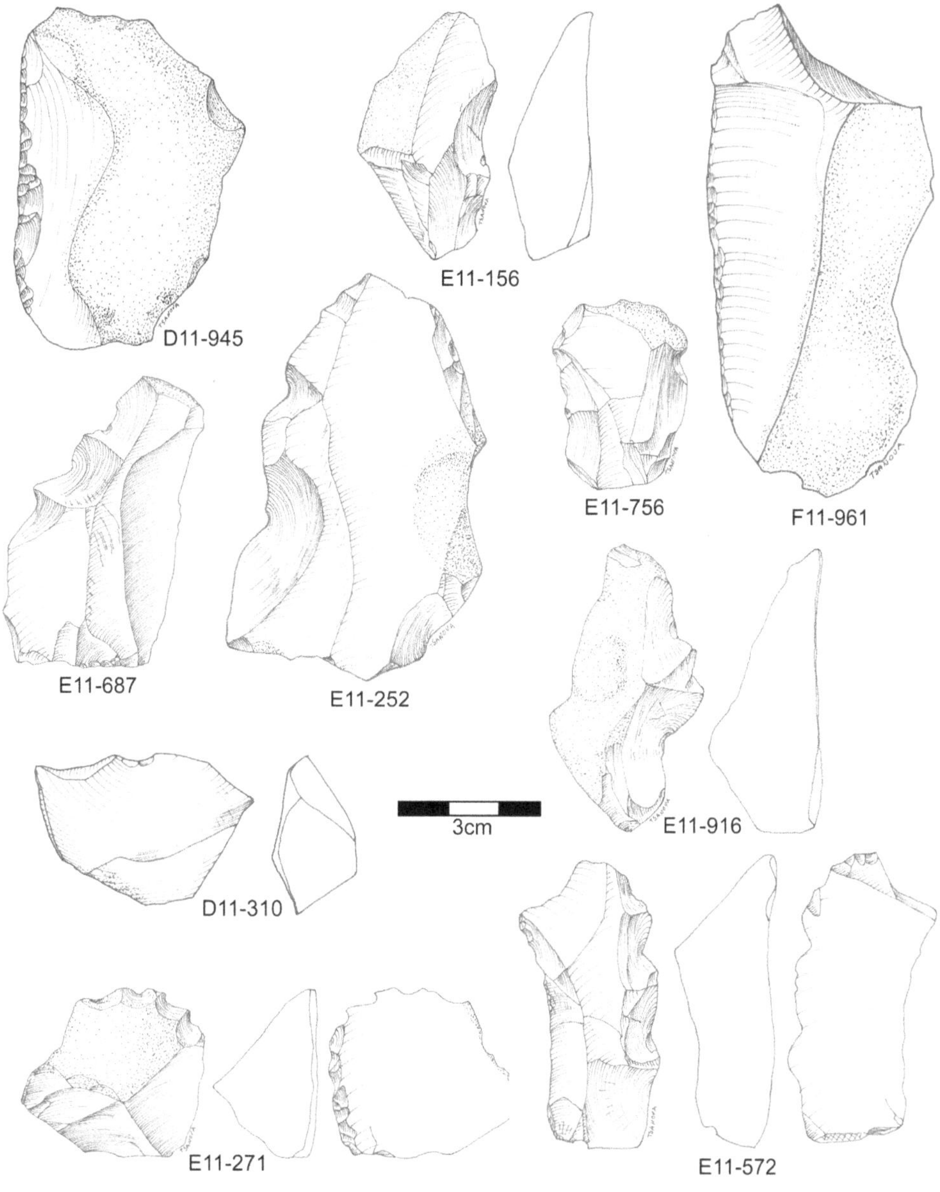


**Fig. 6.49** Retouched pieces from Combined Unit I-A. D11-945: Single scraper; E11-156: Denticulate; F11-961: Single scraper; E11-687: Denticulate; E11-252: Notch; E11-756: Denticulate; D11-310: Clactonian notch; E11-916: Denticulate; E11-271: Denticulate; E11-572: Denticulate

this case. In terms of the number of flakes to cores, the Unit IV assemblages are actually quite low in comparison with the other layers. This said the number of cores is also quite high because any flake can be converted into a core. Note too that in this statistic, truncated-facetted artifacts are not counted as cores. If they were included, the flake to core ratio would drop further still. So if we look at it another way to avoid this issue of counting cores and instead look at the number of blanks extracted from 1 kg of material, on average, Unit IV falls in the middle of the Pech IV assemblages. If we also look at tool production, Unit IV has the lowest number of tools per blank meaning the production of flakes was not driven by the need to produce more tools. One is tempted to conclude that during this phase of the site,

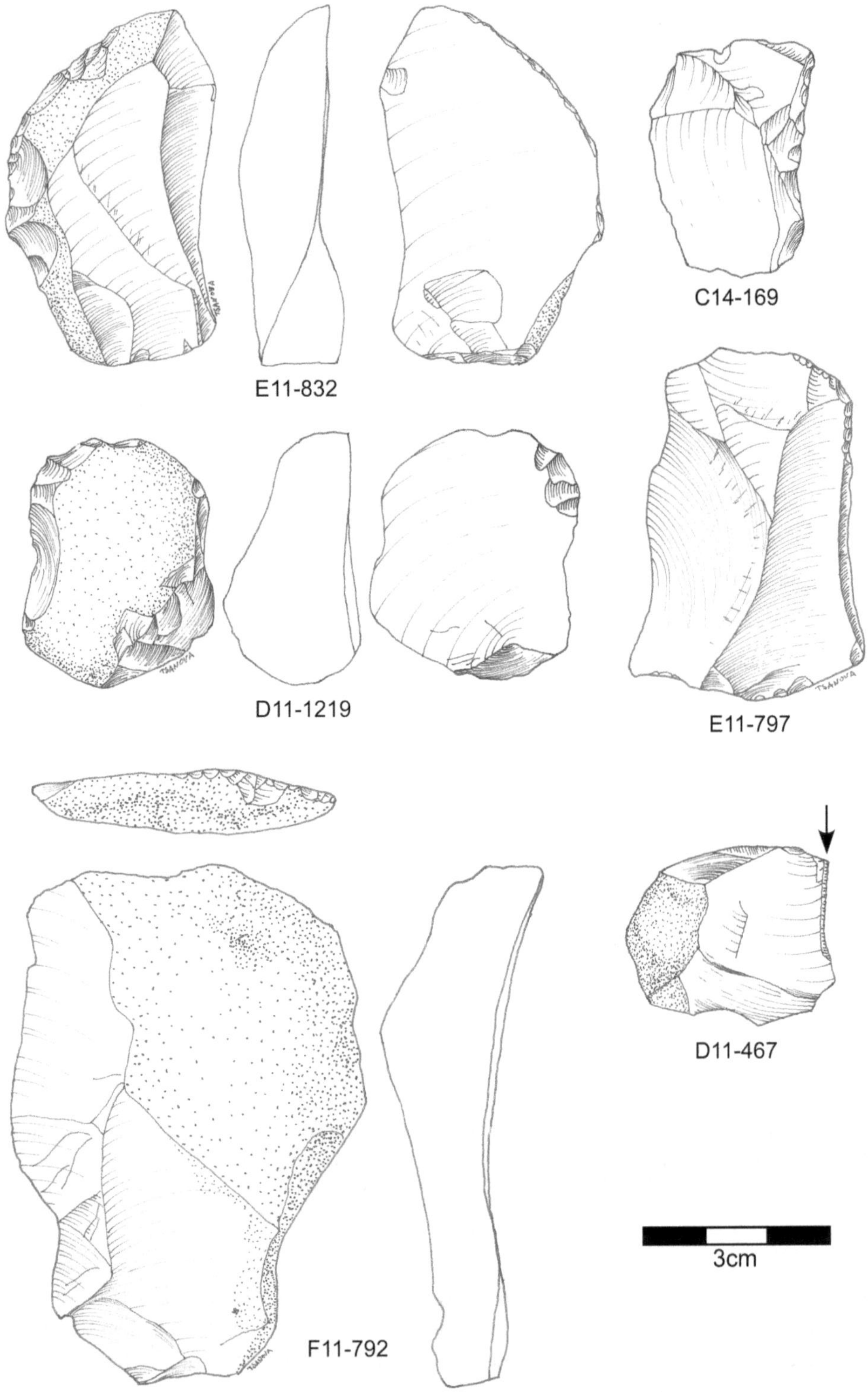


**Fig. 6.50** Retouched pieces from Combined Unit I-A. E11-832: Backed knife; C14-169: Backed knife; D11-1219: Endscraper; F11-792: Backed knife; D11-467: Burin

**Table 6.92** Summary technological and typological data

| Unit | Group II | Group III | Group IV | Il | Ifs | T-F | Kombewa | Core edge | Nat. backed | Type 5s |
|---|---|---|---|---|---|---|---|---|---|---|
| I-A | 14.5 | 11.76 | 61.97 | 3.15 | 9.96 | 5.04 | 0.14 | 4.09 | 8.92 | 10.2 |
| I-B | 29.37 | 6.91 | 51.73 | 6.59 | 13.2 | 2.05 | 0.18 | 2.31 | 7.02 | 4.5 |
| II-A | 69 | 3.5 | 21 | 4.79 | 15.19 | 2 | 0.14 | 1.09 | 4.1 | 2.4 |
| II-B | 73.24 | 5.63 | 11.27 | 8.42 | 20.06 | 3.52 | 0 | 1.49 | 5.2 | 0.9 |
| II-C | 72.66 | 4.68 | 13.19 | 12.11 | 27.92 | 3.54 | 0.09 | 1.74 | 7.53 | 1.4 |
| III-A | 68.42 | 5.79 | 13.68 | 12.02 | 24.38 | 5.79 | 0.08 | 2.04 | 11.37 | 2.4 |
| III-B | 53.09 | 7.46 | 21.11 | 11.98 | 26.49 | 11.3 | 0.27 | 1.47 | 6.82 | 1.7 |
| IV-A | 15.27 | 6.7 | 35.93 | 10.61 | 23.71 | 32.42 | 0.25 | 1.15 | 5.03 | 1.6 |
| IV-B | 18.7 | 6.78 | 41.19 | 10.33 | 24.03 | 22.22 | 0.2 | 2.01 | 5.87 | 2.9 |
| V | 61.06 | 7.26 | 17.35 | 9.37 | 21.63 | 7.61 | 0.22 | 1.23 | 5.85 | 2.4 |

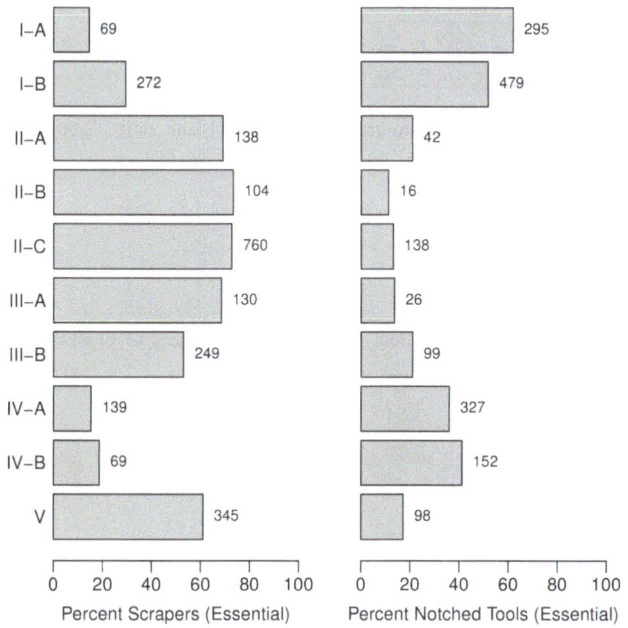


**Fig. 6.51** Percentages of scrapers (*left*) and notched tools (*right*) through the sequence. *Numbers next to each bar* are actual counts

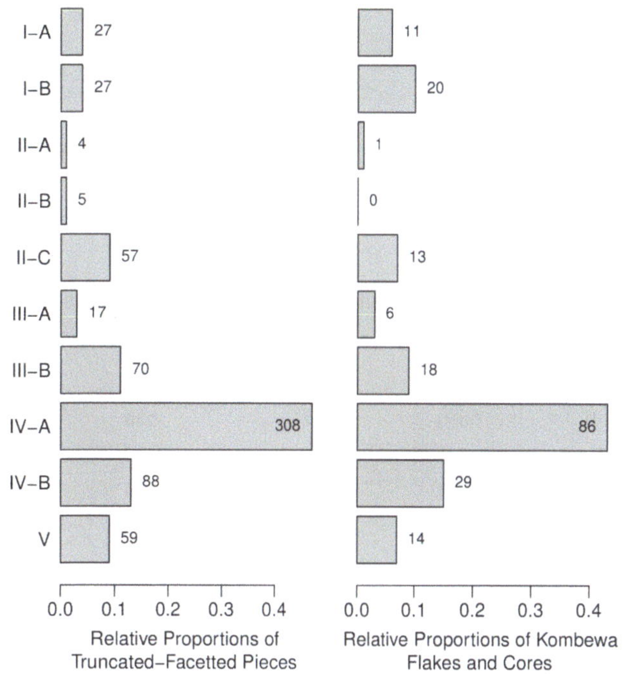


**Fig. 6.52** Relative proportions of truncated-faceted artifacts and Kombewa flakes and cores through the sequence. *Numbers in each bar* are actual counts

Pech IV was used for activities that required small sharp edges over large sharp edges.

Second, we were able to confirm in our excavations the presence of bifacial, discoidal, and Levallois technologies at the top of the sequence (Unit I), which dates to very near the end of the Mousterian in southwest France. Our Unit I was Bordes' Layer F which he subdivided into four arbitrary levels so that he could better see changes within them (see discussion in McPherron et al. 2005). We were able to confirm this subdivision, though we divided the layer into

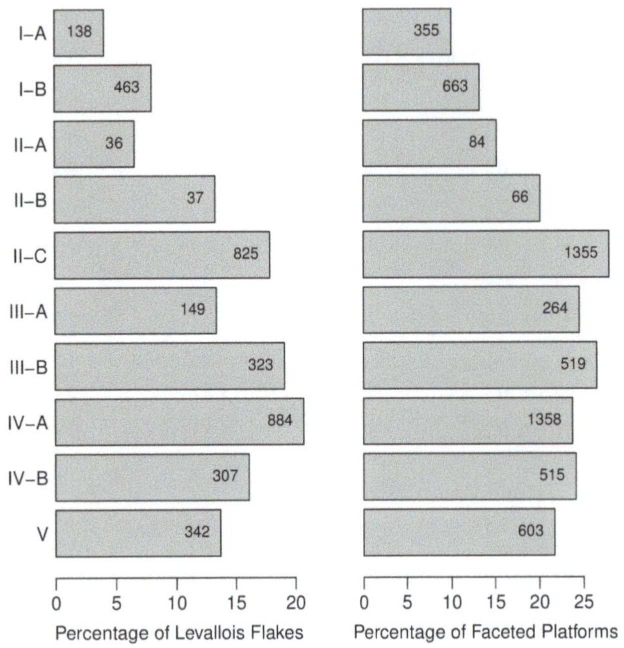


**Fig. 6.53** Percentage of Levallois flakes and faceted platforms through the Pech IV sequence. *Numbers in each bar* are actual counts

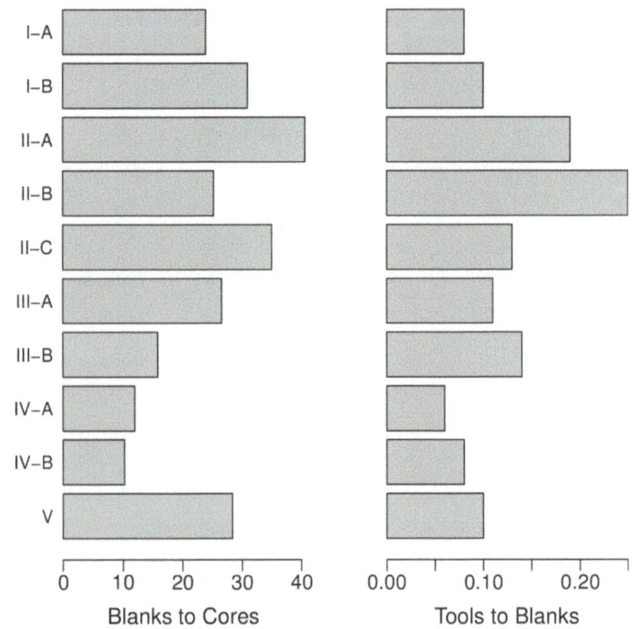


**Fig. 6.54** Blank to core ratio and tool to blank ratio through the Pech IV sequence

**Table 6.93** Blank, tool, and core ratios (platforms and complete cores only)

| Unit | N_blanks | N_cores | N_tools | Blanks_to_cores | Tools_to_cores | Tools_to_blanks |
|---|---|---|---|---|---|---|
| I-A | 4350 | 182 | 355 | 23.9 | 1.95 | 0.08 |
| I-B | 6680 | 216 | 656 | 30.93 | 3.04 | 0.1 |
| II-A | 731 | 18 | 139 | 40.61 | 7.72 | 0.19 |
| II-B | 404 | 16 | 103 | 25.25 | 6.44 | 0.25 |
| II-C | 5813 | 166 | 767 | 35.02 | 4.62 | 0.13 |
| III-A | 1223 | 46 | 136 | 26.59 | 2.96 | 0.11 |
| III-B | 2580 | 163 | 359 | 15.83 | 2.2 | 0.14 |
| IV-A | 8258 | 693 | 529 | 11.92 | 0.76 | 0.06 |
| IV-B | 2932 | 287 | 230 | 10.22 | 0.8 | 0.08 |
| V | 3574 | 126 | 360 | 28.37 | 2.86 | 0.1 |

only two units (Unit I-A and I-B). One of the differences between these two is the artifact density. Both are rich, but Unit I-B has the highest density of artifacts and the greatest number of artifacts in the sequence (setting aside the highly disturbed Layer 7). We also see site formation differences between these two units with Unit I-B looking to be minimally disturbed and with Unit I-A showing high isotropy values suggestive of an unusual depositional environment (for instance a toss zone just under the cave wall) or some sort of post-depositional reworking. Moreover, there are differences in the stone tool industries that are compatible with the type of changes once described as distinguishing the

**Table 6.94** Size measures on Levallois and single-surface cores (completes only)

| Unit | N | Length_median | Length_iqr | Width_median | Width_iqr | Thickness_median | Thickness_iqr | Weight_median | Weight_iqr |
|---|---|---|---|---|---|---|---|---|---|
| I-A | 56 | 43.48 | 19.82 | 35.57 | 14.76 | 16.86 | 8.21 | 30.5 | 44.5 |
| I-B | 112 | 47.46 | 12.89 | 37.79 | 11.44 | 16.53 | 8 | 33 | 29.75 |
| II-A | 9 | 52 | 13.94 | 43.87 | 8.3 | 15.93 | 8.56 | 48 | 38 |
| II-B | 10 | 50.65 | 19.3 | 41.01 | 14.06 | 19.6 | 12.47 | 57 | 53.5 |
| II-C | 102 | 50.83 | 18.42 | 39.02 | 9.68 | 18.03 | 7.72 | 43 | 36.5 |
| III-A | 29 | 49.39 | 13.83 | 39.78 | 11.85 | 17.2 | 7.96 | 38 | 31 |
| III-B | 104 | 42.25 | 16.77 | 33.55 | 11.71 | 14.02 | 8.07 | 21.5 | 31.75 |
| IV-A | 450 | 37.97 | 10.47 | 30.89 | 8.07 | 12.94 | 5.8 | 17 | 13 |
| IV-B | 186 | 38.4 | 10.19 | 30.65 | 8.22 | 12.61 | 5.82 | 16 | 12 |
| V | 73 | 44.47 | 12.6 | 35.98 | 10.68 | 13.73 | 6.36 | 24 | 17 |

**Table 6.95** Blank technology (all flakes and tools)

| Unit | *N* | Biface. retouch | Blade | Burin | Clactonian | Discoidal | Kombewa | Levallois | Normal | Other | Retouch. flake |
|---|---|---|---|---|---|---|---|---|---|---|---|
| I-A | 3359 | 0.7 | 0.6 | 0.3 | 1.8 | 0.7 | 0.2 | 4.1 | 90 | 1.7 | 0 |
| I-B | 5777 | 1.9 | 0.3 | 0.2 | 1.2 | 1.1 | 0.2 | 8 | 85.9 | 1.2 | 0.1 |
| II-A | 543 | 7 | 0 | 0.2 | 1.5 | 0.9 | 0.2 | 6.6 | 79.7 | 1.5 | 2.4 |
| II-B | 278 | 0.7 | 0 | 0.4 | 2.5 | 0.4 | 0 | 13.3 | 82.7 | 0 | 0 |
| II-C | 4612 | 1.1 | 1.1 | 0 | 0.5 | 0.6 | 0.1 | 17.9 | 78.2 | 0.5 | 0.1 |
| III-A | 1108 | 1.3 | 2.1 | 0.1 | 0.1 | 0.5 | 0.1 | 13.4 | 81.5 | 0.9 | 0 |
| III-B | 1694 | 0.8 | 1.1 | 0.1 | 0.5 | 0.3 | 0.4 | 19.1 | 76.8 | 0.4 | 0.5 |
| IV-A | 4269 | 0.4 | 0.6 | 0 | 0.3 | 0.3 | 0.5 | 20.7 | 76.4 | 0.8 | 0 |
| IV-B | 1908 | 0.6 | 0.5 | 0.1 | 0.4 | 0.6 | 0.3 | 16.1 | 79.8 | 1.6 | 0.1 |
| V | 2503 | 0.5 | 1.2 | 0.1 | 0.1 | 0.3 | 0.3 | 13.7 | 83.4 | 0.4 | 0 |

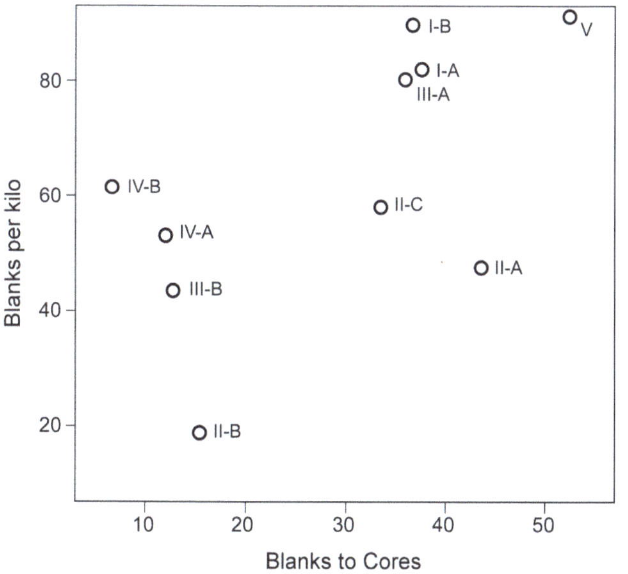


**Fig. 6.55** The number of blanks per unit of raw material (1 kg) versus the number of blanks per core

Mousterian of Acheulian Tradition Type A from Type B (see also McPherron et al. 2005) with, in particular, a peak in backed knives occurring very high in the sequence. In addition to backed knives, we were able to confirm a general pattern of more frequent backed forms in the upper part of the sequence (again, especially Unit I-A) as has been suggested by Soressi (2002, 2004). On the other hand, we were not able to confirm an increase in elongated shapes (Soressi 2005). Rather, at Pech IV, the increase in backed forms is caused in part by an increase in discoidal flakes which are themselves not elongated.

The Pech IV lithic assemblage is now a large, well-dated sample of Neandertal technological behavior within a substantial period of time just prior to the arrival of modern humans into this part of Europe. What remains is to better integrate these date with the growing list of similar high-resolution data sets documenting late Pleistocene Neandertal behavior in southwest France but also more generally in Europe to have a better idea of the Neandertal adaptation.

**Table 6.96** Platforms (all flakes and tools)

| Unit | *N* | Chapeau.g. | Cortical | Dihedral | Faceted | Other | Plain | Punctiform |
|---|---|---|---|---|---|---|---|---|
| I-A | 3562 | 0 | 8.7 | 9.5 | 10 | 0.8 | 67.9 | 3.2 |
| I-B | 5013 | 0.1 | 13.3 | 8.2 | 13.2 | 0.4 | 61.9 | 2.9 |
| II-A | 552 | 0.2 | 12.9 | 6 | 15.2 | 1.4 | 60.1 | 4.2 |
| II-B | 329 | 0.3 | 13.4 | 6.1 | 20.1 | 0.9 | 58.1 | 1.2 |
| II-C | 4854 | 0.1 | 13 | 8.2 | 27.9 | 0.7 | 46.7 | 3.4 |
| III-A | 1076 | 0.5 | 10.8 | 10.2 | 24.5 | 1 | 49.9 | 3.1 |
| III-B | 1958 | 0 | 13.7 | 7.3 | 26.5 | 0.7 | 49.3 | 2.6 |
| IV-A | 5724 | 0.1 | 11.3 | 8.8 | 23.7 | 0.7 | 52.6 | 2.8 |
| IV-B | 2139 | 0.2 | 11.1 | 9.4 | 24.1 | 1.1 | 51.4 | 2.8 |
| V | 2784 | 0.4 | 8.5 | 10.6 | 21.7 | 0.6 | 55.1 | 3.1 |

**Table 6.97** Blank form (all flakes and tools)

| Unit | *N* | Angular | Blade | Broad | Burin | Core. edge | Expanding | Crested. blade | Naturally. backed | Normal | Overshot | Point | Pustule |
|---|---|---|---|---|---|---|---|---|---|---|---|---|---|
| I-A | 3522 | 14.5 | 3.9 | 4.3 | 0.1 | 5.2 | 3.4 | 0.6 | 11.4 | 52.1 | 0.8 | 1.8 | 2.1 |
| I-B | 6044 | 6.9 | 4.4 | 3.5 | 0.1 | 2.7 | 3.6 | 0.1 | 8.4 | 67.2 | 0.8 | 1.3 | 1.2 |
| II-A | 572 | 7 | 1.9 | 3.1 | 0.2 | 1.4 | 6.6 | 0.3 | 5.8 | 69.6 | 0.7 | 1.9 | 1.4 |
| II-B | 272 | 5.1 | 5.1 | 6.6 | 0.4 | 2.2 | 2.6 | 0 | 8.8 | 67.3 | 0.4 | 1.5 | 0 |
| II-C | 4592 | 4.4 | 7.3 | 2.4 | 0 | 2.3 | 2.4 | 0.2 | 10.8 | 65.1 | 1.1 | 2.5 | 1.5 |
| III-A | 1289 | 4.3 | 7.1 | 3.7 | 0.1 | 2.2 | 3.9 | 0.3 | 11.9 | 62.3 | 0.9 | 1.7 | 1.5 |
| III-B | 1748 | 8.1 | 4.9 | 2.6 | 0.1 | 2.2 | 3.2 | 0.1 | 11.1 | 64.4 | 1 | 1.1 | 1.3 |
| IV-A | 4603 | 6.1 | 4.3 | 2.8 | 0 | 2.1 | 3 | 0.1 | 9.9 | 68 | 1.4 | 1.1 | 1.1 |
| IV-B | 2227 | 6.6 | 3.1 | 2.7 | 0 | 2.9 | 2.2 | 0.1 | 9.5 | 69.2 | 1.4 | 0.7 | 1.5 |
| V | 2893 | 5.8 | 7.1 | 4.3 | 0.1 | 1.7 | 3.1 | 0.2 | 8.1 | 64.6 | 1 | 3 | 1 |

**Table 6.98** Flake elongation by layer

| Unit | *N* | Elongation_mean | Elongation_sd | Elongation_cv |
|---|---|---|---|---|
| I-A | 2898 | 1.53 | 0.51 | 3 |
| I-B | 3963 | 1.5 | 0.52 | 2.86 |
| II-A | 407 | 1.52 | 0.53 | 2.88 |
| II-B | 190 | 1.56 | 0.66 | 2.36 |
| II-C | 3232 | 1.68 | 0.62 | 2.7 |
| III-A | 756 | 1.63 | 0.57 | 2.88 |
| III-B | 1222 | 1.56 | 0.51 | 3.03 |
| IV-A | 3537 | 1.54 | 0.5 | 3.08 |
| IV-B | 1286 | 1.49 | 0.5 | 3.01 |
| V | 1731 | 1.59 | 0.56 | 2.87 |

**Table 6.99** Percentage of backed pieces (Types 5, 36, 37, 38, Naturally backed and Débordant complete tools and flakes)

| Unit | *N* | N_backed | Perc_backed | Percent_type5 | Percent_retback | Percent_natback | Percent_coreedge |
|---|---|---|---|---|---|---|---|
| I-A | 1399 | 256 | 18.3 | 16.41 | 2.34 | 60.55 | 30.86 |
| I-B | 2521 | 301 | 11.94 | 6.98 | 0.66 | 78.74 | 17.28 |
| II-A | 200 | 21 | 10.5 | 14.29 | 0 | 80.95 | 9.52 |
| II-B | 41 | 4 | 9.76 | 0 | 0 | 100 | 0 |
| II-C | 564 | 62 | 10.99 | 9.68 | 0 | 72.58 | 24.19 |
| III-A | 1305 | 169 | 12.95 | 5.33 | 1.18 | 82.25 | 15.98 |
| III-B | 586 | 64 | 10.92 | 3.12 | 0 | 85.94 | 12.5 |
| IV-A | 2027 | 228 | 11.25 | 5.26 | 0.88 | 85.53 | 12.28 |
| IV-B | 1885 | 212 | 11.25 | 10.85 | 0.94 | 71.23 | 20.28 |
| V | 2240 | 171 | 7.63 | 7.02 | 0.58 | 83.63 | 11.11 |

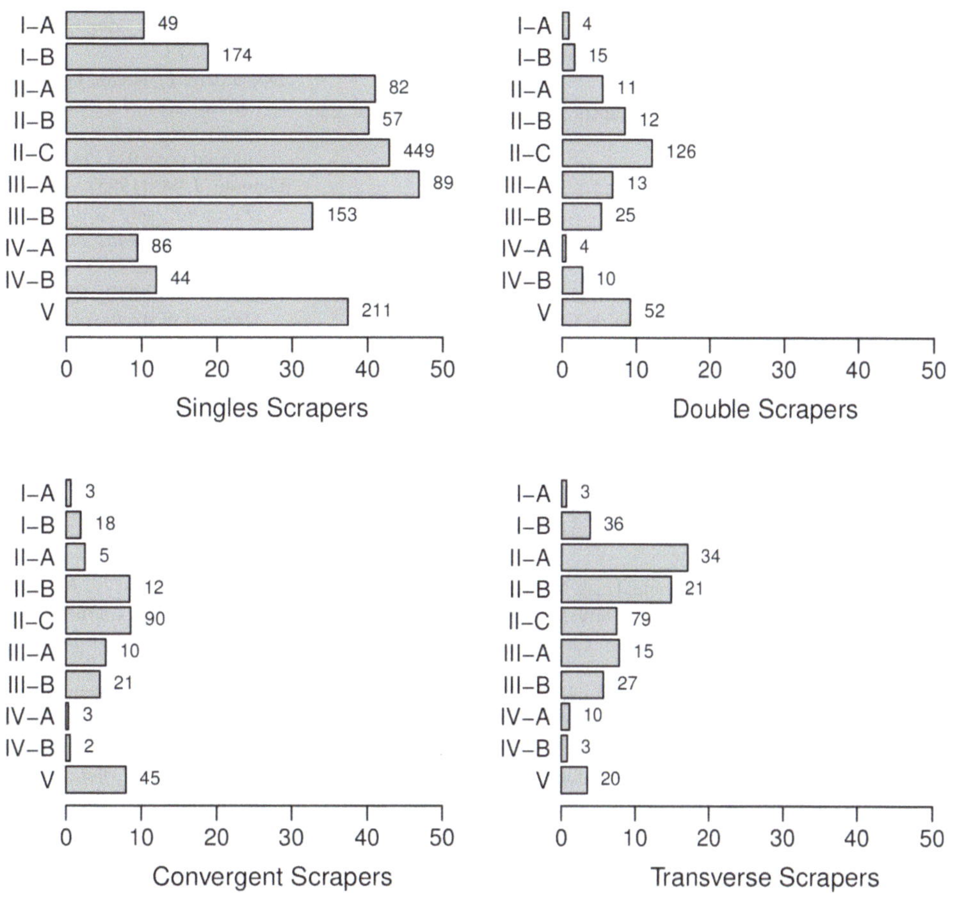


**Fig. 6.56** Proportions of various scrapers types through the Pech IV sequence. *Numbers along each bar* are actual counts

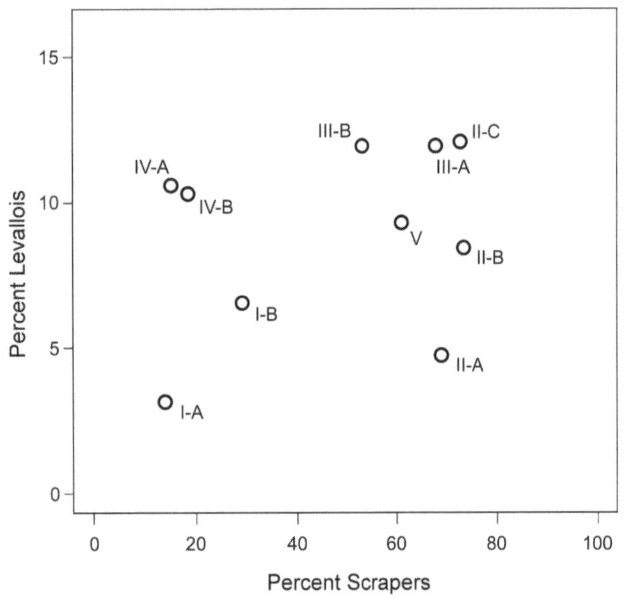


**Fig. 6.57** The percentage of scrapers in each layer versus the frequency of Levallois flakes

## References


Bordes, F. (1961). *Typologie du Paléolithique ancien et moyen*. Cahiers du Quaternaire no 1, Ed. Presses du CNRS, 2 t., (1981).

Bordes, F. (1975). Le gisement du Pech de l'Azé IV. Note Préliminaire. *Bulletin de la Société Préhistorique Française, 72,* 293–308.

Capdeville, J. (1988). *Carte géologique France (1/50 000), feuille Sarlat (808) et notice explicative*. Orléans: Bureau de Recherches Géologiques et Minières.

Chase, P. G., Armand, D., Debénath, A., Dibble, H., & Jelinek, A. J. (1994). Taphonomy and zooarchaeology of a Mousterian faunal assemblage from La Quina, Charente, France. *Journal of Field Archaeology, 21,* 289–305.

Chase, P. G., Debénath, A., Dibble, H., & McPherron, S. P. (2009). *The cave of Fontéchevade: A new investigation of the site and its paleoanthropological implications*. Cambridge: Cambridge University Press.

Cornford, J. M. (1986). La Cotte de St. Brelade 1961–1978: Excavations by C. B. M. McBurney. In P. Callow & J. M. Cornford (Eds.), *La Cotte de St. Brelade 1961–1978: Excavations by C. B. M. McBurney* (pp. 337–352). Norwich: Geo Books.

Debénath, A., & Dibble, H. (1994). *Handbook of paleolithic typology: Lower and middle paleolithic of Europe* (Vol. 1). Philadelphia: University of Pennsylvania Museum.

Demars, P.-Y. (1982). *L'utilisation du silex au Paléolithique supérieur: choix, approvisionnement, circulation: l'exemple du bassin de Brive*: Editions du Centre national de la recherche scientifique (CNRS).

Dibble, H. (1988). Typological aspects of reduction and intensity of utilization of lithic resources in the French Mousterian. In H. Dibble & A. Montet-White (Eds.), *Upper Pleistocene prehistory of western Eurasia* (pp. 181–197). Philadelphia: University of Pennsylvania Museum Press.

Dibble, H. (1995). Biache Saint-Vaast, Level IIA: A comparison of analytical approaches. In H. Dibble & O. Bar-Yosef (Eds.), *The definition and interpretation of Levallois technology* (pp. 93–116). Madison, WI: Prehistory Press Madison.

Dibble, H., & Bar-Yosef, O. (1995). *The definition and interpretation of Levallois technology* (Vol. 23). Madison, WI: Prehistory Press.

Dibble, H., Holdaway, S. J., Lenoir, M., McPherron, S., Roth, B. J., & Sanders-Grey, H. (1995). *Techniques of excavation and analysis the middle Paleolithic Site of Combe-Capelle Bas (France)*. Philadelphia: University Museum Press.

Dibble, H., & McPherron, S. (2006). The missing mousterian. *Current Anthropology, 47,* 777–803.

Dibble, H., & McPherron, S. (2007). Truncated-faceted Pieces: Hafting modification, retouch, or cores? In S. McPherron (Ed.), *Tool v core: New approaches in the analysis of stone tool assemblages*. Cambridge: Cambridge Scholars Publications.

Dibble, H., & Rezek, Z. (2009). Introducing a new experimental design for controlled studies of flake formation: results for exterior platform angle, platform depth, angle of blow, velocity, and force. *Journal of Archaeological Science, 36,* 1945–1954.

Dibble, H., Schurmans, U. A., Iovita, R. P., & McLaughlin, M. V. (2005). The measurement and interpretation of cortex in lithic assemblages. *American Antiquity, 70,* 545–560. doi:10.2307/40035313.

Dogandžić, T., Braun, D. R., & McPherron, S. (2015). Edge length and surface area of a blank: Experimental assessment of measures, size predictions and utility. *PLoS ONE, 10,* e0133984. doi:10.1371/journal.pone.0133984.

Geneste, J.-M. (1985). *Analyse lithique d'industries moustériennes du Périgord: une approche technologique du comportement des groupes humains au Paléolithique moyen*. Bordeaux: University of Bordeaux.

Hiscock, P. (2002). Quantifying the size of artefact assemblages. *Journal of Archaeological Science, 29,* 251–258. doi:10.1006/jasc.2001.0705.

Jaubert, J., Hublin, J.-J., McPherron, S., Soressi, M., Bordes, J.-G., Claud, E., et al. (2008). Paléolithique moyen récent et Paléolithique supérieur ancien a Jonzac (Charente-Maritime): premiers résultats des campagnes 2004–2006. In J. Jaubert, J.-G. Bordes, & I. Ortega (Eds.), *Les Sociétés du Paléolithique dans un Grand Sud-ouest de la France: nouveaux gisements, nouveaux résultats, nouvelles méthodes* (pp. 203–244). Mémoire de la Société Préhistorique Française 47.

Jelinek, A. J. (1977). A preliminary study of flakes from the Tabun Cave, Mount Carmel. In *Eretz-Israel: Archaeological. Historical and Geographical Studies* (pp. 87–96).

Jelinek, A. J. (2013). *Neandertal lithic industries at La Quina*. Tucson: University of Arizona Press.

Lin, S. C., McPherron, S., & Dibble, H. (2015). Establishing statistical confidence in Cortex Ratios within and among lithic assemblages: A case study of the Middle Paleolithic of southwestern France. *Journal of Archaeological Science, 59,* 89–109. doi:10.1016/j.jas.2015.04.004.

Lin, S. C. H., Douglass, M. J., Holdaway, S. J., & Floyd, B. (2010). The application of 3D laser scanning technology to the assessment of ordinal and mechanical cortex quantification in lithic analysis. *Journal of Archaeological Science, 37,* 694–702. doi:10.1016/j.jas.2009.10.030.

McPherron, S. (2003). Typological and technological variability in the Bifaces from Tabun Cave, Israel. In M. Soressi & H. Dibble (Eds.), *From prehistoric bifaces to human behavior: Multiple approaches to the study of bifacial technology* (pp. 55–76). Philadelphia: University of Pennsylvania, Museum Press.

McPherron, S., & Dibble, H. (2000). The lithic assemblages of Pech de l'Azé IV (Dordogne, France). *Préhistoire Européenne, 15,* 9–43.

McPherron, S. P., Dibble, H. L., & Goldberg. P. (2005). "Z." *Geoarchaeology, 20*(3), 243–262. doi:10.1002/gea.20048.

Morala, A. (1983). A propos des matières premières lithiques en Haut-Agenais. Bulletin de la Société Préhistorique Française. *Comptes Rendus des Séances Mensuelles Paris, 80*(6).

Park, S. J. (2007). *Circulations des matières premières et Industries lithiques au Paléolithique moyen récent et final. Approche techno-économique des industries lithiques à partir des dernières fouilles de La Quina, Charente.* Thèse, Université de Paris X-Nanterre.

Roe, D. A. (1969). British lower and middle palaeolithic handaxe groups. *Proceedings of the Prehistoric Society, 34,* 1–82. doi:10.1017/S0079497X00013840.

Séronie-Vivien, M., & Séronie-Vivien, M.-R. (1987). *Les silex du Mésozoïque nord-aquitain: approche géologique de l'étude des silex pour servir à la recherche préhistorique*. Société linnéenne de Bordeaux.

Soressi, M. (2002). *Le Moustérien de tradition acheuléenne du sud-ouest de la France. Discussion sur la signification du faciès à partir de l'étude comparée de quatre sites: Pech-de-l'Azé I, Le Moustier, La Rochette et la Grotte XVI.* Université Bordeaux I.

Soressi, M. (2004). *From Mousterian of acheulian tradition type A to type B: technical tradition, raw material, task, or settlement dynamic changes?* (Vol. 2). Paper presented at the Settlement dynamics of the Middle Paleolithic and Middle Stone Age.

Soressi, M. (2005). Late mousterian lithic technology: Its implications for the pace of the emergence of behavioural modernity and the relationship between behavioural modernity and biological modernity. In L. R. Backwell & F. D'Errico (Eds.), *From tools to symbols* (pp. 389–417). Johanesburg: University of Witswatersand Press.

Turq, A. (2000). *Paléolithique inférieur et moyen entre Dordogne et Lot*. Paléo Supplément No. 2.

Turq, A., Antignac, G., & Roussel, P. (1999). Les silicifications coniaciennes du Sarladais et du Gourdonnais: inventaire et implications archéologiques. *Paléo, 11,* 145–160.

Turq, A., Dibble, H., Goldberg, P., McPherron, S., Sandgathe, D., Jones, H., et al. (2011). Les Fouilles Récentes du Pech de l'Azé IV (Dordogne). *Gallia Préhistoire, 53,* 1–58.

Turq, A., Roebroeks, W., Bourguignon, L., & Faivre, J-. P. (2013). The fragmented character of Middle Palaeolithic stone tool technology. *Journal of Human Evolution, 65*(5), 641–655. doi:10.1016/j.jhevol.2013.07.014.

# Summary and Conclusions

# 7

Harold L. Dibble, Shannon J.P. McPherron, Paul Goldberg, and Dennis M. Sandgathe

The Middle Paleolithic site of Pech de l'Azé IV (Pech IV) is one of the cluster of four Lower and Middle Paleolithic sites located in the Perigord region of southwest France. Although originally thought to be a shelter situated at the foot of the cliff, it is, in fact, a collapsed cave, most likely part of the same karstic system associated with that of Pech I and II. Pech IV was discovered and tested by Bordes in the spring of 1952, during his excavations at the other Pech sites, but it was not until 1970 that he began excavating there. Altogether, his excavation spanned eight seasons and removed approximately 115 $m^3$. The new excavations, focused on the western part of the site, took place over four seasons (2000–2003), and removed just over 15 $m^3$.

H.L. Dibble (✉)
Department of Anthropology, University of Pennsylvania, Philadelphia, PA, USA
e-mail: hdibble@sas.upenn.edu

H.L. Dibble · S.J.P. McPherron
Department of Human Evolution, Max Planck Institute for Evolutionary Anthropology, Leipzig, Germany
e-mail: mcpherron@eva.mpg.de

H.L. Dibble
Institute for Human Origins School of Human Evolution and Social Change, Arizona State University, Tempe, AZ, USA

P. Goldberg
School of Earth and Environmental Sciences, Centre for Archaeological Science, University of Wollongong, Wollongong, NSW 2522, Australia
e-mail: paulberg@bu.edu

P. Goldberg
Institute for Archaeological Sciences, Eberhard Karls Universitat Tübingen, Tübingen, Germany

P. Goldberg
Department of Archaeology, Boston University, Boston, MA, USA

D.M. Sandgathe
Department of Archaeology and Human Evolution Studies Program, Simon Fraser University, Burnaby, BC, Canada
e-mail: dms@sfu.ca

D.M. Sandgathe
Museum of Archaeology and Anthropology, University of Pennsylvania, Philadelphia, PA, USA

## The Pech IV Sequence

Bordes' original description of the deposits encompassed a mixture of lithology and lithic industries within the different levels, sometimes with arbitrary subdivisions based on elevations. The upper series of levels, from top to bottom, were labeled A through J, with A–E being mixed Holocene deposits. Three levels, X, Y, and Z, formed the base of the deposits. The current stratigraphic divisions (see Chap. 2) are based on lithostratigraphy using the criteria of color, composition, texture, grain morphology, internal structure, and lateral and vertical changes. Altogether, eight major layers were identified, along with subdivisions based primarily on geological criteria. As can be seen in Table 7.1, there is a general similarity between our units and Bordes'. Cases of overlap (where one of Bordes' levels spans two of our layers) could be due to lateral (east–west) variation within the site itself. However, based on lithostratigraphy and comparative lithic analysis between his collections and ours, we were able to link the East and West Sections left by Bordes' excavations, thereby showing that contemporaneous occupations occurred throughout the excavated space. Therefore, there is no doubt that Pech IV represents a single depositional sequence.

A major focus of the new excavations was to establish a firm chronological sequence for the site. Four independent dating methods were employed to construct the chronological framework for Pech IV (Chap. 3). The absolute dating techniques used were radiocarbon ($^{14}C$), electron spin resonance (ESR), thermoluminescence (TL), and optically-stimulated luminescence (OSL), and the totality of dated samples makes Pech IV one of or perhaps the most intensively dated Middle Paleolithic sites in this region. These techniques produced notably consistent age estimates, ranging from approximately 100 ka (MIS 5c) at the base of


H.L. Dibble et al. (eds.), *The Middle Paleolithic Site of Pech de l'Azé IV*, Cave and Karst Systems of the World, https://doi.org/10.1007/978-3-319-57524-7_7

**Table 7.1** Stratigraphic correlations between Bordes' levels and the current layers, and the designations given to the combined assemblages

| Bordes level | Dibble/McPherron layer | Combined assemblage units |
|---|---|---|
| A–E | 1, 2 | N/A |
| $F_1$ | 3A | I-A |
| $F_2$ | | |
| $F_3$ | | |
| $F_4$ | 3B | I-B |
| G | 4A | II-A |
| $H_1/H_2$ | 4B | II-B |
| $I_1$ | | |
| $I_2$ | 4C | II-C |
| $J_1$ | 5A | III-A |
| $J_2$ | 5B | III-B |
| $J_3$ | | |
| $J_3$A | 6A | IV-A |
| $J_3$B | | |
| $J_3$C | 6B | IV-B |
| X | 7 | N/A |
| Y | 8 | V |
| Z | | |

the sequence to between 54 and 45 ka (MIS 3) at the top of the Pleistocene deposits. These dates suggest that Pech IV was occupied during varying paleoclimatic conditions, from temperate conditions at the base to generally colder periods near the top; this view is also supported through analysis of the faunal remains.

Given such a long sequence, it is not surprising that the various lithic assemblages of Pech IV showed variability as well. Bordes interpreted most of these as representing his "Typical Mousterian" facies, with moderate percentages of both scrapers and notched pieces and varying frequencies of Levallois, with assemblages representing "Mousterian of Acheulian Tradition", that is, with bifaces, at the top. He also defined a new facies, the "Asinipodian", in his Level J, which yielded relatively high percentages of extremely small Levallois flakes and cores, Kombewa flakes and cores, and, as identified by us, numerous truncated-faceted pieces. The frequency of these elements has been later interpreted as representing an emphasis on the production of small flakes by a variety of techniques, which is not unique to these levels nor to Pech IV itself. Most of the lithic assemblages are composed of locally available raw materials.

## The New Project

In the years leading up to the start of the new excavation, a major effort was expended on finishing processing the material excavated by Bordes. This involved making digital scans of his field notebooks and entering data from them into a database. A large portion of the material was washed and labeled, and roughly ten percent of the lithic objects were digitally photographed. All objects (lithics and fauna) were assigned to the level from which they came, based either on information contained in the notebooks or by their location ($X$, $Y$, and $Z$ coordinates) within the site. The entire lithic collection was then analyzed and published.

The new excavations at Pech IV, which ran for four seasons from 2000 through 2003, concentrated on the western part of the site, moving back Bordes' western profile by 1 m and removing most of a bench (roughly 9 $m^2$) of deposits that contained the lower part of the sequence. We applied a number of excavation techniques designed to increase both the accuracy and efficiency of recording and to maintain strict standards or protocols on how the excavation proceeded. All provenienced objects (any object greater than 2.5 cm in maximum dimension, plus all teeth) and samples were recorded with a total station, following the same 1 m grid system used by Bordes at the site. While most provenienced objects were recorded with a single $X$, $Y$, and $Z$ coordinate, elongated objects were recorded with a point at each end, which allowed for more detailed studies of their orientations in three-dimensional space. Larger rocks were recorded with multiple shots. All of the sediments removed during excavation were put into 7-L buckets, and the area from which the sediment was derived was also provenienced. Those buckets of sediment were then wet-screened through two mesh sizes (6 and 2 mm), and various materials (e.g., lithics, fauna, minerals) were sorted and bagged separately. A relational database was constructed to integrate the provenience information of objects and sediments, along with analytical data that were subsequently taken. Finally, a large number of digital photographs (more than 100,000) were taken from the site during excavation and of the lithic objects recovered. Chapter 1 details our work on the older collections and the methods used for the new excavation.

In effect, there are now three collections of archaeological materials from Pech IV. The first is the one excavated by Bordes (what we call "Levels", with letters indicating various strata) and the second is that recovered by the new project ("Layers", using Arabic numbers). Because it was possible to correlate our layers with Bordes' levels, the third collection is the combination of both of these, using a new set of labels ("Units" designated with Roman numerals) for the combined strata. Chapter 6 presents basic descriptions of the lithic material from the new excavations, along with an overview of all of the combined collections. An analysis of the faunal material from two of Bordes' levels are presented in Chap. 4, while the totality of the faunal assemblages from the new excavations are presented in Chap. 5.

## Summary of the Pech IV Sequence

### Layer 8

Bordes originally identified three, mostly anthropogenic, basal levels (X, Y, and Z) above the bedrock, which were similar in archaeological composition. We have grouped the lower two (Y and Z) into our Layer 8, while his Level X is represented by our Layer 7. He described this ensemble as having been affected by burning, and indeed it contains many combustion features. However, the very dark color in this layer reflects not so much the presence of charcoal, but rather char—gelified organic matter or fat from bones and meat formed by burning, which also gives a 'greasy' feel to the deposits. There are some darker bands in this layer that are capped by thin, cm-thick bands of ash, probably representing the remains of discrete fire events, but there is also clear evidence for the raking out of hearths, sometimes accompanied by trampling. These processes spread out previously combusted materials before the next burning event took place, though did not change the displacement of artifacts to a major extent, that is, more than a few cm vertically and perhaps a bit more horizontally. At the time that this layer was being deposited, it is likely that much of the original ceiling of the cave was still present, with an opening toward the south due to the earlier erosion of the bedrock.

Layer 8 has been dated by both single-grain OSL and TL. The former method produced dates on six samples, ranging in age from 98.6 $\pm$ 11.1 ($1\sigma$) to 91.5 $\pm$ 7.6 ka, with a weighted mean age of 95 $\pm$ 4 ka. The TL age estimates based on six samples of burned flint range from 106 $\pm$ 12 to 89 $\pm$ 9 ka, with a weighted mean age of 96 $\pm$ 5 ka.

These dates place Layer 8 into the MIS 5c interglacial period, which is supported on the basis of the recovered fauna. In this layer, red deer were the predominant prey, with smaller numbers of boar, roe deer, ibex, reindeer, hare, horse, and beaver, all of which indicate temperate conditions in a forested landscape. Seasonality data indicate that the cave was occupied repeatedly during most or all of the year, suggesting that at least red deer were available locally throughout the year. Among the more notable specimens is a third phalanx from the talon of a medium-sized raptor, which has cutmarks on the articular surface, and one of the few carnivore (wolf) remains found in the site.

Red deer remains include at least three fetal individuals, and among the adults, canine teeth indicate six females and one male, suggesting a predominance of females. Both red deer and roe deer are represented by similar percentages of heads and postcranial elements, suggesting that entire carcasses were transported to the cave for processing and consumption. However, cutmarks on these two species were less frequent, both in Layers 8 and 6, as compared to the upper layers. Cutmarks were recognized on the lateral aspect of the second phalanx of a beaver specimen, perhaps for preparing the animal for consumption or for removing its skin.

Given the abundance of fire evidence in this layer, it is not surprising that the faunal assemblage was significantly affected by burning, though mostly at lower temperatures. However, the data from the faunal remains and analysis of the thin sections (including the identification of char) suggest that bone was used as fuel, although cooking or incidental exposure to heat cannot be ruled out as another possibility. Analysis of bone breakage also supports the micromorphological studies showing a high degree of breakage due to trampling.

In terms of lithics, Layer 8 is one of the richer layers. Scrapers dominate, and of these, simple scrapers are most frequent. Notches are more common than denticulates, though some of these types may also be the result of trampling. There are some truncated-facetted artifacts and small Levallois cores and flakes similar to what will become more prevalent in the overlying Layer 6. Levallois blanks are common, and there is a relatively high percentage of prepared platforms. Likewise, the core types are heavily dominated by single-surface and/or Levallois types. This layer has by far the highest percentage of exotic raw materials: 13.8% compared to a mean of 4.3% for all layers. However, analysis of the cortex ratios shows no strong evidence for artifact transport.

### Layer 7

The deposits of Layer 7 experienced a very high degree of disturbance due to cryoturbation, a result of extremely cold conditions. Because of its generally geogenic character, lack of organic matter, and its lighter color, it is lithologically distinct from Bordes' Levels Y and Z. This is reflected in the lithic artifacts, which exhibit a high degree of edge damage and breakage, but also by high percentages of notched pieces and flakes with an abrupt and alternating retouch (roughly 80% of the assemblage); there is also an increase in small lithic elements due to breakage. Likewise, while faunal remains were relatively rare in this layer, those that were present were also severely damaged by post-depositional processes. There are no chronometric dates for this layer, but the evidence for cryoturbation and the significant roof fall event that caps this layer most probably means that it was deposited during the cooler conditions of the MIS 5b. This, in turn, suggests that this layer is closer in time to that of Layer 8 than it is to Layer 6B. Nonetheless, the lithic assemblage appears to represent something intermediate between Layer 8 and the overlying Layer 6B, with intermediate frequencies of truncated-faceted pieces and Kombewa flakes and cores.

### Layer 6

This layer clearly correlates with Bordes' levels J3a, J3b, and J3c. In our view, however, there are only two subdivisions: the basal part, Layer 6B, is characterized by large, flat-lying blocks of limestone roof fall lying directly on the upper surface of Layer 7, and the upper part, Layer 6A, which contains many fewer large blocks of bedrock. Up to now, Layer 6 has been dated only using single-grain OSL from a series of five samples taken from Layers 6A and 6B and the boundary between them. The age estimates were all found to be internally consistent, with a weighted mean age of 77 ± 4 ka or during the MIS 5a interstadial. Again, these dates are consistent with the faunal evidence that includes temperate taxa such as roe deer, boar, and beaver—all indicating warmer, wooded environments—in addition to red deer and some horse. This is very similar to what was present in Layer 8.

Layer 6 yielded the largest faunal assemblage from the site. Seasonal information is based on fetal red deer (hunted during spring and winter) and boar (indicating death in spring/early summer), though cementochronology results indicate that both red deer and roe deer were also hunted during the fall. This means that occupations occurred throughout most, if not all, of the year. There is a predominance of red deer prime adults, with just a few juveniles and old adults, and both females and males are represented. As was the case in Layer 8, cutmarks are not as frequent as in the overlying layers, but there is clear evidence of marrow exploitation.

The lithic assemblages from Layer 6 correspond to Bordes' "Asinipodian" facies, which, in fact, reflects an emphasis on the production of small flakes. These were removed either from small Levallois cores, or from larger flakes, from the interior surfaces (Kombewa technique) or from exterior or interior removals reflected in truncated-faceted pieces. The core types are single-surface and, to a lesser extent, Levallois and Kombewa; especially in Layer 6B, these cores, and their flake products, are the smallest in the site. At first glance, it would be tempting to conclude that the emphasis on small flakes reflected a higher intensity of lithic reduction (due, for example, to lack of raw materials), but this interpretation is not supported by the low levels of flake and tool production. Analysis of cortex ratios shows no evidence for artifact transport. A higher proportion of broken and damaged pieces in Layer 6B may reflect the roof fall that occurred and perhaps, though less likely, some mixing with the underlying Layer 7 material.

Fire use, as evidenced by burned flint or bone or through micromorphological analyses, decreases during this time relative to that seen during Layer 8 times, but it is still much more frequent than in subsequent layers.

### Layer 5

The lower subdivision of this layer, Layer 5B, consists of numerous and typically rounded lithoclasts within a yellowish-red silty sandy matrix; there is evidence of some solifluction. Analysis of artifact orientations, breakage, and edge damage is consistent with this interpretation but also shows it to be spatially limited to a portion of the layer. Layer 5A is somewhat similar but is characterized by flatter, well-bedded limestone clasts in a sandy matrix, and it contains numerous generally angular decimeter-size blocks of limestone roof fall. This represents the beginning of the collapse of the overhang that was the roof of the original cave, and its more open nature meant that the surface of the deposits was exposed to rain and snow, which encouraged weathering of the material on the surface, leading to the formation of red iron-rich dusty clay coatings and cappings seen on many of the grains.

Four single-grain OSL samples produced a weighted mean age of 76 ± 4 ka. Similarly, four heated flints produced a weighted mean age of 74 ± 5 ka. This suggests that this layer was deposited near the end of the MIS 5a interstadial, although there is a clear dominance of reindeer, which suggests somewhat cooler conditions, perhaps corresponding to the beginning of MIS 4. However, the faunal assemblage also includes red deer, bison, horse, and roe deer thus suggesting a gradual shift in climate conditions over the course of the deposition of Layer 5. Seasonality data suggests that occupations took place during the winter and spring.

Unfortunately, this layer is remarkably poor in fauna, perhaps reflecting a distinct change in subsistence and/or site use. As in the overlying Layer 4, cutmarks are present in higher frequency, which may reflect more intensive processing of the faunal elements.

Likewise, Layer 5B has a low density of lithic artifacts, though it increases in Layer 5A. Both assemblages show a shift in the lithic assemblages. While Levallois blank production remains important and there are some Kombewa elements and truncated-facetted artifacts, these drop back to the frequency seen in Layer 8, and, likewise, lightly retouched scrapers again become dominant. This is accompanied by a greater emphasis on tool production. Identifiable cores tend to be single-surface varieties, and core size also increases. A higher cortex ratio indicates an over representation of cortical pieces, which could reflect the transport of non-cortical artifacts from the site after manufacture. The higher degree of edge-damaged pieces in Layer 5B is consistent with the interpretation of solifluction having affected this layer.

There is little direct evidence for the use of fire during Layer 5 times, and frequency of burned lithics and bone decreases to a very low percentage.

**Layer 4**

The sediments of Layer 4 are generally fine grained but there are also, especially in Layer 4B, numerous blocks of roof fall. This represents, then, one of the main retreats of the roof, resulting in the cave being considerably more open. Also, because the blocks fell more on the southern portion of the site (at the limit of the overhang), they effectively formed a wall that served to trap sediments between it and the northern cliff face. Eleven OSL samples produced weighted mean age estimates of $68 \pm 4$ ka for Layer 4C, $62 \pm 3$ ka for Layer 4B, and $57 \pm 3$ ka for Layer 4A. The date for Layer 4C is consistent with that obtained for two TL samples, which gave individual ages estimates of $71.8 \pm 6.7$ and $68.5 \pm 6.6$ ka. Thus, Layer 4 was deposited over a protracted period of time, from 72 to 54 ka ago. However, the occupations during this long time span were not homogeneous, but rather occur as more or less discrete lensing of artifacts separated by relatively sterile sediments.

Based on these dates, it is reasonable to conclude that most of the deposition of Layer 4 took place during the cooler climatic conditions of MIS 4, though the deposition of Layer 4A probably extended into MIS 3. This is supported by the faunal remains, which are overwhelmingly represented by reindeer.

This reindeer-dominated faunal spectrum marks all three sublayers of Layer 4, along with a few horse specimens and two elements identified as wolf (from Layer 4C). Both juveniles and prime adults are represented. Cutmark frequencies are high in Layer 4 (along with Layer 5A), which may reflect more intensive processing of elements for their food yield, and the general lack of cranial remains shows differential transport of elements into the site. Seasonality data, which was relatively abundant for Layer 4, shows that the occupations during this time occurred primarily during the spring and perhaps early summer, which is also the time of reindeer migration. It seems likely that the occupations at the site and the hunting of *Rangifer* occurred in seasonally restricted timeframes, presumably when this prey species was available locally as it migrated through the area. Along with the discontinuous deposition, or lensing, of artifacts generally, this suggests shorter durations of occupations. There is no direct evidence of fire (e.g., charcoal, ash), and the percentages of burned lithics and bone are among the lowest at the site.

The density of lithic artifacts is also quite low, especially in Layers 4A and 4B, though they are not distributed uniformly in the deposits but rather as discrete lenses or occupation horizons. All three sublayers are heavily dominated by scrapers (>75%) and, in Layer 4A, some of these show Quina retouch and transverse types on relatively thick flakes. Thus, there seems to be a shift in Layer 4 from an emphasis on Levallois blank production in Layer 5 to more of a Quina-type blank production in Layer 4A, as well as a number of Clactonian notches, a low number of cores, and a high tool to flake ratio. All of these are features characteristic of Quina-type assemblages.

The lithics from Layer 4B exhibit some features that suggest some effects due to water flow. These features include a relatively low frequency of small flakes and debris and a rather high level of linearity in the artifact orientations. On the other hand, there is no apparent evidence for running water that is observable in the thin sections. It is more probable that the size distribution resulted from an importation of larger pieces that were originally manufactured elsewhere, which is supported by the fact that this layer has a very high percentage of nonlocal raw materials, and a general low level of on-site blank production and tool manufacture, both activities that generate large amounts of small debris.

**Layer 3**

Layer 3 is lithologically quite distinct from the Layer 4 complex, being more compact and cemented and with fewer limestone fragments; it consists of gritty, compact to cemented quartz and limestone sand. The contact between 4 and 3 is sharp, though irregular, and it is inclined toward the south–southwest. Layer 3B is somewhat coarser and richer in flints and exhibits clearer bedding than is seen in Layer 3A. At this point in the morphology of the shelter, most of the remaining overlying roof collapsed. This is likely why the artifacts from this layer have a relatively high average plunge angle.

This layer is the only one to be dated by all four chronometric dating techniques. OSL yielded a weighted mean age of about 51 ka. Two TL ages of 54.6 $\pm$ 5.0 and 45.7 $\pm$ 3.9 ka were obtained for sublayer 3B. Five ESR age estimates were obtained on teeth, producing weighted means of 50 $\pm$ 3 (using an early uptake model) and 52 $\pm$ 5 ka (assuming linear uptake). Calibrated $^{14}C$ dates (seven samples) fall between 46.2 $\pm$ 1.6 and 42.3 $\pm$ 0.8 ka cal BP and two other ages are too old to be calibrated. All told, there is remarkable agreement among all of the age estimates obtained for Layer 3 using the four independent absolute dating techniques, all within the interval of MIS 3. The faunal assemblages, consisting primarily of reindeer and *Bos/Bison* but also some roe deer, suggest that the occupations took place during the intermittent warmer periods. Evidence of a slight increase in fire use is documented both by the presence of occasional charcoal fragments and slightly higher percentages of burned bone and flint.

Because of poor preservation, indeterminate long bone shaft fragments were most common. A single left femur shaft from a fetal horse was identified in Layer 3B, and its age-at-death estimates suggest a winter occupation. Unfortunately, the low abundance of faunal remains does not allow us to draw broader conclusions for this layer.

With Layer 3B lithic artifact densities rise significantly (in comparison to Layer 4), but drop somewhat in Layer 3A. Overall, notched artifacts are dominant, and some of the denticulates show a very regular, saw-tooth type pattern of complex notches and are made on thin flakes. In Layer 3A, Upper Paleolithic-type tools, especially backed knives, reach their highest value in the sequence. The Levallois technique is poorly represented, especially in Layer 3A where it is the lowest in the sequence. Single-surface cores are the most frequent with a scattering of other types throughout Layer 3. There is also a very high percentage of pseudo-Levallois points (Type 5), and generally there is a shift toward discoidal flake production. Bifacial technology is also present. There is a statistically significant over representation of cortex.

### Layers 1 and 2

Layer 2 occurs only in a limited part of the West Section, partially because of truncation by Layer 1 in historic times. It appears to represent a small depression produced by water cascading from the former brow of the cave. The Layer 1 complex is a slope and roof fall deposit that truncated many of the underlying deposits, especially toward the southern part of the site. This layer was deposited following the final collapse of the roof of the shelter, thus resulting in today's appearance of the site being situated at the foot of a cliff.

## Final Summary

When occupation of Pech IV began, the morphology of the site was more consistent with a cave than a rockshelter. The earliest deposition was primarily anthropogenic and activities took place directly on the exposed surface of the bedrock. Through time the roof of the former cave began retreating toward the northern cliff face. This collapse, which began at the onset of Layer 6B, was completed just following the Pleistocene occupations of the site, giving today's appearance of its being a rockshelter.

The deposits of Pech IV date to between ~99 and 45 ka. As such, the site was occupied during the alternating warm and cold periods of the Late Pleistocene. These changes in paleoclimate are confirmed by the presence of more temperate species at certain times (especially during the earliest deposits but again at the very top of the sequence) and colder species dominating the middle range of deposits, especially during Layer 4. The limited seasonality data available suggests that occupations during the temperate periods took place during all four seasons of the year, but primarily during spring and summer months during Layer 4. Use of the site is most frequent in the basal layers of the site (especially in Layer 8), and appears to drop significantly during most of the deposition until rising again slightly in Layer 3. Occupation at the site appears to be more or less continuous, except during the deposition of Layer 4, where there is clear lensing of artifactual material within an otherwise sterile sediment deposition.

Only a couple of layers show some substantial post-depositional disturbances (particularly Layer 7 and parts of Layer 5), but these appear to have modified the assemblages very locally and essentially in place and they did not likely result in substantial artifact transport. In the layers thus affected, it would be more accurate to say that sediments moved but the artifacts did not. However, trampling was a common feature during the deposition of Layer 8. Both geological and artifact orientation data were used to evaluate these post-depositional processes, and except for chemical weathering that affected the faunal assemblages, the Pech IV deposits are remarkably intact. The orientations of the lithics showed, for the most part, general alignments toward the south (along the slope of the bedrock) or southeast (along the axis of the cave system).

The faunal sequence was clearly dominated by large cervids (reindeer or red deer), with other ungulates represented by much smaller numbers. There are only a couple of examples of wolf, and small mammals and birds are infrequent in this assemblage. Faunal preservation was not uniform, which limits overall comparisons of how the various prey species were exploited. Relative abundances of the high-survival set of skeletal elements (i.e., the long bones and heads) are consistently higher among all main prey taxa and across the Pech IV assemblages overall. Even the somewhat robust (but still considered low-survival) scapulae and pelves are poorly represented, indicating that these elements were rarely transported into the site. Carpals, tarsals, and patellae are rare. Since their underrepresentation is likely not taphonomic, this may relate to anthropogenic factors, namely that limbs were disarticulated and defleshed before transport to the cave. This corroborates evidence for the layers in which we see the selective transport of the richest marrow bones. Distinctive patterns in cutmark abundance are seen across the faunal sequence. In general, the reindeer-dominated assemblages show the highest frequencies. Because articular ends of long bones are generally scarce throughout the sequence, the majority of cutmarks are found on the shaft portions, indicating the removal of meat and tissue.

The new excavations allowed for additional insights on the Pech IV lithic assemblages beyond what had been described by Bordes. First, it was possible to get a better understanding of Layer 4. It is now clear that Layer 4A shows Quina-type blank production techniques that are quite different from those of the rest of the sequence. Second, the particular nature of small flake production in Unit IV (Layer 6) was confirmed. It has already been shown that small flake production and their associated products—small cores and flakes, Kombewa cores and flakes, and truncated-faceted pieces—are part of many, if not most, Mousterian and Middle Paleolithic assemblages. So, in spite of it being quite intensively expressed in this particular layer, it is not clear that it warrants being considered as a distinct variant of French Mousterian assemblage variability. In fact, it is becoming increasingly clear that Bordes' so-called Mousterian facies are not particularly useful for interpreting assemblage variability, which is why such designations have been avoided here. Third, the presence of bifacial, discoidal, and Levallois technologies at the top of the sequence was confirmed, and at Pech IV it dates to very near the end of the Mousterian in southwest France. While Bordes made arbitrary subdivisions of his Layer F, we were able to show that only two subdivisions (Layers 3A and 3B) are warranted based on lithology or other criteria. On the other hand, changes in the stone tool industries from these two layers are somewhat compatible with the progression of changes he described as distinguishing his Mousterian of Acheulian Tradition Type A to Type B facies. Thus, there is a clear increase in the frequency of backed pieces, although there is not an increase in elongated shapes. Again, given that this reflects continuous change through time, it does not appear to support a notion that these are discrete industrial variants.

In sum, there is a tremendous amount of lithic variability in the Pech IV sequence. Different technologies (Levallois, discoidal, bifacial) become more or less dominant in different parts of the sequence and different retouched types change in terms of their relative frequencies. It is clear that we are still far from understanding what is behind this variability, but it is equally clear that thus far it does not relate to obvious external factors. For example, the lithic assemblages from Layer 6 are very distinct from those of Layer 8, but they both occur in temperate conditions and there are few differences in terms of the fauna represented and only a small decrease in the use of fire. Aside from changes through time, another avenue of research that deserves more attention at Pech IV is whether there are detectable differences in the use of space. We know that the configuration of the site changed through time with the retreat of the cave, but how this impacted artifact discard and, therefore, the assemblages we studied, is less clear for the moment.

One of the more interesting aspects of the Pech IV sequence does have to do with the evidence for the use of fire. This evidence is both direct (the remains of charcoal, ash, and char) and indirect (the burning of bones and lithics). Clearly some evidence of fire (or burning) is present in all layers, but by comparing percentages of burned objects found throughout the sequence it is possible to quantify it beyond more presence or absence. Thus, in this way it is clear that there is much more evidence for fire in the lower part of the sequence, very little in the middle part, and a slight increase in the upper Layer 3. What is behind this variation is still unclear, but it is somewhat ironic that the evidence for fire is much more abundant in warmer climates than in colder ones.

Of course, no one site will ever be able to answer all questions surrounding Neandertal behavior. However, the two excavations at Pech IV have produced a tremendous dataset—not just in terms of quantity of objects recovered, but also in terms of their context, including their stratigraphic context, their chronology, and their spatial distribution. The results of these projects, then, provide a tremendous resource for future research.

# Appendix A: Manganese-rich Lumps from Pech de l'Azé IV: A Preliminary Analysis

Francesco d'Errico and Marie Soressi

In the last few years, the utilization of manganese oxides by Mousterian and Châtelperronian Neanderthals has been the subject of a number of studies (Soressi et al. 2008; Heyes et al. 2016; Salomon 2009; Dayet et al. 2014). Pech de l'Azé IV (Pech IV) is one of the Mousterian sites from the Dordogne region that has yielded a substantial collection of such objects. They consist of dark gray to brown soft stones, often bearing clear traces of modification. Although not analyzed yet for their elemental and mineral composition they are in all likelihood, by comparison to similar fragments found at other Mousterian sites, mainly composed of manganese-rich oxide minerals. They were recovered during both Bordes' and the recent excavations reported on here. In 2003, we examined this material and provided an inventory and a first description, included in the 2003 excavation report (d'Errico and Soressi 2003). Here we summarize the results of that preliminary assessment, which can be of use for future study on this material.

## Inventory

The Pech IV collection of manganese artifacts is composed of 26 pieces, weighing 392 g in total. They include 25 lumps of black mineral ore, likely rich in manganese dioxide, and one small fragment of a ferruginous sandstone (Fig. A.1; Table A.1). Seventeen of these pieces were recovered during Bordes' excavations; the other nine come from the more recent excavations. Three fragments were refitted into three larger blocks, which reduce the size of the collection to 23 pieces. Most of the blocks were found in squares D-G/11-13, with a maximum concentration in squares D11, E12, and F11.

## Description

The weight of the pieces varies between 1 and 226 g, with very few pieces being heavier than 10 g (Table A.1). Maximum dimensions range between 11 and 67 mm with a mean of 30 mm. The blocks are mostly irregular in shape and relatively light (Fig. A.1). However, they seem on average, denser than those from Pech de l'Azé I. The surface of the blocks is irregular and porous. Traces of abrasion creating distinct facets are observed on 14 pieces. A single facet was found on 10 pieces, 2 facets on 3 pieces and 3 facets on 1 piece. The abrasion facets are systematically elongated (Fig. A.3), and their length/width ratio shows a remarkable degree of correlation (Fig. A.2). Use-wear suggesting use posterior to grinding or present on objects not modified by grinding was detected on 16 pieces (Table A.1).

Fourteen facets are covered by fine parallel striations, usually oriented along the main axis of the facet or, less often, slightly oblique to it (Fig. A.3). This indicates that the facets were produced, as observed at Pech de l'Azé I, by gently displacing the object back-and-forth along its main axis on a fine-grained grindstone.

The facets without striations show a polish that may have been produced by rubbing the piece on a soft material such as human or animal skin (d'Errico and Soressi 2002).

Two pieces differ from the others. One has a higher content of quartz grains. The other (D14-3090) is substantially larger and more heavily modified than the others in the collection (Fig. A.4). Two shallow wide grooves extending over the whole surface show at their bottom-deep striations produced by scraping the block with a stone tool. Facets due to abrasion are present on the remainder of the surface.



H.L. Dibble et al. (eds.), *The Middle Paleolithic Site of Pech de l'Azé IV*,
Cave and Karst Systems of the World, https://doi.org/10.1007/978-3-319-57524-7

**Fig. A.1** Manganese oxide-rich lumps uncovered at Pech de l'Azé IV during Bordes' and the recent excavations (*Photo* d'Errico and Soressi)

## Research Perspectives

The excellent quality of the contextual data available for this material, the proximity of the Pech de l'Azé I site, which has yielded the richest known collection of Mousterian manganese oxides (Soressi et al. 2008; Heyes et al. 2016), and the good state of preservation of the traces of modification make this collection an essential point of comparison for gaining a better understanding of how and why manganese-rich rocks were used by Neanderthals and to document changes in source use and technology through time. The inventory and preliminary descriptions presented are the necessary prerequisites before more detailed surface, elemental and mineralogical analyses are conducted on this interesting material.

**Table A.1** Contextual, taphonomic, technological, and dimensional information on mineral ore blocks from Pech de l'Azé IV (Bordes' and the recent excavations)

| Excavation | Layer | Square | Number | Material | Fragmentation | Abrasion facets (*N*) | Use-wear | Length (mm) | Width (mm) | Thickness (mm) | Weight (g) |
|---|---|---|---|---|---|---|---|---|---|---|---|
| D&McPh | 3B | F11 | 1214 | MO | Complete | 2 | * | 32.8 | 15.4 | 6.9 | 6.4 |
| D&McPh | 3A | D11 | 1667 | MO | Complete | 1 | * | 39.6 | 24.8 | 17.1 | 20.1 |
| D&McPh | 3A | D11 | 1563 | MO | Complete | | *Conjoins with D11-1667* | | | | |
| D&McPh | 3A | D11 | 1480 | MO | Broken after use | 1 | * | 32.6 | 22.9 | na | 11.3 |
| D&McPh | 3B | F11 | 1153 | MO | Complete | | * | 19.1 | 12.4 | 10.9 | 6.6 |
| Bordes | I1 | G13 | 386 | MO | Complete | | | 22.7 | 19.9 | 10.9 | 8.6 |
| D&McPh | 3B | F11 | 1259 | MO | Complete | | | 27.4 | 19.9 | 11.6 | 10.2 |
| D&McPh | 4A | G11 | 831 | MO | Complete | | * | 30.8 | 23.9 | 9.4 | 13.4 |
| D&McPh | 3A | E11 | 1136 | MO | Indet | | | 28.8 | 19.6 | 12.6 | 9.5 |
| Bordes | I1 | G13 | 387 | MO | Indet | | | 30.5 | 23.9 | 16.4 | 10.7 |
| Bordes | I2 | D14 | 3090 | MO | Complete | 2 | * | 67.1 | 53.1 | na | 225.5 |
| Bordes | J3B | F13 | 1536 | MO | Complete | 1 | * | 34.5 | 17 | 4.5 | 5.2 |
| Bordes | I2 | F12 | 1477 | MO | Broken | | *Conjoins with F13-1536* | | | | |
| Bordes | I2 | D15 | 1368 | MO | Complete | 1 | * | 25 | 19.9 | 5.8 | 2.8 |
| Bordes | IND | IND | IND | MO | Broken after use | 1 | * | na | na | na | 4.5 |
| Bordes | J2 | D15 | 2408 | MO | Complete | 1 | * | 50.1 | 26.2 | 23.9 | 26.2 |
| Bordes | F4 | E12 | 1397 | MO | Complete | | | 18.4 | 14.2 | 9.4 | 3.1 |
| Bordes | F4 | E12 | 1332 | MO | Complete | 1 | * | 27.3 | 12.5 | 10.9 | 5.1 |
| Bordes | F2 | E12 | 370 | MO | Complete | | | 11.4 | 11.2 | 5.3 | 1.1 |
| Bordes | F4 | D12 | 925 | MO | Complete | 3 | * | 22.6 | 11.1 | 5.8 | 2.4 |
| Bordes | IND | IND | IND | MO | Complete | 1 | * | 15.1 | 11.7 | 10.8 | 3.4 |
| Bordes | IND | IND | IND | MO | Complete | 1 | * | 26.6 | 17.6 | 15.2 | 8.2 |
| Bordes | I1 | E13 | 1616 | MO | Complete | 2 | * | 40.9 | 26.4 | 9.7 | 9.9 |
| Bordes | I1 | E13 | 1617 | MO | Complete | | *Conjoins with E13-1616* | | | | |
| Bordes | F | E13/F13 | 305 | MO | Complete | 1 | * | 20.6 | 10.5 | 10.3 | 4.1 |
| Bordes | IND | IND | IND | FS | Indet | | | na | na | na | 0.7 |

The inferred use-wear was visible to a naked eye or assessed using a hand lens. *MO* Manganese oxides and *FS* Ferruginous sandstone

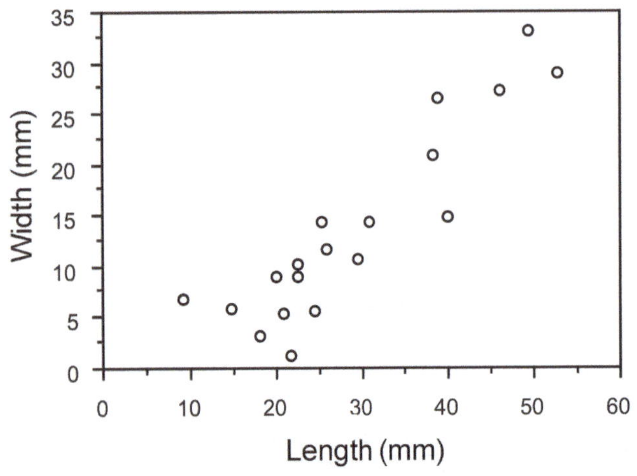


**Fig. A.2** Correlation between the length and the width of factes produced by grinding ($n$ = 19)

**Fig. A.3** Manganese-rich lumps from Pech de l'Azé IV showing facets covered by thin longitudinal striations. *Top left* conjoin of pieces D11-1563 and D11-1667 with a close-up view of the main facet on the *right* of the figure. *Bottom left* object from Bordes' excavations with no provenance information (*Photo* d'Errico and Soressi)

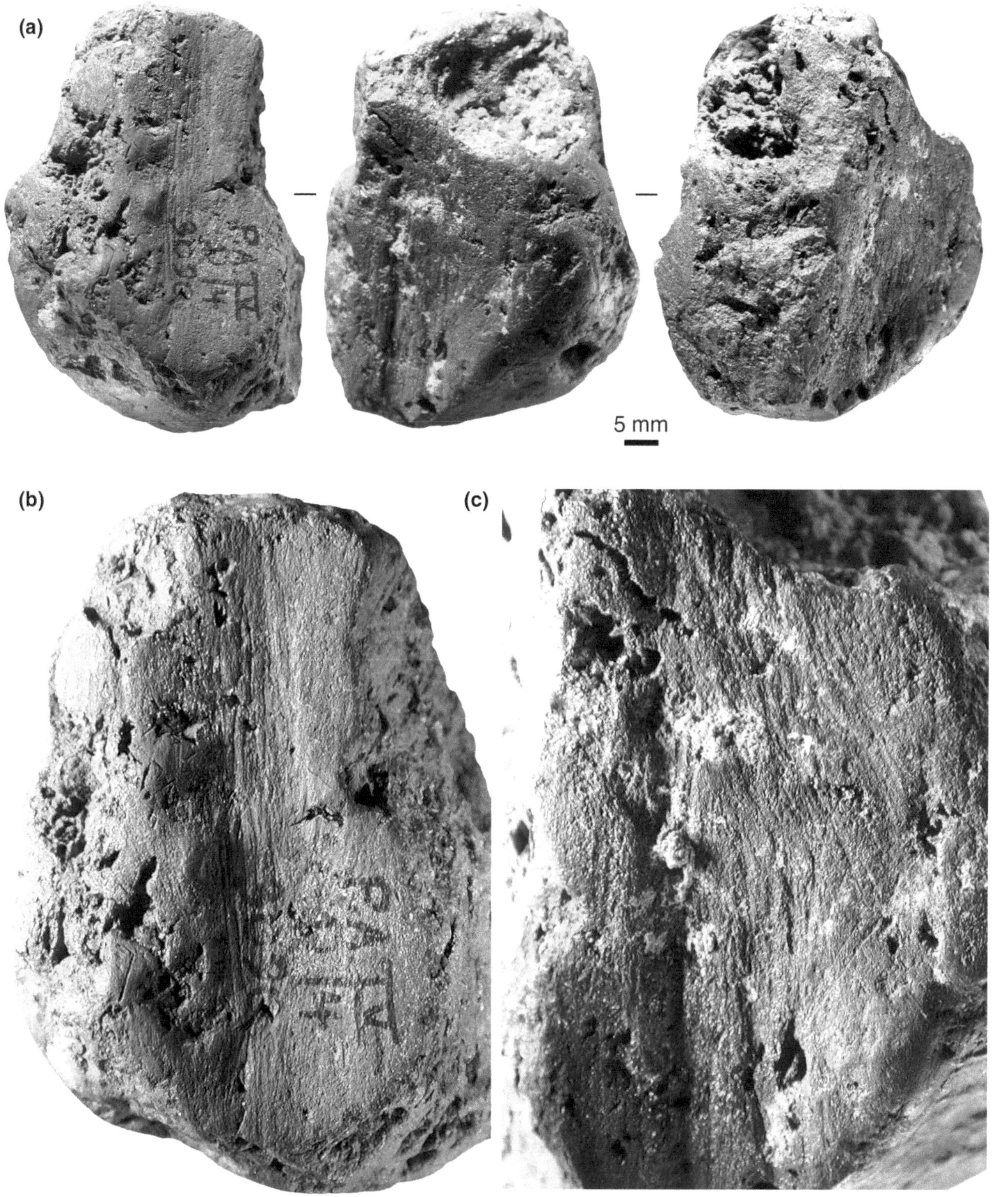


Fig. A.4 **a** Three aspects of the manganese oxide block D14-3090; **b–c** close-up views of areas covered by striations produced by scraping the object with a stone tool (*Photo* D'Errico and Soressi)

## References


Dayet, L., d'Errico, F., & Garcia-Moreno, R. (2014). Searching for consistencies in Châtelperronian pigment use. *Journal of Archaeological Science*, *44*, 180–193.

d'Errico, F., & Soressi, M. (2002). Systematic use of manganese pigment by the Pech-de-l'Azé Neanderthals: Implications for the origin of behavioral modernity. Paleoanthropology Society meeting, Denver. *Journal of Human Evolution, 42*(3), 13

d'Errico, F., & Soressi, M. (2003). Les colorants des fouilles Bordes et Dibble-McPherron. In H. Dibble, S. McPherron, P. Goldberg, & A. Turq (Eds.), *Pech de l'Aze IV (Carsac, Dordogne): Rapport d'operation pour les années 2000–2003*. Report prepared for Direction Régionale des Affaires Culturelles de l'Aquitaine et Service Régional de l'Archéologie.

Heyes, P. J., Anastasakis, K., de Jong, W., van Hoesel, A., Roebroeks, W., & Soressi, M. (2016). Selection and use of manganese dioxide by Neanderthals. *Scientific Reports*, *6*, 22159.

Salomon, H. (2009). *Les matières colorantes au début du Paléolithique supérieur: sources, transformations et fonctions*. Doctoral dissertation, Bordeaux 1.

Soressi, M., Rendu, W., Texier, J.-P., Claud, E., Daulny, L., D'Errico, F., et al. (2008). Pech-de-l'Azé I (Dordogne, France): nouveau regard sur un gisement moustérien de tradition acheuléenne connu depuis le XIXe siècle. In J. Jaubert, J.-G. Bordes, & I. Ortega (Eds.), *Les sociétés Paléolithiques d'un grand Sud-Ouest: nouveaux gisements, nouvelles méthodes, nouveaux résultats.* Société préhistorique française, Paris, Mémoire 47, pp. 95–132.

# Appendix B: The Unique Hominin Remain: A Fragment of an Upper Permanent Molar Germ

Bruno Maureille

This piece, F11-4490, was discovered on June 29, 2002, during the sieving process of the fine sediments from a 7 L bucket. The sediments come from the F11 square and Layer 5A (global Layer III-A; see volume). This layer is the equivalent of Bordes' Layer J1 which yielded a Typical Mousterian with large scrapers (Bordes 1975) (Fig. B.1).

The find is a tooth fragment which preserves a part of the occlusal face with a notable complexity in its relief (numerous grooves and crests). One principal cusp is complete. This morphology is without any doubt a molar one (Hillson 2002; Lautrou 1997). The preserved part of the root is not eroded, but present a small postmortem break. Below the cervix, the height of the calcified root is 2.5 mm, which is useful for the age determination. Considering its apical fringe, it corresponds to the E stage of Demirjian et al. (1973). Its enamel thickness, size, and the complexity of the occlusal surface argue for a permanent molar germ. Considering the apical view, on the junction of the dentine and the pulp chamber roof, we do not observe the depression corresponding to the cervical extremity of the dental pulp related to each cusp. Such a trait characterizes lower molars more than upper ones (personal observations). Then, considering the development of the high main cusp and the curvature between the parts of the two poorly preserved crown faces, this cusp could correspond to a distolingual one (the hypocone) of an upper right permanent molar germ (URM?), more probably a first or a second than a third one.

Then, with such a root calcification and following Alqahtani et al. (2010), it could belong to a 4–5 (hypothesis # 1) or a 8–9-year-old child (second hypothesis # 2) or a 15-year-old child if a URM3.

The morphology of the preserved part of the crown does not allow us to identify any peculiar trait (Turner et al. 1991). The disto-cusp is well individualized which suggests a URM1 or URM2.

As there are no derived Neandertal traits on such a fragment, a comparison with Neandertal teeth with no or light occlusal abrasion allows us to underline some crown morphological similarities (complexity of the reliefs) with the URM1 of the Le Moustier 1 Neandertal teenager (personal observations). Then, unfortunately, the morphology of this fragment of tooth does not help to precise the taxonomic status of the fossil. Nevertheless, as it is easy to see on its natural breaks, the dentine is deep and the enamel is almost 1 mm thick at the level of the hypocone cusp. As the germ has been micro-scanned at the Grenoble synchrotron (McPherron personal com.), future investigations may lead to its taxonomical determination (Macchiarelli et al. 2006).

The preserved part of this germ is not useful for any metrical crown diameter comparison (Maureille 2001), and we cannot compare its scores relative to Neandertal or to the extant human variability.

There is no evidence for pathological/stress (hypoplasia, enamel defect, pit, carious lesion) or erosion on the fragment. This tooth was never functional.

To conclude, we cannot discuss any hypothesis explaining the presence of this hominin fragment within the deposit, but, for sure, it cannot be related to any natural ante-mortem process.


H.L. Dibble et al. (eds.), *The Middle Paleolithic Site of Pech de l'Azé IV*,
Cave and Karst Systems of the World, https://doi.org/10.1007/978-3-319-57524-7

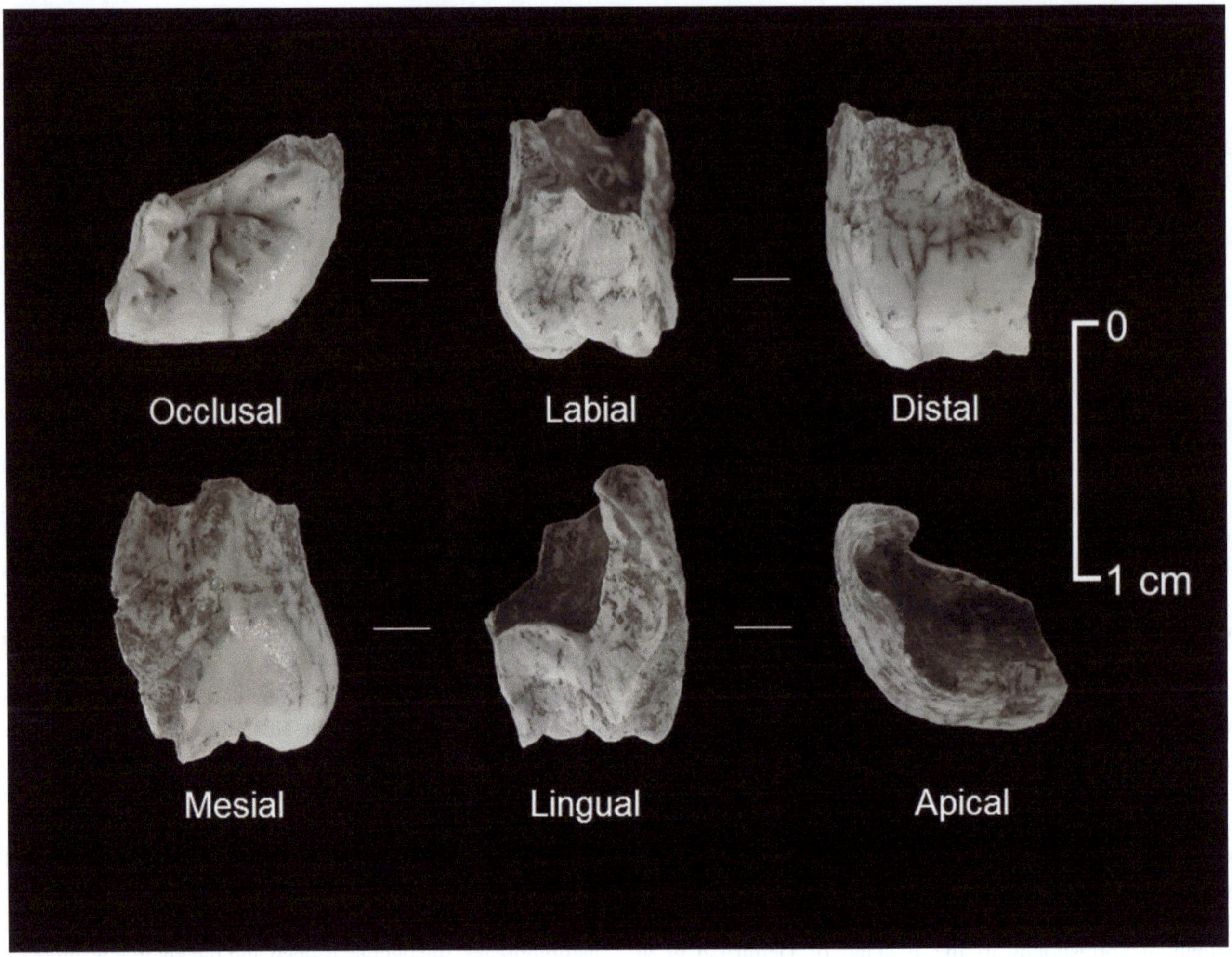


Fig. B.1 The different views of the Pech de l'Azé IV fragment of an *upper right* permanent molar germ

**Acknowledgements**

We want to thank Jean-Jacques Cleyet-Merle, head of the Musée national de Préhistoire (Les Eyzies-de-Tayac) and Amult Hoffmann curator at the *Museum für vor- und Frühgeschichte archäologie Europas* for the authorization to study the collections conserved at their institutions. This research was supported by the project «Transitions» convention 20051403003AB of *Région Aquitaine*, the ANR (French National Research Agency) project: LabEx Sciences archéologiques de Bordeaux, n° ANR-10-LABX-52, sub-project *NéMo* and the Research Program of the *Nouvelle Région d'Aquitaine*: Neanderthalenses Aquitanensis*: Territoires, Chronologie, Humanité,* convention n° 2016-1R40204-00007349-00007350.

## References

Alqahtani, S. J., Hector, M. P., & Liversidge, H. M. (2010). Brief communication: The London atlas of human tooth development and eruption. *American Journal of Physical Anthropology, 142*, 481–490.

Bordes, F. (1975). Le gisement du Pech-de-l'Azé IV: note préliminaire. *Bulletin de la Société préhistorique française, 2*, 293–308.

Demirjian, A., Goldstein, H., & Taner, J. M. (1973). A new system of dental age assessment. *Human Biology, 45*, 211–227.

Dibble, H. L., Raczek, T. P., & McPherron, S. (2005). Excavator bias at the site of Pech de l"Azé IV, France. *Journal of Field Archaeology, 30*, 317–328.

Hillson, S. (2002). *Dental anthropology*. Cambridge: Cambridge University Press, 373 p.

Lautrou, A. (1997). *Anatomie dentaire*. Paris: Masson, 264 p.

Macchiarelli, R., Bondioli, L., Debénath, A., Mazurier, A., Tournepiche, J.-F., Birch, W. (2006). How Neanderthal molar teeth grew. *Nature, 444*, 748–751.

Maureille, B. (2001). *Variabilité dans le genre Homo: les mensurations des couronnes dentaires déciduales et permanentes.* Synthèse de l'activité scientifique pour l'obtention de l'habilitation à diriger des recherches. UMR 5809, Laboratoire d'Anthropologie des populations du Passé: Université Bordeaux 1, T. I (text), 168 p, T. II (annex), 124 p.

Turner, C. G., Nichol, C. R., & Scott, G. R. (1991). Scoring procedures for key morphological traits of the permanent dentition: the Arizona State University dental anthropology system. In M. A. Kelley & C. Spenser Larsen (Ed.), *Advances in dental anthropology.* New York: Wiley-Liss, pp. 13–31.

MIX
Papier aus verantwortungsvollen Quellen
Paper from responsible sources
FSC® C105338

If you have any concerns about our products,
you can contact us on
**ProductSafety@springernature.com**

In case Publisher is established outside the EU,
the EU authorized representative is:
**Springer Nature Customer Service Center GmbH**
**Europaplatz 3, 69115 Heidelberg, Germany**

Printed by Libri Plureos GmbH
in Hamburg, Germany